Reproductive Laws for the 1990s

Contemporary Issues in Biomedicine, Ethics, and Society

Reproductive Laws for the 1990s, edited by ***Sherrill Cohen and Nadine Taub,*** 1989
The Nature of Clinical Ethics, edited by ***Barry Hoffmaster, Benjamin Freedman, and Gwen Fraser,*** 1988
What Is a Person?, edited by ***Michael F. Goodman,*** 1988
Advocacy in Health Care, edited by ***Joan H. Marks,*** 1986
Which Babies Shall Live?, edited by ***Thomas H. Murray and Arthur L. Caplan,*** 1985
Feeling Good and Doing Better, edited by ***Thomas H. Murray, Willard Gaylin and Ruth Macklin,*** 1984
Ethics and Animals, edited by ***Harlan B. Miller and William H. Williams,*** 1983
Profits and Professions, edited by ***Wade L. Robison, Michael S. Pritchard, and Joseph Ellin,*** 1983
Visions of Women, edited by ***Linda A. Bell,*** 1983
Medical Genetics Casebook, by ***Colleen Clements,*** 1982
Who Decides?, edited by ***Nora K. Bell,*** 1982
The Custom-Made Child?, edited by ***Helen B. Holmes, Betty B. Hoskins, and Michael Gross,*** 1981
Birth Contol and Controlling Birth, edited by ***Helen B. Holmes, Betty B. Hoskins, and Michael Gross,*** 1980
Medical Responsibility, edited by ***Wade L. Robison and Michael S. Pritchard,*** 1979
Contemporary Issues in Biomedical Ethics, edited by ***John W. Davis, Barry Hoffmaster, and Sarah Shorten,*** 1979.

Reproductive Laws for the 1990s

Edited by

Sherrill Cohen

and

Nadine Taub

Women's Rights Litigation Clinic
Rutgers Law School–Newark

Humana Press • Clifton, New Jersey

Library of Congress Cataloging in Publication Data
Main entry under title:

Reproductive laws for the 1990s/edited by Sherrill Cohen and Nadine Taub.
472 pp. cm.–(Contemporary issues in biomedicine, ethics, and society)
Includes index.
ISBN 0-89603-157-8
1. Human reproduction–Laws and legislation–United States.
2. Human reproduction–Technological innovations–Moral and ethical aspects. I. Cohen, Sherrill. II. Taub, Nadine. III. Series.
KF3771.R46 1988
344.73'0419–dc19
[347.304419] 88-25867
CIP

Nadine Taub, Director
Women's Rights Litigation Clinic
School of Law–Newark
Rutgers, The State University
15 Washington Street
Newark, NJ 07102

PREFACE

The Project on Reproductive Laws for the 1990s began in 1985 with the realization that reports of scientific developments and new technologies were stimulating debates and discussions among bioethicists and policymakers, and that women had little part in those discussions either as participants or as a group with interests to be considered. With the help of a planning grant from the Rutgers University Institute for Research on Women, the Women's Rights Litigation Clinic at Rutgers University Law School-Newark held a planning meeting that June attended by approximately 20 theorists and activists in the area of reproductive rights. Project purposes, methods, and general shape took form at the meeting. Two goals have characterized the Project's work since then: first, to generate discussion, debate, and, where possible, consensus among those committed to reproductive autonomy and gender equality as to how best to respond to the questions raised by reported advances in reproductive and neonatal technology and new modes of reproduction; and second, to ensure that those shaping reproductive law and policy appreciate the ramifications of these developments for gender equality. In meeting this twofold agenda, the Project focused on six areas: time limits on abortion; prenatal screening; fetus as patient; reproductive hazards in the workplace; interference with reproductive choice; and alternative modes of reproduction. The Project identified individuals to take responsibility for drafting model legislation and position papers in the six areas (for the drafters, see the Appendix). It obtained funding support from the ADCO Foundation, the Ford Foundation, the Fund for New Jersey, the General Service Foundation, the Huber Foundation, the Ms. Foundation, the Muskiwinni Foundation, the Norman Foundation, the Pettus Crowe Foundation, the Robert Sterling Clark Foundation, the Rockefeller Foundation, and the 777 Fund.

At the same time, a working group of 25 academic and activist feminists was formed to discuss the crosscutting issues and develop the theoretical framework for evaluating the proposals (see Working Group in the Appendix). Two additional background papers--on the impact of Project issues on the disabled and on poor women of color--were commissioned.

The following year, with the help of twelve different funders, the working group met with the drafters. At the close of this initial drafting phase, the proposals were presented for discussion and critique to representatives of the pro-choice and women's rights communities. Held on the Douglass Campus in New Brunswick in the fall of 1986, the working conference involved a wonderfully geographically, ethnically, and occupationally diverse group in intense and respectful conversation of visions, goals, and strategies. Although the position papers presented here remain the responsibility of their drafters, they clearly reflect and have benefited from each stage of the Project's collective work. In some cases, such as the question of workplace hazards, that work has produced consensus. In other cases, such as the question of surrogacy arrangements, it has not. Perhaps more important than the revisions of individual proposals that resulted from these discussions was the overwhelming understanding that these six areas could not be considered in isolation from broader social, political, and economic issues.

The revised position papers were presented to the larger bioethics community at a Forum on Reproductive Laws for the 1990s held in New York City on May 4, 1987 and attended by over 400 people. Co-sponsored by the New York Academy of Sciences' Section on Science and Public Policy and Committee for Women in Science, the Forum included reaction and commentary by recognized experts in the fields of law, medicine, and public policy. The Forum resulted in an appreciation of the necessity of considering the demands of gender equality when structuring laws and regulations regarding the new reproductive technologies. The exchange of viewpoints that occurred at the Forum advances our attempts, as a society, to make optimal policies concerning reproduction. This volume contains both the Forum proceedings and important contextual essays prepared for the Project.

A number of people, in addition to the Project's drafters, working group, and funders, have contributed significantly to making this book possible. Janet Gallagher was particularly important in alerting women to the need to make their voices heard in policymaking concerning new modes of reproduction. The members of the Project's advisory committee provided key information and critical readings of the first drafts of the position papers (see

Advisory Committee in the Appendix). At Rutgers University Law School in Newark, consultant to the Project, Edith Jaffe, offered invaluable advice and aid concerning the Project's conferences and publications. Roberta Francis also served as a consultant and carefully oversaw the final phases of production of this manuscript. On the secretarial and word-processing staff at the law school, Joyce Brown, Arlene Woodyard, Elizabeth Urbanowicz, and Gwen Ausby provided assistance throughout the life of the Project. A special debt is owed those affiliated with the Project's co-sponsor, the Institute for Research on Women: Catharine Stimpson, the former Director of the Institute; Carol Smith, the Institute's current Director; Ferris Olin, its Executive Officer; and Arlene Nora, its Principal Secretary, who ably did the Project's accounting. At the New York Academy of Sciences, Anne Briscoe, chairperson of the Committee for Women in Science; Frank Fischer and Sally Guttmacher, co-chairpersons of the Section on Science and Public Policy; and Ann Collins, Public Relations Director, lent their generous efforts in helping to arrange the Forum. Humana Press' president, Thomas Lanigan, and copyeditor, Susan Hannum, have shown sustained enthusiasm and support for the Project, and made the publication of this book an easy and pleasant endeavor. Finally, many thanks are due Elizabeth Bocknak for her editing assistance and Helaine Randerson of Fastidious Word Processing for her keen skills in typesetting this book and useful editing and production suggestions.

Nadine Taub and Sherrill Cohen

CONTENTS

I. THE SOCIAL CONTEXT FOR REPRODUCTIVE LAWS

INTRODUCTION

Working Group of the Project on Reproductive Laws for the 1990s

Much of the media coverage of the new reproductive technologies--techniques like in vivo and in vitro fertilization, embryo transfer, prenatal genetic screening, and fetal surgery--describes these scientific advances as major medical breakthroughs. Yet it is important to assess technological impact in terms of its context, and who controls and deploys its use. A critical perspective on the new reproductive technologies reveals potential drawbacks as well as potential liberation.

American culture has a predilection for a quick fix--a seemingly simple, technological solution to complex social and personal problems. History holds many examples of new and seemingly innovative technologies that served to replicate and worsen existing social problems, rather than to transform them. Some medical "revolutions" quickly turn into disasters. X-rays, for example, were hailed right after World War II as a potent and positive tool in assessing problematic pregnancies, predicting the need for cesarean sections, and diagnosing multiple births. Two decades later, British epidemiologists revealed the dramatic increase in childhood leukemia and other cancers that accompanied the use of X-rays in the first trimester of pregnancy.

Reproductive technologies and alternative modes of reproduction are being introduced into a society permeated by social and economic inequalities. These developments threaten to exacerbate present social and gender inequalities. They may divert resources from less exciting but more needed basic prenatal services, and may focus attention exclusively on women, instead of women and men, in the effort to achieve optimal reproductive outcomes. Thus, although the new reproductive developments hold tremendous promise, there is reason for caution in evaluating their relative merits and dangers.

The members of the Project's working group, united by their common commitment to reproductive freedom and gender equality, came together to assess these new technological developments and consider appropriate legislative and policy recommendations. The 25 participants, including lawyers, doctors, social scientists, and community activists, addressed six areas of reproductive law in which technological progress appeared to be dramatically transforming social problems and practices:

1. Time Limits on Abortion - What limits on a woman's right to choose abortion may be appropriate in light of actual and projected developments in fetal technology?

2. Prenatal Screening - What limits on prenatal screening leading to selective abortion are appropriate in light of the pressures on prospective parents created by advances in genetic testing capability and societal attitudes toward the disabled?

3. Fetus as Patient - By whom and how should claims of the fetus as patient be resolved?

4. Reproductive Hazards in the Workplace - How should reproductive hazards in the workplace be controlled to maximize reproductive choice and equal employment opportunity?

5. Interference with Reproductive Choice - Under what circumstances should civil or criminal sanctions be imposed on those who interfere with reproductive choice? This interference ranges from sterilization abuse and workplace hazards to abortion clinic violence and harm done to a pregnant woman that causes fetal death.

6. Alternative Modes of Reproduction - What controls should be placed on alternative modes of reproduction? This topic includes artificial insemination by donor, surrogate pregnancy, and new technologies such as in vitro fertilization and embryo transfer.

The group began with the working assumption that our thinking about reproductive law in the coming decades would necessarily be future-oriented. Yet as we worked, we were constantly struck by the connection between scientific advances and the past. The new technologies have emerged out of old social relations, and often play on old notions about a woman's place. For example, as scientific and technological advances have yielded more information about the fetus at various stages of development and made possible fetal surgery and other interventions during pregnancy, calls to recognize the separate claims of the fetus have been heard increasingly. These claims have, in turn, prompted calls to impose behavioral restrictions on pregnant women. Such calls are troubling, because at the same time that they raise important concerns, they also imply that women are incapable of acting responsibly on their own and must always subordinate their needs to those of others.

In American society today, securing full reproductive freedom is key to achieving gender equality. Women must be able to determine if, when, and how they will bear children if they are ever to gain ground and participate as equals in social, economic, and political life. At the heart of gender relations in our society lies the social construction of reproductive functions. Reproduction is tightly tied to our cultural definitions of who women are, the importance of children, and the meaning of family life. The fact of being able to reproduce--or not being able to, in the case of the development of many new reproductive technologies as cures for infertility--has a powerful influence on the way girls are raised to become women in America. Thus, the availability of highly advanced techniques such as in vitro fertilization and simpler means of conception through surrogate parents, as well as access to abortion, may well shape sexual roles in years to come.

Similarly, the way girls are raised to become women has a powerful influence on their reproductive activities. We have babies for complex reasons: for the joy and continuity of reaching out to the future, for the sensuous pleasures of intimate connection to children, for the proof of our own adulthood and its connection to our families and communities. We also have babies because we feel pressured to do so by normative gender roles, because we want to "give a baby" to a lover or a parent, or because we hope the child will give us the things we lack in our own lives. The complex motives each woman feels in having children make it hard to "come clean" on motherhood, to admit that we are not only selfless madonnas, content to give our children everything we can, at whatever personal cost, but also that there are more selfish, or self-centered, reasons for having children. In short, women are both agents and victims of our reproductive capacities, and our "individual choices" in having or not having children reflect the larger social world in which we gain and lose a sense of our own self-worth and standing vis-à-vis others.

In attempting to assess the impact of the new developments, the group discussion returned frequently to the conditions that currently circumscribe individual choices regarding reproduction. In addition to the gender considerations just discussed, these include socioeconomic circumstances, the health care delivery system, the legal culture, racial bias, and attitudes toward the disabled. In light of these circumstances, the group found that broader policy initiatives as well as specific legal proposals were necessary to effectuate reproductive choice and promote gender equality.

The socioeconomic structures of American society represent an important set of constraints on reproductive "choice." The vast socioeconomic inequities that rift our society very simply deny low-income women the same kind of access to choice that other sectors of the population enjoy. We as a society debate the morality of late abortion now that advances in neonatology and prenatal treatment reveal the complexity of fetal development. However, we cannot discuss the ethics of late abortion without recognizing that over 90 percent of all abortions now occur in the first trimester, and the remainder are "chosen" by women who are disproportionately young, poor, and from minority groups. The conditions of their lives are not changing enough to permit the "choice" of early abortions, no matter how much information on fetal development the medical profession expounds. Likewise, access to prenatal diagnosis, especially to amniocentesis, varies greatly, depending on where a woman lives and her financial resources. By the same token, a woman's ability to resist an employer's pressure to be sterilized to retain her job depends to a very great extent on her economic circumstances.

The nature of the present health care delivery system in the United States compounds these problems. In a system dominated by the fee-for-service principle and characterized by fragmented and inadequate medical care programs, great numbers of Americans lack the resources necessary to obtain basic services. Those who cannot afford to go to private practitioners must seek their medical care at large teaching hospitals, where they often become objects of teaching needs, including the tendency of such hospitals to perform a high rate of surgical and experimental procedures. Women, particularly poor and vulnerable women, who are persuaded to undergo unnecessary cesarean sections and hysterectomies have their right to individual reproductive choice abrogated.

The current health care delivery system has other consequences as well. For example, the 123-plus clinics offering "revolutionary" in vitro fertilization that have sprung up in the last few years in the United States must be evaluated in light of their low success rates and freedom from customary ethical constraints. Because the clinics tend to be funded by venture capitalists and are freestanding, rather than attached to established medical centers, procedures that ordinary hospital ethics committees might deem experimental pass as normal without further discussion. Also, some clinic directors have made their screening preferences for "acceptable" infertility patients baldly clear--no poor people, unmarried couples, single mothers, disabled women, or lesbians need apply. Under the weight of such judgments, even "revolutionary" technologies may be used to perpetuate only the most traditional of family arrangements. Even as we acknowledge the exciting possibilities inherent in many new technologies, we need to step back and ask a classic public health question: who benefits from them and at what cost? When in vitro fertilization costs, on average, $7000 per cycle, and black Americans have infertility rates at least one and a half times those of white Americans, why is more not being done about the structural causes of the problem, rather than developing a high-tech solution for a small number of infertile couples who can, sometimes with great hardship, pay the price? The appearance and spread of the new reproductive technologies pose many such taxing questions about the ranking of priorities in allocating societal resources.

With health care in general, biological reproduction has become highly medicalized, limiting choice even further. Increasingly, women in America derive their understanding of what it is to be a woman, become (or avoid becoming) pregnant, have a baby, experience infertility, and deal with "problem" children, in the language of medicine. The exaltation of the medical "expert" in our culture diminishes the standing of all patients, but especially that of minorities and women, who frequently are perceived by health professionals as incapable of making proper choices on their own. Physicians, and sometimes other health professionals, too easily fall into an omniscient and authoritative manner, too readily accepted by patients.

Legal pressures on health care providers have resulted in additional constraints on choice. Despite desire on the part of individual providers to be diligent and sensitive to patients' needs, the necessity of avoiding malpractice and product liability suits leads all too often to health care being delivered defensively. Doctors, concerned with skyrocketing insurance costs, may feel compelled, for example, to persuade patients to submit to amniocentesis and consequently to terminate a pregnancy involving a fetal anomaly, to minimize the risks of litigation. Likewise, genetic counselors may find that they are engaged not so much in counseling to bring about a truly "informed consent" as in delivering information whose purpose is to protect the doctors and hospitals for whom they work from later suits. Fear of product liability suits may also restrict choice by making certain reproductive technologies less available, as in the recent decision by manufacturers to cease production of the Lippes loop and Copper-T IUDs in the wake of a product liability judgment against the makers of the Dalkon Shield IUD. It is imperative that pharmaceutical companies continue research and development to produce a range of contraceptive options, but they must institute far more careful testing procedures in order to protect women's health. Concerns about legal liability similarly undermine respect for confidentiality. How can the confidentiality of patients be protected, when genetic screening may reveal factors that employers and insurers want to use to limit their liability? Concerns about legal liability have also been used to excuse discriminatory policies such as those excluding fertile women from certain workplaces.

Even apart from its persistent connection with income disparities, race is a factor that limits reproductive choice. Both past and present eugenics efforts in this country have been accompanied by an impulse to restrict the reproduction of minority racial groups. For example, sterilization techniques were tried on poor blacks in the American south in the 1940s. Experiments with the birth control pill were conducted among women in Puerto Rico in the 1950s. At times, the racism has been deliberate; at other times, simply unthinking. Because the provision of today's high-tech reproductive services is so geared toward the consumer demands of the majoritarian white population, there has not been adequate sensitivity to the needs of minority groups seeking these services, and if great care is not taken, a similar bias may affect their delivery. For example, genetic counselors and other providers have not been sufficiently attuned to the different cultural values, family forms, and political histories shaping the concerns of minority patients, thus lessening the latitude that minority men and women have for exercising their reproductive autonomy. Similarly, the prospect of embryo transfer raises the specter of women of color being used as a breeder class to gestate the fetuses of the dominant race.

Another set of cultural factors influencing reproductive choices in our society is the attitudes many Americans have toward disabilities. We hold our breath as geneticists use gene-splicing techniques to isolate the genes that cause diseases like Huntington's Chorea or Tay-Sachs disease. However, we rarely engage in public discussion of the prejudices and institutional barriers that disabled children and their families confront, and how our attitudes about "perfect children" influence our feelings about prenatal diagnosis. Just as our attitudes toward disability inform our views on the children we envision raising, they also inform our behavior toward disabled individuals and the choices they face concerning their own procreation. Disabilities vary enormously, and it would make little sense to group together blindness, cerebral palsy, Down syndrome, and spina bifida, except that all of these conditions ensure that the people who have them will confront institutional, medical, and cultural prejudice in many arenas of American life.

In other words, what makes disabled people an "unexpected minority" is their political responses to the negative attitudes and institutional barriers they too often face. Ironically, this is particularly true in the fields of reproductive medicine and law, where concepts like "the quality of life" and "wrongful life" often prevail. Such concepts describe a eugenic edge to much of medically current thinking about prenatal screening and late abortion--that is, fetuses who would become disabled people and the fetuses of the poor are often viewed as having only a minimal potential for a high quality of life. Prenatal screening for genetic disabilities raises many possible scenarios: the sad but understandable choice of women to protect themselves and other family members from the problems of raising a disabled child; the prospect of bearing a child whom parents fear will lead a lifetime of pain; and the eugenic specter of mothers as "stewards of quality control," aborting fetuses who would become socially unacceptable, stigmatized people. The members of the Project's working group support any and every woman's right to decide about abortion in the full range of circumstances. We especially note that the "choice" to have a disabled child in a society that offers so little support to aid children and their families is a very hard choice indeed. However, we want to point out that the dice are loaded in the attitudes many medical professionals hold toward what constitutes a "serious" disability. Until we demedicalize popular understandings of disabilities, and educate the public at large, we will continue to use a "universal" medical language to describe the complex ethical choices that different communities of women, and their supporters, confront.

The Project's working group believes that, ultimately, one of the most pressing concerns is the trade-off between maximizing individual reproductive autonomy and allocating societal resources in an equitable way. There is always a tension between seeking to guarantee the widest and most democratic access to state-of-the-art medical care, and asking, "At what cost?" The group believes that a national health plan would help to allocate resources more equitably. Additionally, by providing greater preventive care, a national health plan may enhance the total amount of resources available for health needs. Yet even if such a reform is implemented, it may not be possible to avoid hard choices totally. As a society, we will need to opt for fewer intensive care nurseries and prenatal DNA screens, while providing universal, high-quality prenatal care that would guarantee, on average, many more healthy mothers and babies. We may, as a society, need to clean up our workplaces, making it a crime to threaten the reproductive health of all workers, rather than offer genetic screening to those who have been exposed to teratogenic agents.

The members of the Project's working group are all troubled by the nature of the necessary trade-offs. In grappling with these issues, we have identified more agonizing questions than clear-cut solutions. However, the one conclusion upon which we all agree is that allocations of societal resources must be determined in ways that do not deepen the present socioeconomic inequities and existing discrimination in terms of race, class, gender, sexual preference, and disability. The criteria for deciding which consumers will benefit from reproductive technologies must be factors other than those characteristics. Legislators and policymakers formulating regulatory measures must be alert to the wider ripples of social ramifications that their measures are certain to have. The legislative proposals and policy recommendations that have emerged from the Project on Reproductive Laws for the 1990s are thus accompanied by broad discussions addressing our concerns about such social effects.

II. REPRODUCTION AND ACCESS TO HEALTH CARE: A LEGISLATOR'S VIEW

REPRODUCTION AND ACCESS TO HEALTH CARE

George Miller

The issues involved in formulating reproductive laws are among the most sensitive legal and policy questions confronting the courts, the Congress, and the American people. They affect emerging medical and scientific technology, and raise complex questions about civil rights, labor protections, and public health policy. Also, of course, they touch deeply on many highly sensitive questions of morality, religion, and personal values. Establishing policy on reproductive issues will undoubtedly be very difficult. Yet medical technology and social trends compel us to confront these issues.

I would like to discuss some of the most basic of all "reproductive rights":

- the fundamental right to health care, nutrition, and other services that increase chances for a healthy pregnancy
- the right to family planning services
- and the essential right of *all* workers--male and female alike--to protection from occupational hazards that endanger their ability to enjoy healthy offspring and long lives.

For some people concerned about reproductive rights, the means for confronting these questions are through the courts; for others, through administrative agencies and the medical community. I play my role as a member of Congress, and particularly in my position as chairman of the House Select Committee on Children, Youth and Families.

One of the committee's major responsibilities--indeed, one of the reasons I sought its creation in 1983--was to provide a congressional forum for the examination of such controversial and emerging issues. Because we are a nonlegislating committee, we have the unusual luxury of considering topical and controversial issues, not in the context of a particular statute, but as policy questions that demand future enlightened consideration. We have focused on highly controversial issues: teen pregnancy, school-based health clinics, and now, reproductive policy. Based on that record, we have affected numerous statutes ranging from child abuse to nutrition and health policy.

In 1986, we secured inclusion of a "Children's Initiative" in the congressional budget resolution. Our Initiative targeted supplemental funds to effective, high-priority programs that serve children and pregnant women, programs like maternal and child health, childhood immunization, Medicaid, the Women-Infants-and-Children's (WIC) nutrition program, and child welfare services. As a result, although virtually all other domestic programs were frozen or cut, funding for programs included in the Children's Initiative actually *rose* by more than $350 million. In the budget resolution just passed by the House, we more than doubled support for these proven programs, and we added an additional $100 million for the Social Services Block Grant (Title XX) to address one of the most urgent problems confronting families across the income spectrum: the lack of adequate child care.

I mention the success of the Children's Initiative in order to illustrate a promising trend in congressional responses to developmental and health-related issues, even in an era of budget deficits and spending constraints. Winning bipartisan support for the Children's Initiative under these political conditions is the result of our ability to shift the public debate--from one about *increased* costs associated with government interference, to one about budget *savings* through timely intervention. The national preoccupation with deficit reduction has inadvertently focused attention on programs that are cost effective, and that save children and money simultaneously. We have won support for WIC by demonstrating the investment payback of more than three dollars in averted medical costs for every dollar spent on the program. We have won more dollars for maternal health programs by focusing on the eleven-to-one lifetime savings associated with prenatal examinations, increased birth weight, and maternal care.

Yet although we have won congressional support for particular programs, we have not established a *legal* right to a healthy baby, to adequate nutrition, or to proper health care. So, the avoidable tragedies continue:

■ More than 350,000 poor children in America lost their access to health care because of program changes and funding reductions proposed by President Reagan and enacted by the Congress in 1981.

■ Since 1981, we have cut support for programs that aid the poor, including health care, social services, nutrition, housing, and legal services, by nearly $100 billion. At the same time, the President has increased military spending to the obscene level of $800 million a day.

■ We have thrown three million children into poverty since 1981. Today, one-fifth of all children--13 million Americans under the age of 18--are growing up poor, with all of poverty's attendant health, educational, and developmental consequences.

■ By destroying our public housing and social service support programs, the Administration has created within our midst a nation of the homeless, the largest number of whom are families with children.

■ Although reports of child abuse have risen over 50 percent in the last four years, resources to combat this tragedy have risen by just 2 percent. Through our failure to address the issues of maternal and child health, family planning, and adolescent sexuality, we are sowing the seeds for decades of financial and human turmoil.

■ Nearly 25 percent of all children are born to women who received *no* prenatal health care whatsoever.

■ Over 40 percent of preschoolers are not immunized against one or more of the preventable childhood diseases.

■ Our progress in reducing the infant mortality rate has slowed, leaving us with a rate far higher than that of many developing nations.

Even those efforts that are highly effective, like maternal and child health services, reach only a fraction of the legally eligible, medically needy population.

In addition to these longstanding concerns, policymakers must recognize that even more challenging and controversial issues demand attention. From an economic, educational, and social standpoint, there is no more devastating an impediment to the fulfillment of a teenager's potential than an unwanted or unplanned pregnancy. Today in America, more than 12 percent of all teenage girls are already mothers. Each year, more than one million teenagers will become pregnant--one in four sexually active young women. Pregnancy is now the leading reason why young girls drop out of school. Four-fifths of all teen mothers *never* finish high school, and are twice as likely to depend on welfare. In fact, 60 percent of all welfare spending goes to women who first gave birth as teenagers. If just 10 percent more teen mothers completed their education, we could save $53 million in welfare expenditures every year.

Many of the children of teenage parents, like millions of other American infants, will be born into a system that fails to provide adequate health care. Only 13 states offer publicly supported maternal health care, and only *one* was providing funding sufficient to meet the needs of the full target population. Twenty-eight states have admitted turning away uninsured pregnant women in labor, and another 23 admit that they refuse to arrange in advance for delivery services for uninsured women who cannot afford a preadmission deposit.

Not surprisingly, we as a nation will live with the products of these shortcomings for decades to come. Twenty-two of the 34 states with measurable black infant deaths in 1984 will not meet the target figure for reducing mortality in 1990, including New Jersey, Connecticut, the District of Columbia, and Delaware. Thirteen states will not meet the overall objective for diminishing infant mortality. Forty-four states with a sufficient number of postneonatal deaths to measure progress will fail to meet their goals in reducing the number of those deaths, and 24 are actually getting worse! At the current rate of progress, *no state* will meet the target for prenatal care, and 10 states are getting worse, including Connecticut, Pennsylvania, the District of Columbia, and Maryland.

Even if teenage pregnancy is averted, ominous new dangers confront sexually active teens, as well as the rest of the population. In some areas of the country, as many as 8 of every 1000 young adults are testing positively for the AIDS virus, and more than 100 cases of AIDS have already been diagnosed among 13 to 19 year olds, the age group with the highest overall rate of sexually transmitted diseases among all Americans. Nor can we be insensitive to the related issue of abortions among teens. Forty percent of teenage pregnancies--400,000 cases each year--end in abortion, and teenagers account for more than a quarter of all abortions in this country annually. Regardless of our feelings about the right of women to have full access to medically safe abortions, no one should be willing to suggest that nearly half a million teenage abortions constitutes an appropriate response to adolescent pregnancy.

In light of the sorry statistics of preventable tragedy, what are we to make of the President's proposals for $1.3 billion in cuts in health services for the poor; of freezing the Maternal and Child Health program at a time when nine million poor and near poor children have no health care; or of reducing family planning assistance below the already inadequate level? The responses of the political structure of this country have been worse than misguided. They have been ignorant, they have been malicious, and they must be changed.

Access to decent, affordable, and necessary health care cannot be treated like a medieval morality play at which policymakers blanch whenever the words "sex education," "family planning," or "contraception" are mentioned. We have tried ignorance as a policy. It has failed miserably. And let us remember; the price is being paid not by the President and members of Congress, all of whom have health care coverage, but by the pregnant teens, the low-birth-weight infants, and the developmentally disabled children. Ignoring the issues has not cut costs or reformed sexual behavior: but it has filled the abortion clinics with frightened teenagers, filled the delivery rooms with low-birth-weight premature babies born to teenage mothers, and added billions of dollars in future costs to our health, education, rehabilitation, and welfare obligations. The time has come--actually, it has long since passed--for us to treat these issues as legal and health concerns rather than as matters for religious moralizing. Let us get the evangelicals back to their pulpits and television studios, and the health care professionals into the community where they are desperately needed.

The willingness of some in the Congress to confront the need for a far more aggressive national policy on health and reproductive issues comes at a critical time in the national debate. During the course of the next 19 months, this nation will be engaged in a broad debate over our post-Reagan political agenda. In that debate, the kinds of health issues that I have discussed must take a central role.

First, we must make a national commitment to provide, before the end of the next Administration, a guarantee of full access to quality health care for all Americans, regardless of age, of economic condition, or of where they live in this country. In particular, we must focus attention on special groups in our population whose failure to secure quality health care has immediate and tragic results: especially infants and children, the poor, workers in high-risk occupations, the disabled, and adolescents.

We additionally must provide a range of support programs--including school-based health clinics, family planning clinics, community health and mental health facilities, and expanded child care and respite care services--to the mostly underserved and highly vulnerable populations.

Lastly, we clearly must have a more enlightened perspective on the very serious subject of occupational health hazards. Recently, we won a bittersweet victory when the Department of Labor, after a struggle of 15 years, finally issued regulations giving farm workers the right to drinking water, toilets, and sanitation facilities. The fact that it took until the end of the twentieth century to issue these Victorian-era safeguards sadly demonstrates how far we are from addressing workplace safety hazards, like the elevated miscarriage rates reported among women in the computer chip industry.

As chairman of the Subcommittee on Labor Standards for four years, I know well the scientific and medical complexities of these issues, as well as the vigorous opposition of the business and insurance industries to regulation and compensation legislation. We cannot, however, be deterred from developing national policies that protect workers from serious health and safety hazards on the job. The trade bill just passed stipulates that denial of safe working conditions abroad, including occupational safety and health protections, constitutes "unfair trade practices" that justify penalties. If hazardous workplaces are unacceptable in Kenya, Colombia, and Korea, how can elected officials in this country allow them to continue to exist in Cleveland, Chicago, and Kalamazoo?

The recent tragic loss of two dozen lives in a collapsed building in Connecticut has earned headlines in every newspaper around the nation. Yet every year, 100,000 people die from occupationally related diseases, the equivalent of a full 747 crashing *every day of the year*. Are these workers' deaths less tragic or less significant because many are the victims of slow poisoning? Millions more are subjected to dangerous substances in the workplace. Over the course of their lives, the exposure will have drastic effects on their health, on that of their offspring, and even on their ability to conceive healthy children.

Nationwide, only 15 of the 500 largest companies have comprehensive policies covering reproductive hazards on the job. The growing threat of reproductive hazards on the job was recently recognized when the National Institute of Occupational Safety and

Health placed reproductive hazards on its list of "top ten" work-related diseases and disabilities. The federal government estimates, in fact, that as many as 20 million workers are exposed to substances that may cause damage to their reproductive systems, or to their offspring. Yet to date, just 6 percent of the tens of thousands of chemicals in common use have been tested for reproductive effects, and only 9 percent for genetic effects. And only *four* substances are currently banned because of the hazards they represent to reproduction (ionizing radiation, lead, ETO, and DBCP).

As with the farmworker regulations, the Reagan Administration has obstructed efforts to address reproductive hazards in the workplace. When scientists from the National Institute of Occupational Safety and Health developed a study of women workers at Bell-South, the President's Office of Management and Budget (OMB) at the company's behest threatened to terminate funding unless questions about worker fertility and stress were deleted. This is no isolated incident of obstruction. A study by Harvard and Mt. Sinai researchers last year concluded that OMB has "delayed, impeded and thwarted governmental research" on occupational issues.

Even where the danger is recognized, many employers prefer to banish fertile women workers from the dangerous worksite rather than remove the hazardous agent. The essential principle of workplace safety legislation, however, has always been to make the worksite safe, and that means for *all* workers, fertile or not, men as well as women. Indeed, the practice of workplace banishment, as practiced by some in the chemical and computer industries, raises serious legal questions. However noble the intent of such "fetal protection policies," the net result is a discriminatory and irrational rule that forces workers to choose between being employed and having a healthy baby.

Finally, I have little doubt that Congress and the state legislatures will begin to focus on many of the emerging and complex issues that are the subject of this conference: the innumerable permutations of human reproduction now possible through high technology. With infertility affecting 15 percent of all couples in the United States, and a growing interest in families among non-married people, the developments in reproductive technology will inevitably accelerate and become more complex, both from a medical and legal standpoint. Surrogate parenthood, embryo transfer, artificial insemination, and in vitro fertilization will prove vexing questions to legislators who are overwhelmed by such comparatively simple issues as school-based health clinics. Yet these issues raise difficult questions about parental rights and responsibilities that, in all likelihood, will require legislative standardization.

I view the movement of government into these delicate areas with much trepidation. Congress does not fare well when considering such issues, as we have seen in the ongoing debates about freedom of choice, funding for international family planning, and sex education in public schools. Some of my colleagues sometimes show an unfortunate willingness to decide other peoples' morality and legal rights, particularly when those people are impoverished and without the means to defend their own interests. That is one of the reasons why I am pleased by the growing involvement of the legal community, through the Women's Rights Litigation Clinic and similar organizations. As we have seen in the field sanitation case, the *Baby M* case, and many others, the courts often play a crucial role in the resolution of these issues, and sometimes can act more decisively than the lengthy legislative process.

I hope to expand on the consideration of these issues during our upcoming hearings on reproductive policy. As we work in the Congress on the wide range of issues touching on reproductive rights, I am very hopeful that I will continue to benefit from the insights, advice, and expertise of all of you who are involved with these concerns.

III. SOCIETY AND REPRODUCTION

REPRODUCTIVE LAWS, WOMEN OF COLOR, AND LOW-INCOME WOMEN

Laurie Nsiah-Jefferson

Introduction

Reproductive rights, like other rights, are not just a matter of abstract theory. How these rights can be exercised and which segments of the population will be allowed to exercise them must be considered in light of existing social and economic conditions. Therefore, concerns about the effects of race, sex, and poverty, as well as law and technology must be actively integrated into all work and discussions addressing reproductive health policy.[1]

This essay concerns the six areas identified by the Project on Reproductive Laws for the 1990s[2] as they affect low-income and women of color.[3] Many, though not all, women of color are poor. Women of color are not all one group, just as women of color and poor women are not one group. They have different needs, behaviors, and cultural and social norms. One thing they do share is having been left out of the decision-making process concerning reproductive rights. Although my experience is as a black woman, I will attempt to identify issues that appear to be nearly universal to both women of color and poor women, and point out instances where their perspectives might differ.

There is little information available about the reproductive needs of women of color.[4] In general, the demographic data about non-Caucasian women are clustered together under the heading "nonwhite" as if there were only two racial groups, white and non-white. For example, published abortion statistics are broken down only into two ethnic categories--white and black. As a result of this dichotomization, understanding of the experience of specific groups such as Native American, Asian/Pacific Islander, and Latina women is inadequate. This dichotomization is itself evidence of the pressing need for more precise data gathering on issues concerning women of color.[5] The information that is available generally fails to consider the obvious cultural and social differences related to differences in ethnicity and national heritage. In many cases, this has made it difficult to define and address particular problems and to make recommendations about their solutions.

For many women of color, taking control over their reproduction is a new step, and involves issues never before considered. One reason for this is that women of color have not always had access to the pro-choice movement. In the past, it has been difficult for many middle-class white feminists to understand and include the different perspectives and experiences of poor and minority women. Thus, it is particularly important that adequate information on the needs and experiences of all women be made available now.

The broader economic and political structures of society impose objective limitations on reproductive choice, that is, decisions as to when, whether, and under what conditions to have a child. Very simply, women of color and poor women have fewer choices than other women. Basic health needs often go unmet in these communities. Poor women and women of color have a continuing history of negative experiences concerning reproduction, including their use of birth control pills, the IUD, and contraceptive injections of Depo-Provera;[6] sterilization abuse;[7] impeded access to abortion;[8] coercive birthing procedures and hysterectomy;[9] and exposure to workplace hazards.[10]

Thus, the primary reproductive rights issues for poor women and women of color include: access to health services and information, and the ability to give informed consent or informed refusal; access to financial resources; an end to discrimination relating to class and race, which creates the potential for abuse of the new technology; development of new policies and programs geared toward their needs; medical experimentation; and the need to explore and promote the extended family concept and alternative family structures. Given the history and circumstances of these groups, there are two overarching concerns. One is the desire to make reproductive services, including new technologies, broadly accessible. The other is the need to safeguard against abuse.

After considering each of the six topics, this essay makes policy recommendations relating to the needs of poor women and women of color. These recommendations are designed to ensure:

1. access to quality prenatal care
2. the birth of healthy, wanted children
3. protection against sterilization abuse
4. protection against occupational and environmental conditions harmful to fertility and health
5. protection from pharmaceutical experimentation and unnecessary medical procedures
6. access to accurate information about sex, conception, and contraception and
7. access to safe, affordable abortion

In light of the structural nature of the limitations on the exercise of reproductive choice by poor women and women of color, the recommendations often focus on affirmative policy initiatives, rather than legal restraints.

Time Limits on Abortion

Poor women and women of color often live under circumstances that make it difficult for them to obtain early abortions. For instance, in 1971, nearly one in three nonwhite women of reproductive age lived below the poverty level.[11] It is therefore important to develop affirmative programs that improve access to early procedures and, even more importantly, that reduce the risk of unwanted pregnancy. Unfortunately, however, such affirmative programs cannot totally obviate the need for late abortions. Thus, it is important to understand that laws restricting late abortions will continue to have a particular impact on poor women and women of color.

The Disproportionate Need for Post-First-Trimester Abortions

A significantly higher percentage of nonwhite women who get abortions do so after the first trimester, or first 12 weeks, of pregnancy. Of all abortions obtained by white women in 1983, 8.6 percent took place in the 13th week or later, but 12.0 percent of nonwhite women having abortions obtained them in that period.[12] These figures represent the numbers of women who actually succeeded in obtaining post-first-trimester procedures, and they may seriously understate actual demand. Financial, geographical, and other barriers to access are likely to have a greater impact on nonwhite women, whose overall abortion rate is more than twice that of whites.[13]

There is little information directly concerning very late abortions. Available data on women who obtain abortions after the first trimester, however, demonstrate that financial factors are very important. The enactment and implementation of the Hyde Amendment terminating federal Medicaid funding for abortions has caused many poor women to delay having abortions while they raise the necessary funds. A study of a St. Louis clinic, for example, showed that, in 1982, 38 percent of the Medicaid-eligible women interviewed who sought abortions after the 10th week attributed the delay between receiving the results of their pregnancy tests and obtaining their abortions to financial problems.[14] Yet Medicaid-eligible women were not significantly later in obtaining abortions than other women before the Hyde Amendment went into effect.[15] Even where state Medicaid funding is in theory still available for abortions, it is often not available in practice. Welfare workers and other state officials do not always inform Medicaid recipients of their right to obtain Medicaid-funded abortions.[16] Not all abortion providers are aware that reimbursement is available from Medicaid.[17] Some providers who are aware are unwilling to accept Medicaid, in part because doctors are reluctant to assert that the abortions they perform fall within the particular categories being funded in their states[18] and in part because Medicaid reimbursement rates are so low.[19]

Difficulty in locating abortion services also causes delay. In 1984, there were no abortion providers identified in 82 percent of the counties in the United States--that is, where 30 percent of all women of reproductive age lived.[20] The availability of abortion services also varies considerably by state.[21] Because abortion facilities are concentrated in metropolitan areas, access to abortion services is particularly difficult for rural women. In 1984, 79 percent of all nonmetropolitan women lived in counties that had no abortion facilities.[22] Although geographic access may not pose a significant problem for women of color from northern states who are concentrated in inner cities, it is a concern for women of color in southern states.

Not only are Native American women who live on reservations denied federal funding for abortions, but no Indian Health Service clinics or hospitals may perform abortions even when payment for those procedures is made privately.[23] The Indian Health Service may be the only health care provider within hundreds of miles of the reservation, and as a result the impact of the regulations can be quite severe.

Women in prison, who are disproportionately poor and of color, may also have great difficulty in gaining access to abortion facilities. Abortion services are rarely available at the prison, and prison authorities are unwilling to release inmates for treatment.[24] Recently adopted federal regulations specifically deny abortion services to federal prisoners.[25]

Even where abortion services exist, lack of information about them deters early abortion. Language barriers and the absence of culturally sensitive bilingual counselors and educational materials make gaining information about abortion services a special problem for Asian/Pacific and Hispanic women.[26] This information gap would be severely exacerbated by the Reagan Administration's proposed new Title I regulations, which would prohibit family planning services receiving federal monies under the Title X program from giving any information about the abortion option.[27]

Three factors have been identified as especially important in accounting for very late abortions: youth, medical conditions, and fetal anomalies. At least two of these, youth and medical problems, are likely to have disproportionate significance in the case of women of color. The significance of the problem of fetal anomalies for poor women and women of color is discussed below in the section on prenatal screening.

In 1981 (the latest year for which data are available), 43 percent of all abortions performed after the 20th week of pregnancy were performed on teenagers.[28] Women under 15 years of age are most likely to obtain the latest abortions (those at 21 weeks or more gestation).[29] Their delay is understandable in terms of the difficulties very young women experience in obtaining abortions. These difficulties include the parental notice and consent requirements in effect in some states,[30] as well as the financial and information problems already discussed. Teenagers of color often have particular difficulty in obtaining an abortion. One study found that 4 out of 10 black teenagers were unable to obtain a desired abortion, as compared to 2 out of 10 white teenagers.[31]

Medical problems are also a factor in late abortions, including very late abortions. A major reason for very late abortions is the onset or worsening of certain diseases. Given the nature of their health problems, poor women and women of color are particularly vulnerable to such developments. For example, black women have higher rates of diabetes, cardiovascular disease, cervical cancer, and high blood pressure[32] than other women, and may therefore be in greater need of late abortions. Similarly, the lack of prenatal and general health care that results from poverty may mean that serious health problems arise during pregnancy for many poor women.

Different Forms of Time Limits

Time limits on abortion may be imposed by various laws. Currently, there is concern about statutes that impose prohibitions on post-viability abortions or seek to compel the use of the method most likely to preserve fetal life unless the woman's health would be jeopardized. Poor women and women of color bear the brunt of such laws because women with money and power can find ways to circumvent the law, just as they did prior to the legalization of abortion. Affluent women can either travel to a place where a procedure is legal or find a doctor who will certify that their health is at stake. Poor women who do not have such options are denied autonomy because, as the experience with Medicaid provisions allowing reimbursement only for health-threatening situations suggests, few doctors are willing to risk prosecution under these statutes.

Time limits on abortion may result from a provider's decision not to perform procedures past a certain point in pregnancy. Poor women and women of color today have limited access to facilities that provide abortions after the first trimester.[33] Public hospitals are a major source of health care for poor women, yet only 17 percent of all public hospitals report performing abortions in 1985.[34] Even where the lack of access does not result in an outright denial of abortion, it may cause women further delay that subjects them to increased health risks.[35]

Because most poor women must get abortions where they can find them, they may be severely limited in their choice of method. Although abortions done by the dilatation and evacuation (D & E) technique are safer and less upsetting for women, D & Es are not universally available.[36] To obtain a D & E, a woman may be required to pay for a private gynecologist or travel to a facility where the procedure is done.[37] The problem of obtaining an abortion after the 20th week is even more acute. Because such a limited number of providers perform this procedure, locating a facility, scheduling the procedure, and traveling can all impose serious burdens on poor women.[38]

The question of abortions very late in pregnancy pits the well-being of the pregnant woman and other people against that of the unborn fetus. Although there is no consensus among poor women and women of color that the woman's interests are paramount, there is widespread appreciation of the circumstances that bring women to late abortions and a general sense that the state must not make the decision for the woman. Compelling the use of abortion methods that lead to fetal survival raises serious questions. How would the fate of a surviving fetus be determined? If a fetus were born alive, who would be responsible for its care? What if the mother did not want it? Who would be responsible for financial support? Where would the unwanted fetus be sent? Could it be experimented on? Given their economic circumstances and their history of being subjected to experimentation, poor women and women of color have valid fears about the intentions of the state toward an unwanted fetus.

Family Planning and Life Choices

The number of abortions needed can be drastically reduced by teaching men and women how to prevent unintended pregnancy, but the process may not be simple. When members of a community are denied their rights, how can they know what those rights are, much less learn to assert them? To be effective, family planning services must present information and services in culturally appropriate ways, involving bilingual materials and personnel. Family planning programs must also take account of cultural attitudes and biases about birth control. Some women of color have been unwilling to limit their reproduction in order to redress past population decreases that resulted from war, famine, infant mortality, or genocide. Thus, such programs must make women of color aware of how the ability to take control of reproductive decisions will benefit their lives.

Another important aspect of providing family planning services is helping teenagers make life-enhancing decisions despite the many barriers for young people in our society today. Many teenagers, faced with an empty future, believe that becoming a parent will stabilize their lives. Teenagers need information services, decision-making skills, opportunities for success, and help in building their skills and interests regarding both school and work. They also need family life and life-planning education, and adolescent health services staffed by concerned adults.[39]

Recommendations

Family Planning

1. Information must be made available to young people and adults, on sex, pregnancy, contraception, and abortion, and on how to make choices about them in ways that are culturally appropriate and targeted to the needs of specific communities. Interpreters should be available where necessary. Television, magazines, newspapers, and radio should help provide this information in a variety of languages.

2. Comprehensive job-skill development programs for young people and adults should be available in schools and community programs. In addition to providing needed job training and workplace skills, this type of training can build self-confidence and encourage men and women to make appropriate childbearing choices.

3. Expanded funding should be available to enable sexually active youngsters and teenagers to obtain family planning services. If more young people and adults learned how to prevent unwanted pregnancies, there would be savings in the Aid to Families with Dependent Children and Medicaid programs. Knowledge about spacing pregnancies and education about prenatal care could also reduce the incidence of low-birth-weight babies and associated medical costs.

4. Pro-choice groups should develop stronger alliances with those concerned about teenage pregnancy.

5. Statistical data should be gathered regarding Latina, Asian, and Native American, as well as black and white, populations.

6. The Hyde Amendment should be repealed.

7. In states funding abortions, Medicaid should offer more realistic and prompter reimbursement to encourage more providers to accept Medicaid patients without insisting on cash payments.

8. Where abortion funding is available, information clarifying abortion payment policies should be disseminated to health care providers. Welfare workers and hospital and clinic staff should be trained to know what Medicaid pays for. Community-based non-governmental organizations should assist in disseminating information and in monitoring the information provided by public agencies.

9. Family planning services must be able to provide abortion information and referrals.

10. Adequate services must be available at all stages of gestation.

Postviability Abortions

11. There should be no laws compelling completion of a pregnancy under any circumstances.
12. Responsibility for determining the fate of a live-born fetus must lay with the woman who bore it.
13. Fetal health should be secondary to that of the mother.

Prenatal Screening

Prenatal screening offers women the opportunity to obtain limited information about the status of the fetus they are carrying, i.e., whether or not it is likely that the fetus is "affected" in certain respects. Of the many social, economic, and political issues that the use of this technology poses, questions of access, cultural and class differences, informed consent, confidentiality, and eugenics are of particular concern to poor women and women of color. All of these concerns bear on their rights to choose whether to undergo prenatal screening and whether to continue a pregnancy once a problem is identified.

Financial, cultural, social, and geographic factors all affect access to services. A particularly important factor for women of color and poor women is the cost of many prenatal screening procedures. For example, estimates on the cost of amniocentesis range from $400 to $1000.[40] Amniocentesis for genetic purposes should be performed between the 16th and 20th weeks of pregnancy.[41] The federal government directly supports providers of genetic services on a very limited basis.[42] Most funds are disbursed at the state level, where they are subject to local political pressures. Amniocentesis is available through Medicaid in all states (Hawaii imposes some limitations), and through the Maternal and Child Health Program (MCHP) in 36 states.[43] The quality and extent of genetic counseling with (or without) amniocentesis can vary considerably.[44] However, more critically, abortion is available through Medicaid in only 14 states.[45] A woman dependent upon Medicaid or the MCHP may be able to learn the physical condition of her fetus but be unable to afford an abortion, the only "treatment" alternative in almost every case. For the many women who do not have private health insurance[46] and who are ineligible for Medicaid or the MCHP, as well as those who live in states that do not provide comprehensive funding, the combined cost of amniocentesis and a possible abortion can be prohibitive.[47] Thus, many pregnant women may be faced with the "choice" of learning nothing about the health of the fetus or going without some other necessity to raise the money to pay for amniocentesis and a possible abortion as well.

Educational level and geography also influence access to prenatal screening techniques. Health professionals believe,[48] and data confirm,[49] that low-income women and women of color would use amniocentesis if it were publicized and made available to them. For example, a study conducted in Atlanta found that, after education about amniocentesis, the acceptance rate for the procedure rose considerably.[50] However, the availability of services varies widely, both by geographical area and populations to be served. For example, only three hospitals in Harlem offer counseling and screening services. Similarly, significant numbers of Hispanic people are born with sickle cell anemia, but little effort is made to publicize and provide amniocentesis to the Latino community.[51] Even where communities are aware of specific genetic risk, there may be only a few facilities that provide amniocentesis services. Counseling is not routinely available to Hispanic and Asian women in their native languages.[52]

Access to prenatal screening is also related to access to general health care. Many low-income women are unable to avail themselves of prenatal screening because they begin prenatal care too late or receive none at all. Some poor women and women of color request screening as late as 20 weeks into their pregnancies--too late to schedule counseling, undergo the procedure, obtain the results, and have further counseling on the decision of whether to continue the pregnancy.[53] To ensure that poor women have access to prenatal screening, as a practical matter, outreach methods must take into consideration the lack of prenatal care for poor women. More prenatal services must be provided. Increased outreach and education can encourage such women to seek out prenatal care earlier in their pregnancies.[54]

Cultural and Socioeconomic Factors and Prenatal Screening

Culture, religion, and their connection with childbearing and abortion are key to decision making regarding prenatal screening. Communities thus vary in their attitudes toward particular disabilities. In general, white women, regardless of socioeconomic status, appear to use amniocentesis more often than black women. Urban women--black and white--also appear to seek out amniocentesis more often than their rural sisters. For example, a 1980 study of districts outside Atlanta, found that only 0.8 percent of black women used amniocentesis as compared to 19.4 percent in the city. Other important predictors of the acceptance of amniocentesis are level of education and family history. Ninety-two percent of black women with an education level of 12th grade or higher accepted amniocentesis; 62 percent of those with an education level below the 12th grade accepted it.[55] Black women who had a family member with a congenital disability were also found more likely to use prenatal screening than those who were urged to do so by their doctors because of age.[56]

In general, middle-class women appear to be more concerned about mental capacity, and poor women and women of color are more concerned about physical impairments.[57] For the latter groups, the prospect of added hospital visits and medical disruptions seems more critical than lack of educational success. Thus, preliminary evidence suggests, for example, that middle-class women are more likely to abort after chromosomal anomalies such as Down syndrome are diagnosed than are poor women or women of color.

The determinants of attitudes towards prenatal screening are not always clear. For example, black couples in which both partners carry the sickle cell anemia trait have a one in four chance of producing a child with sickle cell anemia. Thus, they may be appropriate candidates for amniocentesis. Yet acceptance of amniocentesis is not high among the members of this group.[58] Possible explanations include fear of genocide, lack of financial resources, and an awareness that sickle cell anemia may not always be very serious, particularly as the quality of life for children with sickle cell has improved with medical advances in the treatment of the condition.

Unfortunately, little information is available concerning attitudes toward and acceptance of prenatal screening by Hispanics, Native Americans, and Asians. Language has been a major barrier for many Hispanic and Asian women.[59] Many Hispanic and Native American women may be suspicious of any prenatal screening because of the high rate of involuntary sterilization in their communities.

Southeast Asians are particularly at risk for thalassemia, yet they are unlikely to use genetic screening or other reproductive health services if the information is not provided to them in a culturally appropriate manner.[60] Many Southeast Asians believe that amniocentesis will endanger the well-being of the fetus. In addition, some believe the procedure will interfere with natural selection, which is seen as sacred by some Southeast Asian populations.[61]

Genetic Counseling and Informed Consent

Free choice in matters of prenatal screening and abortion requires that women be given the opportunity to obtain all the pertinent information about their own situation and the procedures involved. Genetic counseling can thus play an important role in ensuring informed consent. Ideally, genetic counseling is a process of communication that educates the person being counseled about possible etiology, prognosis, management, recurrence, risks, options, and resources relating to genetic diseases.[62] As noted above, access to such counseling is a serious problem for many women. However, even when such counseling takes place, problems may occur. For example, a survey of counseling programs for sickle cell anemia recently found that almost half of the clients studied were screened without their informed consent, and that many facilities were deficient in providing education and postscreening counseling.[63]

Because the ethics, language, attitudes, religious beliefs, and ethnicity of both the counselor and counselee come into play,[64] the counseling process may become directive and judgmental. There is also the possibility that the information given poor women and women of color will be incomplete, and that a patient's financial situation and educational level will unduly influence the counselor's style and emphasis. Communication may break down because the parties misinterpret each other's verbal and nonverbal responses. For example, prospective parents from particular cultures may be unwilling to show lack of understanding or disagreement with an authority figure. At times, the counseling process may fail to acknowledge decision makers who are essential within a particular culture. As a result, some families will have children they would not have had if they had been counseled appropriately about the nature of genetic diseases, possible outcomes and options, and available assistance, and some women may undergo abortions they would not have had had they been counseled appropriately.

Genuine communication and nondirective counseling are more likely to occur if counselors learn to appreciate the cultural and other factors that shape their clients' beliefs, and influence their way of interpreting and responding to the counseling process. Unfortunately, although the importance of such cross-cultural understanding has been recognized, there has been a lack of cross-cultural training for genetic counselors.[65] Starting in 1985, however, the National Society of Genetics Counselors began to address this problem with workshops at its national meetings.[66]

Confidentiality

Confidentiality in the prenatal screening process is an extremely important issue for people of color who have experienced adverse consequences when intimate information is revealed to third parties. For example, when employers have been given access to information concerning individuals who have the sickle cell trait, they have used it to justify refusals to hire, promote, or retain the employees.[67] Likewise, some insurance companies have refused sickle cell carriers health and life insurance or inflated the cost of their premiums, although there is no evidence that the carriers have a higher risk of disease or a shorter life span.[68]

AIDS screening presents special problems. The Centers for Disease Control have suggested that all fertile women at high risk for contracting AIDS or AIDS-Related-Complex be tested for HIV antibodies. This would include prostitutes, hemophiliacs, intravenous drug users, Haitians, and sex partners of men in high risk groups.[69] To date, such testing has not been made mandatory. Although some pregnant women are anxious to find out whether they test positively or negatively for the disease, fear of job loss and ostracism as well as fear of the deadly consequences of the disease itself may prevent other women from seeking needed prenatal care if they know that AIDS screening is part of the treatment. Assurances that test results will be kept confidential should help address the first fear.

A related problem involves the need to assure confidentiality in the identification and testing of prospective parents required for prenatal screening. Where children are conceived outside the bonds of matrimony, it may be harder to get both parties tested. The woman may be unable or unwilling to contact the male partner. Moreover, the possibility that the male partner's identity will be revealed to social workers and other public officials mandated to collect child support from fathers often makes the male unwilling to come forward. In addition, teenage prospective parents may be fearful that their parents will learn of their sexual activity as a result of testing.

Eugenics

For many people, prenatal screening brings to mind the eugenics movement with its history of abuse. By definition, eugenics is a search for "bad genes," and, at least since the turn of the century, it has been associated with attitudes of racial superiority. In the name of eugenics, blacks and other minorities have been forcibly sterilized, urged to reduce the size of their families, and targeted as carriers of sickle cell and similar traits. Prenatal screening has such overtones when it stigmatizes those with genetic diseases such as sickle cell anemia by implying that they should not have been born.[70] Moreover, couples may be pressured to abort in order to rid the human race of a carrier of a "bad gene," even though the gene may not inevitably lead to debilitating disease.

Recommendations

1. All women should have access to prenatal screening, counseling, and abortion services regardless of ability to pay or geographic location. Prenatal screening and education should be available in community health centers in low-income urban and rural communities.
2. Prenatal screening should not be mandatory.
3. Prenatal diagnostic procedures, counseling, and abortion should be paid for by private insurers and, for Medicaid-eligible women, by Medicaid. The state should be the funder of last resort for the near poor, who are not covered by Medicaid or private insurance.
4. The availability and purpose of amniocentesis should be publicized in the print and broadcast media, and by community organizations. Public maternal and child health care programs should also inform women of the availability, purpose, and possible value of prenatal screening. All publicity should emphasize that use of prenatal screening does not require subsequent action.
5. Test results, particularly those involving sensitive areas, should be confidential and not a part of the patient's regular medical record. Legal provisions guaranteeing confidentiality should be strengthened.
6. Minority prenatal counselors should be recruited, and minority physicians, social workers, and public health nurses should be encouraged to learn about genetic counseling. Clients of a prenatal care facility should be able to select a counselor who

shares their culture, religion, and language, or who shares their attitudes toward death, disease, abortion, and disability.

7. Graduate genetic counseling curricula should include classes on counseling culturally diverse populations. Graduates of past programs should be informed of the availability of continuing education. Where such courses are not available, they should be instituted. Licensing agencies and institutions should emphasize the need for these courses.

8. Representatives of the minority groups to be served by particular prenatal diagnostic programs should be involved in the design, development, and operation of the programs from the outset.

9. Legislative protections against discrimination on the basis of hereditary traits should be strengthened.

10. Informed consent forms should be required and written in simple, understandable terms. Forms and informational materials should be available in all languages appropriate to the area served by the health facility.

Fetus as Patient

The topic of fetus as patient involves attempts by medical and legal authorities to compel women to follow doctors' orders, and accept particular medical procedures while pregnant and when they give birth. For example, doctors and hospitals may seek court orders forcing women to undergo surgery on the fetus or to submit to cesarean sections rather than to give birth vaginally. Women may also be subject to criminal prosecution for "fetal abuse" or to civil suit by their children for their behavior while pregnant.

Medical and legal actions in the name of fetal rights raise many issues for poor women and women of color. A basic question is whether it is right to hold individual women responsible for poor outcomes at birth when many women are not able to live under healthful conditions. This topic thus implicates the general socioeconomic conditions poor women and women of color experience that result in their lack of access to basic prenatal care, and advanced prenatal, perinatal, and neonatal technologies. Holding individual women responsible under present circumstances is morally unjust, and it diverts attention from the need to correct the serious inequities that permeate today's society.

Liability for Poor Reproductive Outcomes

There is good reason to believe that poor women and women of color will be especially vulnerable to prosecutors' attempts to hold mothers responsible for bad reproductive outcomes. As a general matter, their children experience greater rates of infant mortality and low birth weight, which can result in physical and neurological illness.[71] Infant mortality and morbidity among mothers who live below the poverty line are greatly increased, sometimes to as much as twice the rate experienced by other women.[72]

Although the data differentiated by racial and ethnic group are sparse and not standardized, they generally show that infant mortality rates for minority groups are disproportionately high. In 1982, for example, infant mortality rates for black infants were almost twice those of white infants.[73] The infant mortality rates for Native Americans are also extremely high.[74] Hispanics present a complex picture. Puerto Ricans generally have the highest infant mortality rates of any Hispanic group.[75] Although the neonatal mortality rate for Mexican-Americans is considered low by some analysts, most studies suggest that the low death rate is the result of underreporting. Recent studies have shown that Mexican-Americans have a higher neonatal mortality rate in all birth-weight categories than do blacks.[76] Cuban-Americans have low infant mortality and high birth weights compared to other Hispanics. This is not surprising, given the higher socioeconomic status of Cuban-Americans compared to the other groups. The Asian population in the United States is quite diverse, and available data are inadequate. In general, perinatal outcomes for Asians in the United States are good, with relatively low incidence of low birth weight. Southeast Asian refugees, however, present a different picture with respect to perinatal outcomes, as a result of lower economic status and early childbearing.[77]

Socioeconomic conditions are an important element in these poor reproductive outcomes. Low-income women and women of color lack access to prenatal and neonatal care. In addition, many suffer from general ill health, broken families, and lack of social supports. They are more likely to be exposed to environmental hazards where they live[78] or work.[79] When poor women and women of color lack the resources necessary to help them bring healthy babies into the world, it does not make sense to hold them responsible for poor reproductive outcomes. Is it fair, for example, to say that an indigent woman is responsible for the consequences of deficiencies in her diet when Medicaid does not pay for vitamins? Similarly, is it fair to say an indigent woman is responsible for bearing a disabled fetus if Medicaid does not pay for abortion? It may be more just morally, if less feasible legally and politically, to hold the state responsible for the high incidence of infant mortality and disability among the babies born to low-income women and women of color.

Compulsory High-Tech Procedures

Recent evidence suggests that hospital authorities' efforts to force pregnant women to accept high-tech procedures will be aimed disproportionately at low-income women and women of color. In 1987, the *New England Journal of Medicine* published a report on the incidence of court-ordered obstetrical interventions, including forced cesarean sections and intrauterine transfusions. The report revealed that 81 percent of the women subjected to such court orders were black, Hispanic, or Asian; 44 percent were not married; 24 percent were not native English speakers; and none were private patients.[80] Attempts to compel submission to procedures such as cesarean section, fetal monitoring, and other technologies presuppose that they have been adequately explained and that the pregnant woman has no good reason for refusing the procedure. Neither assumption may be warranted.

Health professionals report that most women, irrespective of color or education, do not question a doctor's orders. Indeed they stress that the major problem is unquestioning acceptance rather than rejection of prescribed procedures, particularly among low-income women. Some women who do question high-tech procedures may do so because doctors have not been able to clearly explain the risks and benefits. Others may refuse because they have personally had related negative experiences in the past or heard of others' bad experiences.[81] Despite their failure to question the authority of a physician, poor women and women of color might have good reason to do so. They have been the subjects of experimentation in public hospitals and public health care services.[82] In teaching hospitals, unnecessary procedures are known to have been performed to give experience to doctors-in-training.[83] Individual legal actions directed at women who do resist doctor's orders may divert attention from these problems, and encourage other women to submit to unnecessary and risky procedures. Genuine informed consent could be an important tool in addressing these problems. Women need relevant information in a form they can understand and a supportive environment in which to consider it. It is questionable whether our informed consent laws concerning these technologies and procedures work now. What can informed consent mean today when the informer and the person being informed are on the opposite sides of education, class, race, gender, language, and culture lines? We must develop mechanisms that will really allow women to decide what treatment they want, and protect women against being pressured into accepting tests and procedures they either do not want or whose implications they do not understand.

Technology and Resource Allocation

Overuse of sophisticated technology has inflated the cost of providing routine obstetrical care for all women. Perinatal regionalization schemes, with other high-cost equipment and personnel, focus on end-stage care for mothers and babies with medical complications. Little or no attention is paid to organizing a system that ensures that every pregnant woman will receive basic prenatal care in her community and an adequate diet--an investment in preventing complicated pregnancies. More children are likely to benefit from prenatal care than from high-tech therapies. Although a greater emphasis on preventive care is important for all segments of the population, it is especially important for the traditionally disadvantaged. Those concerned with the fetus as patient should focus on these needs rather than question the behavior of individual women.

A change in focus from end-stage high-tech procedures aimed at individuals to broadly aimed basic prenatal care programs will make existing resources go further. When good prenatal care and other health and social interventions are not available, the results are more difficult deliveries and more low-birth-weight babies needing expensive technologies. With fewer pregnancy complications, it should be easier to arrange for all those who need high-tech services to get them.

Recommendations

1. State and local record keeping relating to prenatal care and reproductive outcomes for all women of color should be improved by maintaining separate statistics for black, Hispanic, Asian, and Native American women.

2. Private insurance coverage of maternity benefits should be mandated, and all payment caps should be removed. Where insur-

ance is employment-related, costs should be shared by employers and employees.

3. States should make every effort to enroll all eligible pregnant low-income women in prenatal programs funded by Medicaid. Eligibility standards should be modified to make more low-income women eligible for Medicaid. States should establish a payor of last resort system for situations where neither Medicaid nor private insurance provide maternity coverage.

4. Services available to low-income women should be increased by expanding existing programs for women, children, and families in underserved areas. Such services should be culturally appropriate and multilingual.

5. States should continue efforts to increase the numbers of obstetricians, gynecologists, family practitioners, and mid-level health professionals accepting Medicaid patients by use of incentive programs or legal mandate, if necessary.

6. Medicaid recipients should have the opportunity to use mid-level health professionals such as midwives, nurse practitioners, and physicians' assistants who offer cost-effective prenatal and infant care.

7. Legislation ensuring informed consent regarding the use of fetal monitoring, cesarean sections, ultrasound and similar procedures, and certain drugs should be enacted. Such legislation should be modeled on the present federal and state sterilization regulations, which are designed to ensure that the patient has adequate knowledge and is not making her decision under pressure.

8. Legal remedies should be available for overuse of technology just as malpractice suits currently result in recoveries for underuse of technology.

9. Attempts should be made to identify and prohibit experimental procedures that are potentially harmful. All other experimentation should have rigorous standards of informed consent.

10. Legislation should be enacted to make more resources available for prenatal care by regulating the amount of resources spent on high-tech care.

11. Statistical information regarding the frequency of use of high-tech procedures, including the races and income levels of the recipients, should be published for each health care facility.

Reproductive Hazards in the Workplace

The reproductive health of minority and poor women may be impaired directly, through job-related hazards, or indirectly, as a consequence of having low-paying jobs without benefits. Thus, their reproductive health, like their general health, is affected by their status as workers, as members of a minority group, and as women. Women of color and poor women often have the most hazardous jobs, risking physical, chemical, and psychological injury.[84] Their low income may restrict their access to health care, and force them to live in neighborhoods contaminated by environmental pollutants and to exist on inadequate diets. Many work in positions with low pay and long hours, without benefits such as health insurance, maternity leave, vacation time, or sick pay.[85] Moreover, poverty and discrimination increase stress. Women who are heads of households are particularly likely to suffer hardships.[86]

Poor women and women of color generally have limited recourse when their rights are violated. They have been excluded from trade unions that could have improved their circumstances in the past, and they are afraid to unionize now for fear of losing their jobs.

Large numbers of low-income women and women of color are employed in the health, textile, and apparel industries, and in cleaning services. For example, 30 percent of all ancillary, auxiliary, and service workers in the health service industry are female, and of this 30 percent, 84 percent are black.[87] Women working in low-income jobs in the health field are exposed to heavy lifting and to chemical hazards such as sterilizing gases, anesthetic gases, X-rays, and drugs.[88] As a result, black hospital workers suffer an even higher rate of primary and secondary infertility than black people generally. Similarly, although little research has been done specifically on reproductive hazards encountered by minority or other hospital workers, nonprofessional hospital workers may be at elevated risk for certain types of cancers (especially breast cancer) because of exposure to radiation and various chemical agents.[89] Cancer-causing agents usually also cause spontaneous abortion.[90]

The textile industry is another source of danger to poor women and women of color. For example, in 1980, nationally over 20 percent of all operatives were black women.[91] In New York City, where the bulk of the garment industry is located, approximately 25 percent are Puerto Rican.[92] Sweatshops located in Chinatown and staffed overwhelmingly by Asian women are responsible for a significant share of production.[93] Workers in this industry often work in high-dust areas, spaces in which picking and carding operations take place. They are exposed to chemicals, dyes, arsenic, heat, cold, inadequate ventilation, and excessive noise, all of which affect women's general reproductive health as well as the health of a fetus.[94] Most sweatshops are located in dilapidated storefronts or badly ventilated lofts to the detriment of the women's reproductive and general health. There are approximately 500 sweatshops in New York City, with unsafe and unhealthy conditions.[95] There are no benefits, and the compensation is too low to allow women to pay for or take time off for prenatal care. Most women who work in such jobs are afraid to complain for fear of being fired or reported to immigration authorities as illegal aliens.[96]

Women of color are also found in laundry and cleaning establishments. In 1980, 40 percent of all clothing ironers and pressers, and 23 percent of all laundry and dry cleaning operatives were black.[97] Jobs in this sector also pose serious health risks. The National Cancer Institute found a higher mortality rate among laundry and dry cleaning workers than among the general population as a whole, and found that women in these jobs, particularly women of color, contracted cancer of the lung, cervix, uterus, and skin at excessive rates.[98]

Many minorities, especially blacks and Chicanos, work in agriculture. Of the estimated five million migrant and seasonal workers, 75 percent are Chicano, and 20 percent are black.[99] These workers are exposed to pesticides that cause liver, renal, and reproductive damage.[100]

For some poor women and women of color, the financial precariousness of their work poses the greatest hazard. Women in low-paying positions, whether in agriculture or as domestics in private homes, tend to have no health or other benefits, such as sick leave or vacation. As a result, many women are forced to work throughout their pregnancies and to return to work immediately after giving birth irrespective of the risks to their health. For example, some jobs require women to stand on their feet all day, although continuous standing can cause complications during pregnancy.[101] Moreover, many of these jobs pay just enough to prevent women from being eligible for Medicaid and the prenatal care services it covers.

Employer Policies and Discrimination

Although there is a definite need to protect women from reproductive and other health hazards in the workplace, there is also a danger that protection will take the form of denying them their jobs. Some companies will exclude women of reproductive age from the workplace rather than make working conditions safe. Others may offer a woman another job, usually at reduced pay. Employer policies of this type have a severe impact on poor women and women of color who may be forced to "choose" to be sterilized rather than give up a desperately needed job.[102]

Occupational hazards faced by men can also jeopardize reproduction. This may either come about directly through damage to the reproductive functions of the male partner or indirectly through transmission of toxins to the woman or fetus. Because many minority and poor men work in the least desirable jobs and are exposed to a variety of toxins, this is a serious concern for their partners.[103]

A related area of particular concern for people of color is discrimination based on susceptibility to certain toxins because of genetic traits such as sickle cell. As previously noted, workers may be terminated from their jobs on the grounds that they or their future children are especially likely to be damaged by the work environment, often in the absence of any specific evidence that such harm is likely.[104]

Inadequate Legal Protections

Laws and regulations now on the books at the federal and state levels are supposed to protect workers from hazards and discrimination in the workplace. These laws theoretically guarantee workers the right to know about their working environment and protect those who speak out against hazards.[105] Unfortunately, these laws are rarely enforced adequately.[106] In addition, domestic and agricultural workers, who are disproportionately poor and minority, are excluded from their coverage. Moreover, no legislation addresses the conditions that make workers particularly vulnerable to reproductive hazards, i.e., their lack of health insurance, medical leaves, vacation time, and sick pay.

Recommendations

1. Existing federal and state laws and regulations, including worker compensation laws, health and safety laws, state and federal right-to-know laws, and civil rights laws, should be enforced more vigorously. Employers should be required to clean up workplaces, use substitutes for unsafe substances, provide protective equipment, and be accountable for violations.

2. Criminal sanctions should be imposed on those who violate occupational health and safety standards that can affect reproduction.

3. Existing laws and regulations should be amended to cover all workers, including domestic and agricultural workers. Legislation should also be enacted guaranteeing workers' benefits such as health insurance, medical leave, parental leave, and sick pay.

4. Employers and unions should be required to provide worker education programs about on-the-job risks and about protection from these hazards.

5. All workers should be paid at least the minimum wage.

6. No one should be subject to discriminatory genetic screening practices.

7. Unions should form minority caucuses to address issues of concern to minorities.

Interference with Reproductive Choice

Reproductive choice involves much more than a woman's theoretical right to conceive or not to conceive. Reproductive choice for poor women and women of color is hindered by numerous realities. These include: unsafe working conditions; lack of access to prenatal and postnatal care; lack of quality gynecological services; unnecessary reproductive surgery, including hysterectomy and other procedures resulting in sterility; coerced involvement with experimental contraceptives and medical technology; passage of the Hyde Amendment and other limitations on abortion funding; inadequate informed consent to procedures; lack of information about sex, contraception, and health; lack of culturally appropriate health services; anti-abortion clinic violence; and domestic violence[107] that may lead to unwanted pregnancies as well as abuse during pregnancy.

Although interference can be the product of outright violence, as in the case of abortion clinic violence, it may also take more subtle forms as in the case of nonconsensual procedures and lack of services. It is apparent, however, that interference with reproductive choice is a fact of life for most poor women and women of color. With social services being slashed at every turn and the quality of services declining because of the cutbacks, this constituency's few remaining reproductive health care rights are increasingly under attack.[108]

Sterilization Abuse

The problem of sterilization abuse illustrates the range of ways poor women and women of color experience interference with their reproductive choice. At times, poor women and women of color have been subjected to blatant coercion; at other times, their "choice" of sterilization has been based on inadequate or no informed consent, the effects of poverty, differential government funding schemes, and lack of birth control information.

Blatant sterilization abuse was exposed in the 1970s. Public assistance officials tricked illiterate black welfare recipients into consenting to the sterilization of their teenage daughters.[109] Native American women under 21 years of age were subjected to radical hysterectomies, and informed consent procedures were ignored.[110] Doctors agreed to deliver the babies of black Medicaid patients on the condition that the women be sterilized.[111] Doctors have also conditioned the performance of abortions on "consent" to sterilization.[112]

But one must question how voluntary the choice of sterilization is in other cases as well. Complete information is crucial to voluntary choice, yet many women elect sterilization under the mistaken belief that the procedure is reversible. Medical personnel often encourage that belief by referring to the procedure as "tying the tubes"; many women assume that what can be tied, can be untied later.[113]

Following the revelations of the 1970s, specific regulatory procedures were adopted for ensuring that informed consent was obtained in cases where federal funds subsidized sterilization procedures. It is difficult to determine the effectiveness of these regulations. The Department of Health and Human Services, the federal agency involved, does not directly audit service providers for compliance. It only evaluates the effectiveness of state computer systems for monitoring nonreimbursable sterilization procedures (i.e., those not in compliance with federal regulations). Not even these data are published or made public. No monies have been available for large-scale studies of effectiveness. Without adequate data, there is no basis for future evaluation or policy formation.[114]

Physicians' attitudes are one reason why some poor women and women of color may still be subject to involuntary sterilization. Some doctors, oblivious to their patients' preferences and cultural differences in attitudes towards family size and legitimacy, regard "excessive" childbearing by poor women and women of color as deviant or inappropriate. Doctors convey these attitudes to their patients, who come to believe that they will not be accepted as patients unless they conform to the medical profession's analysis of their behavior and problems.[115] Classism and racism lead physicians and other health care providers to urge sterilization on patients they believe incapable of using other methods effectively. For example, a Boston clinic serving primarily black clients reported that 45 percent of its black clients "chose" tubal ligation as a method of birth control after their first child was born.[116]

Another reason why poor and minority women are coerced into sterilization is the lack of alternatives to publicly funded health services. Sterilization services are provided by states under the Medicaid program, and the federal government reimburses states for 90 percent of their expenses. Most states do not provide abortion services as part of the Medicaid program, and following the enactment of the Hyde Amendment, federal monies are no longer generally available to those that do. Moreover, additional financial pressures are at work. In understaffed and underfunded agencies, sterilization is seen as freeing resources and cutting case loads because sterilized patients will no longer need contraceptive or obstetric care.[117] As feminists have argued, subtle coercion by care providers may often confirm the view of the welfare patient that sterilization is the only alternative to impersonal, degrading reproductive health care that often denies access to safe, effective contraception or to abortion.[118] The absence of information about and access to other contraceptive techniques provides further pressure.[119]

Sterilization rates as high as 65 percent of Hispanic women have been reported in the northeast United States.[120] A study of women 15 to 44 years old in the United States showed that 34 percent of married black women as compared to 25 percent of white women had been sterilized by 1982.[121] Black women of all marital statuses were shown to be more likely than white women to have used sterilization as a method of contraception.[122] The highest rates of hysterectomy and tubal ligation in the United States are in southern black areas.[123] Native American women fare no better. It is estimated that between 30 percent and 42 percent of all Native Americans have been sterilized, resulting in a steadily declining birth rate.[124]

Unnecessary surgery is another form of sterilization abuse. The number of conditions that require removal of female reproductive organs are relatively few, and fibroid tumors are not generally considered in that category. However, many women have gone to a physician for treatment of fibroids--a condition especially common among black women--and been told that a hysterectomy was the only cure.[125] In other cases, doctors have advocated hysterectomy for women they perceive as having too many children,[126] or to provide interns and residents with experience.[127] Unfortunately, such practices appear to persist despite the restrictions on Medicaid reimbursement of hysterectomies for these purposes in effect since 1975.[128]

Sterilization resulting from surgery performed for other reasons may also be seen as a form of sterilization abuse suffered disproportionately by minority communities.[129] The need for such surgery results at least in part from poor access to quality gynecological care and a lack of preventive programs that educate women about gynecological problems.[130] These problems include sexually transmitted diseases, infections, pelvic inflammatory disease, and potentially harmful contraceptive methods, such as IUDs and Depo-Provera, all of which can lead to infertility or to surgery resulting in infertility.[131] Difficulty in obtaining diagnoses and treatment, poor operative care, inadequate postoperative care, and poor health prior to the surgery can all contribute to infertility.[132]

High unemployment and profound economic insecurity have likewise led to the "choice" of sterilization as a method of contraception. Women who feel they must forego permanently the possibility of having children they would like to have because they cannot afford a child or even the cost of delivery have been deprived of their reproductive choice. Moreover, in communities deprived of many other opportunities and sources of satisfaction, the inability to have children can be particularly painful.

Sterilization abuse, like other types of interference with reproductive choice experienced by poor women and women of color, ranges in form from the blatant to the subtle. Legal provisions, the medical profession, and socioeconomic conditions all play a role. Affirmative action is necessary to combat the racism and classism that contribute to this and other kinds of reproductive abuse, and to ensure the true informed consent and access to information and services that genuine reproductive choice requires.

Recommendations

The recommendations listed under the previous topics all address the issue of interference with reproductive choice. In addition:

1. State and federal sterilization procedure guidelines should be strictly monitored and enforced. Sanctions should be applied to violators.
2. Legislation should be enacted to protect women against unnecessary surgery, including hysterectomy.
3. Women should be given access to information about alternative forms of contraception other than sterilization.
4. Women with communicable diseases or genetic defects should not be pressured to become sterilized or have abortions.
5. Women should have access to quality gynecological care to prevent infertility and other gynecological problems.

Alternative Modes of Reproduction

The medical profession's response to the problem of infertility has focused on high-tech modes of reproduction often involving third parties. Most of these infertility procedures are extremely costly and have a disappointingly low success rate. Nevertheless, research funding and enthusiasm are geared toward helping a small number of women conceive and bear healthy babies in this way. Basic societal and medical problems, which have a much broader impact on women's ability to bear healthy children, are not a priority for the medical or scientific community. Many of these problems could be addressed without recourse to new reproductive technologies. Improved education, health care, nutrition, and working conditions, for example, could dramatically enhance the ability of poor women and women of color to bear healthy children. Therefore, the issue of resource allocation is a fundamental concern for poor women and women of color whenever reproductive technologies are being considered.

Infertility in the Minority Community

Infertility rates vary by race. Black women have an infertility rate one and one-half times higher than that of white women.[133] The risk factors contributing to infertility among minority couples are genetic disorders such as sickle cell anemia, alcohol and drug abuse, nutritional deficiencies, infectious diseases such as gonorrhea and pelvic inflammatory disease (PID) that have gone untreated, and infection after childbirth or subsequent to a poorly performed abortion.[134] For example, a study of PID found that rates of the disease among "nonwhite" women were twice those among white women. The differential may be significantly greater since the study covered only visits to doctors' offices, and outpatient and emergency room visits were not included. It was also shown that about 25 percent of women with one or more PID episodes later experienced chronic pelvic pain, infertility, and ectopic pregnancy (which can also result in infertility).[135]

Other reasons for high infertility among poor women and women of color include sterilization abuse,[136] hysterectomies,[137] IUD and birth control pill usage,[138] lack of access to medical treatment,[139] deleterious environmental and working conditions,[140] and unnecessary surgery and medical experimentation.[141] Nonwhite races, particularly blacks, are at a further disadvantage because of their higher rates of fetal mortality, which were approximately 1.5 times greater than those of whites in 1981. Fetal mortality arises from a number of causes, including congenital anomalies and conditions related to the mother's health status and the adequacy of her prenatal care.[142]

Access to New Reproductive Technologies

Because the health care in the United States is organized on a for-profit basis, the needs of poor and minority women receive little attention generally.[143] Access to reproductive technologies is particularly problematic. For example, each in vitro fertilization procedure costs from $3500 to $5000 per attempt,[144] and most couples make several attempts at achieving pregnancy.[145] Private insurers rarely cover such procedures, so the new technologies are often beyond the reach of even middle-class women.[146] In regard to poor and minority women, governmental concern appears to be focused primarily on reducing fertility rather than improving it. Although family planning services are mandated to provide services to women and their families to aid conception as well as prevent it, little or no infertility care is available from public sources.[147] Nor are government efforts to improve the health of poor and minority women and their children adequate to eliminate the factors described above that cause infertility.

Social criteria for services impose additional barriers to access. Most in vitro clinics are highly selective, accepting only married, heterosexual women with adequate resources.[148] This is a problem for many potential clients since, for example, in 1985, 57.3 percent of black women were not married.[149] Socioeconomic grounds should not play a role in assessing whether a person should be helped to conceive a child any more than they should play a role in determining who is entitled to custody of existing children. Certainly, criteria for determining access to the new reproductive technologies should be no more stringent than criteria for adoption in different states. To allow providers to decide who are "deserving and appropriate parents"[150] is a dangerous practice, particularly in view of this country's experience with eugenics.

Given the difficulty poor women and women of color experience in obtaining access to new reproductive technologies, they are highly vulnerable to "consenting" to experimental procedures in order to gain access. It is imperative that poor women and women of color not be given access only to be used as subjects of experiments, and that technological developments not be tested on poor women and women of color without their genuine consent.[151]

Sociocultural Factors

Childlessness is a very serious concern in communities of color. As a result of cultural norms and restricted opportunities for women to have a profession or a career, motherhood and family life are generally valued very highly. Therefore, losing the option of procreating and parenting can be devastating to a poor woman or a woman of color.[152]

Sociocultural factors make it difficult for some poor women and women of color to obtain treatment for infertility. Shame and fear may make it hard for them to discuss their reproductive difficulties. Moreover, physicians and other health personnel may want to involve the woman's partner in the treatment process and expect him to be knowledgeable about her menstrual cycle, and other aspects of her physical condition. This may present particular problems for some poor women and women of color, who live in cultures in which certain subjects are essentially taboo and distinct roles are assigned to men and women. Some women fear losing their mate if they prove infertile. At the same time, infertility testing may be problematic for men in such cultures, especially when their feelings of masculinity are at least partially based on their ability to father children.

Sociocultural factors also shape alternatives to reproductive technologies as the response to unwanted childlessness. Adoption in poor or minority communities is not always the same as the adoption referred to by the agencies serving the primarily white middle class. In many instances, poor or minority women have not formally adopted children, but have raised the children of other family members who, for a variety of reasons, were unable to care for them. This extended family concept is prevalent in many cultures of color and in some white ethnic communities.[153]

Formal adoption is less common, primarily because, until recently, adoption agencies excluded people of color from the adoption process and imposed other socioeconomic barriers. Minorities believe in the concept of adoption, but may be wary of the bureaucracy of adoption administrators.[154] In the past 10 years, more adoption officials have attempted to recruit minority families to adopt children, and formal adoption is becoming a more viable alternative to childlessness.[155]

Surrogacy

Surrogacy arrangements have stark implications for women of color and poor women.[156] Although poor white women may currently be in demand as surrogate biological mothers, it is likely that poor women of color will run a far greater risk of being recruited as surrogate carriers once the process of transferring embryos comes into greater use. The prospect of women of color carrying white babies is frightening and merits serious and immediate consideration. The idea of using poor and minority women as incubators for the white upper and middle classes raises serious moral and practical questions. It is troubling, just as the phenomenon of whites using blacks and other women of color as wet-nurses and childrearers was troubling. Nor are the psychological effects of surrogacy arrangements known. What will the psychological impact be on a woman who carries the child of another couple for the duration of a pregnancy? What will be the effect on her family? What will be the effects on the couple using a poor or minority woman for this service?

Another concern is how much power a childless couple will exert over the poor or minority woman during pregnancy. There are many cultural differences between white middle-class couples and the members of poor or minority communities. Once she has agreed to carry another couple's child, a woman may be asked to adopt a completely different life-style, not because there is anything wrong with her own, but because the couple is not familiar with it. Thus, she could be asked to turn over her eating, drinking, social, and sexual habits to the control of the couple employing her.

Despite these problems, many poor women may agree to such arrangements, in part, because they have so few other job options, particularly paying $10,000 to $15,000 for nine months' work. It may be difficult for a woman in financial distress to weigh accurately the advantages and disadvantages of surrogacy.

Surrogacy has raised many difficult questions. Should surrogacy be regarded as exploitation or gainful employment for poor or minority women? Should it be banned by statute? In considering the problem, one must ask whether exploitation of poor women is inevitable and whether surrogacy could be regulated to protect the surrogate mothers. What other employment options can be made available for poor women so that they do not have to become surrogates? Is it right to have a law telling a woman what she can and cannot do with her body? Would surrogacy without pay be more or less exploitative? Despite the divergence of views on such questions, there is agreement that improving the underlying life conditions for poor women and women of color would be likely to render many of these questions moot.

Recommendations

1. Develop and publicize health education programs and services geared toward preventing infertility, including improved gynecological care and increased education about sexually transmitted diseases.
2. Increase family planning funding and services to provide culturally appropriate care to infertile women.
3. Redirect funding from high-technology reproductive procedures to the prevention of infant mortality and other crucial social issues.
4. Take legal action to prevent health providers from withholding alternative reproductive services from female clients based on their marital status, sexual orientation, age, or income level.
5. Encourage private insurers to fund alternative reproductive procedures.
6. Develop legislation and regulatory provisions to protect potential surrogate mothers from exploitation.

Conclusion

Poor women and women of color have pressing needs for health services, including reproductive health services. They also have a history of maltreatment by the health care delivery system. For such women, making existing rights a reality and meeting the challenges posed by new modes of reproduction and reported advances in prenatal and perinatal technology are crucially related to these needs and history. Reproductive laws and polices for the 1990s must respond to the concerns of all women. The laws for the next decade must:

1. Widen the dissemination of education and information concerning reproductive health.
2. Augment private and public funding to allow financial access to health services.
3. Bar unnecessary and forced medical and surgical treatments.
4. Prohibit discriminatory and eugenicist bias or practices in health care delivery.
5. Ensure confidentiality in health care records.
6. Protect the patient's right to informed consent and informed refusal.
7. Broaden education about reproductive health hazards in the workplace.
8. Facilitate the delivery of culturally appropriate health services to ensure effective health care.
9. Enhance the recruitment of people of color to train as health care providers.
10. Guarantee equity of access to all new reproductive technologies, accompanied by equal protection against abuses of these technologies.
11. Mandate the collection of data on local, state, and federal levels on the reproductive health status of women of color, including Hispanics, Asians, blacks, and Native Americans.
12. Increase preventive health care measures to counter the health problems caused by structural socioeconomic determinants, so that there will be less need to resort to high-tech therapies as solutions.

To bring about the enactment of such laws, more information regarding the views, the life experiences, and the circumstances of poor women and women of color must be made available. Most importantly, poor women and women of color must be included in the decision-making process, so that more attention will be paid to their needs.

Appendix A

"Recommendations" from "An Asian/Pacific Islander Health Agenda for California - 1985"[157]

Based on the experience gained from various Asian/Pacific Islander communities, we are now beginning to understand some of the essential ingredients for the successful implementation of health services within our communities. Because of the limited resources available in the Asian/Pacific Islander community, much support is still needed from federal, state, and local governments.

Federal Level

1. Collection of data on bilingual/bicultural needs is essential in order to have greater statistical documentation of Asian/Pacifics and to eliminate the invisibility of health needs that exists today.
2. Federal funds should be allocated to implement Asian/Pacific health services which are culturally and linguistically relevant to the community.
3. For incentive purposes, Medicaid should establish reimbursement policies for interpreters and service providers who have a bilingual staff.
4. The federal government, particularly the Department of Health and Human Services, must provide support and incentive to the Asian/Pacific community-based programs currently utilized in the Asian/Pacific community.
5. The federal government should develop mandates to the states, so that federal funds being administered by the states [will] be appropriately allocated for program development in the Asian/Pacific community.
6. Federal funding intended for overall refugee allocations should not be limited to job promotion purposes, but should also be specifically targeted for health care.
7. Federal grant support should be provided to conduct research, with respect to specific problems of each ethnic group, into the feasibility of culturally sensitive Asian/Pacific health services. Also, funds should be allocated for research on diseases more prevalent in the Asian/Pacific community.
8. In the development of prepaid health plans, issues relating to the needs of the Asian/Pacific community should be addressed.
9. On the federal level, representatives of the Asian/Pacific community should be placed in top-level staffing positions, as well

as on special commissions, in order to ensure appropriate advisement relating to health care issues.

10. Federal support should be provided to encourage medical curricula to stress culturally sensitive issues relative to health care delivery.

11. Demonstration centers should be established by federal funds to provide technical assistance to communities with underdeveloped Asian/Pacific health resources.

12. The federal government should inform Asian/Pacific communities of the available training and fellowship opportunities funded by the National Institutes of Health and the Department of Health and Human Services.

13. Federal regulations should be revised to allow flexibility for the special needs of the Asian/Pacific Islander population (e.g., MUA description criteria, reimbursement rates, and allowable service modalities).

State Level

1. Collection of data on bilingual and bicultural needs by ethnicities is essential in order to have greater statistical documentation of Asian/Pacifics and to eliminate the invisibility of health needs that now exists.

2. A commission composed of representatives of Asian/Pacific Islander communities should advise the state administration on regulations, policies, problems, and areas of concern with regard to appropriate and adequate provision of health care services.

3. Representatives of the Asian/Pacific communities should be placed in management-level staffing positions to ensure appropriate advisement and follow-up relating to health care issues.

4. State licensing of foreign-trained medical providers should be examined and revised.

5. State funding should be provided for the implementation of culturally sensitive health care services. Bilingual staffing and translated materials are imperative.

6. As federal funding for refugee services is reduced, state agencies should assume responsibility for the continuation of such services.

The lack of access to health care services is, in effect, denial of such services. The following, which the state should be responsible for funding, are proposed to remedy or alleviate the problem of access to services:

7. Availability of bilingual/bicultural staffing.

8. Availability of interpreters and health education materials in primary languages (culturally sensitive materials).

9. Community-based facilities and mobile clinics to provide direct services, with designated service sites, a primary health care team, and bilingual and bicultural health providers.

10. Initiation and expansion of community education outreach efforts through vernacular (language native to particular ethnicities) media.

11. Cultural relevant treatment modalities.

12. The state should require and monitor funded programs located in or near Asian/Pacific communities to ensure adequate services and the provision of bilingual staffing.

Local Level

1 In order to have greater statistical documentation and to eliminate the invisibility of health needs that exists today, it is essential that data be collected, by ethnicities, on the bilingual/bicultural needs of Asian/Pacifics.

2. A commission, comprised of representatives of Asian/Pacific communities, should advise the county administration on regulations, policies, problems, and areas of concern with regard to appropriate and adequate provision of health care services.

3. Representatives of the Asian/Pacific communities should be placed in top management-level staffing positions to ensure appropriate advisement and follow-up relating to health care issues.

4. Local mandates must require staffing of bilingual personnel and translated materials in health care agencies serving Asian/Pacific clients with limited English language skills.

5. As federal funding for refugee services is reduced, county agencies should assume the responsibility for the continuation of such services.

6. Local agencies must take the initiative to inform the Asian/Pacific communities about the availability of health care services, how to gain access to such services, and how to pay for them.

Public and private health providers located in or near Asian/Pacific communities should provide the following:

7. Bilingual and bicultural staffing.

8. Interpreters and health education materials that are culturally sensitive and available in the primary ethnic languages of the Asian/Pacific communities.

9. Community-based facilities and mobile clinics, with designated service sites, a primary health care team, and bilingual/bicultural health providers, to provide direct services.

10. Initiation and expansion of community education outreach efforts through vernacular media.

11. Culturally relevant treatment modalities.

Notes

1. See Alliance Against Women's Oppression, "Caught in the Crossfire: Minority Women and Reproductive Rights," (Jan. 1983), p. 1.

2. The areas are: time limits on abortion, prenatal screening, fetus as patient, reproductive hazards in the workplace, interference with reproductive choice, and alternative modes of reproduction.

3. For purposes of this paper, women of color will be defined as black, Hispanic, Asian, and Native American women.

4. The information for this paper was obtained through the help of health care providers, physicians administrators, and activists who are themselves women of color. The ideas in this paper are still evolving as our understanding of these issues grows. As we receive feedback from our communities, our conclusions and recommendations must be subject to revision.

Other sources providing information include Department of Health and Human Services data, individual research studies, conference proceedings, and testimony before national and governmental bodies.

5. Urban-rural differences are also important, but again there is little data. This need too must be addressed.

6. Clarke, Adele, "Subtle Forms of Sterilization Abuse: A Reproductive Rights Analysis," in Test-Tube Women, Arditti, R., Klein, R. D., and Minden, S., eds. (1984), p. 199; "Birth Control Blamed for Health Problems," Intern Extra (April 7, 1983), p. 60. The United States Indian Health Service continues to give Depo-Provera to mentally retarded Native American women in this country although it has been banned for contraceptive use in the United States since 1984. "Native Americans Given Depo-Provera," Listen Real Loud 8:1 (Spring 1987), p. A-7.

7. Levin, Judith, and Taub, Nadine, "Reproductive Rights," in Women and the Law, Lefcourt, Carol, ed. (1987), pp. 10A-27-28.

8. Blaine, Elaine, Alan Guttmacher Institute, "The Impact on Women of Color of Restricting Medicaid Funding for Abortion," testimony presented to the National Women's Health Network, 1985.

9. Arnold, Charles B., "Public Health Aspects of Contraceptive Sterilization," in Behavioral-Social Aspects of Contraceptive Sterilization, Newman, S. H., and Klein, Z. E., eds. (1978).

10. Mullings, Leith, "Women of Color and Occupational Health," in Double Exposure, Chavkin, Wendy, ed. (1984).

11. U.S. Department of Commerce, Current Population Reports, Series P-60, No. 152 (June 1986), Table 8, p. 28.

12. Spokesperson for the Alan Guttmacher Institute, telephone conversation, Sept. 23, 1987.

13. Henshaw, Stanley K., "Characteristics of U.S. Women Having Abortions, 1982-1983," Family Planning Perspectives 19:1 (1987). Abortion rate data must be understood in the context of nonwhite women's significantly higher fertility rate, a rate that was, for example, 35 percent higher than that for white women in 1981. Centers for Disease Control, op. cit. below, note 28.

14. In 1982, 50 percent of all Medicaid-eligible women at the clinic had abortions at 10 weeks or later. The abortion rate at 10 weeks or later for women not eligible for (or not needing) Medicaid was only 37 percent. Post-13-week abortions were excluded from this portion of the study. Ibid. at 172.

15. Henshaw, Stanley K., and Wallisch, Lynn S., "Medicaid Cut-Off and Abortion Services for the Poor," Family Planning Perspectives 16:4 (July/August 1984), pp. 170, 178.

16. An unpublished survey of welfare caseworkers in the northern counties of New Jersey indicated that, out of the 42 caseworkers interviewed, only six demonstrated adequate knowledge of the availability of state-funded Medicaid reimbursement for abortion. Cohen, Sherrill, "Welfare Caseworkers and Information about Restored State Medicaid Funding for Abortion: Summary of Sample Survey,"

American Civil Liberties Union of New Jersey, (May 1984). The Asian Pacific Planning Council Health Task Force worked to identify and articlate the needs of that community. For its recommendations, see Appendix A below.

17. Henshaw, Stanley K., Forrest, J. D., and Blaine, Elaine, "Abortion Services in the United States, 1981 and 1982," Family Planning Perspectives 16:3 (May/June 1984), p. 127. Of the providers in five states surveyed (Connecticut, Maryland, Massachusetts, New Jersey, and Pennsylvania), 10 percent were unaware that reimbursement was available.

18. Ibid.

19. Ibid.

20. Henshaw, Stanley K., Forrest, J. D., and Van Vort, J., "Abortion Services in the United States, 1984 and 1985," Family Planning Perspectives 19:2 (March/April 1987), p. 65.

21. "In 17 states, fewer than half of all women of reproductive age live in counties with identified abortion providers. In four states--Kentucky, Mississippi, South Dakota, and West Virginia--less than 30 percent of such women live in counties with any abortion provider." Kentucky and West Virginia, of course, have significant poor populations, and Mississippi has both a large poor and a large black population. Henshaw et al., "Abortion Services in the United States, 1981 and 1982," p. 122.

22. Henshaw et al., "Abortion Services in the United States, 1984 and 1985," p. 65.

23. In 1982, the U.S. Department of Health and Human Services (DHHS) promulgated and implemented regulations designed to bring abortion policy in the Indian Health Service in line with that in other DHHS-administered health programs. Under these regulations, abortions may be performed only when the woman's life is endangered. Alan Guttmacher Institute, Issues in Brief 5:6 (Jan. 1985), p. 12.

24. See Monmouth County Correctional Institution Inmates v. Lanzaro, 834 F.2d 326 (3d Cir. 1987).

25. See 28 C.F.R. § 551.23 (Dec. 30, 1986) providing that during fiscal year 1987 the Bureau of Prisons may pay for an abortion only where the life of the mother would be endangered if the fetus were carried to term or if the pregnancy is the result of rape.

26. Leona Pang, Executive Director of the Asian Health Project in Los Angeles, reports that bilingual and bicultural program staffing, interpreters, culturally sensitive health education materials in primary languages, use of the media, and mobile units at designated service sites would greatly reduce the problem of access to reproductive health care in the Asian/Pacific community. The Asian Pacific Planning Council Health Task Force worked to identify and articulate the needs of that community. For its recommendations, see Appendix A below.

27. "Rule On Abortion Counseling Is Blocked," New York Times, Feb. 17, 1988, p. A10.

28. Centers for Disease Control, Abortion Surveillance 1981 (Nov. 1985), Table 14, p. 37.

29. Ibid.

30. Twenty-three states have some form of legislation mandating parental involvement with a minor's abortion decision. See American Civil Liberties Union, Reproductive Freedom Project, "Fact Sheet," updated December 1987.

31. Alan Guttmacher Institute, Teenage Pregnancy: The Problem That Hasn't Gone Away (1981), (an analysis of data from 1970).

32. U.S. Department of Health and Human Services, Report of Secretary's Task Force on Black and Minority Health, Executive Summary, Vol. I (Aug. 1985), p. 71.

33. Henshaw et al., "Abortion Services in the United States, 1984 and 1985," p. 68.

34. Ibid.

35. Cates, W., Jr., and Grimes, D. A., "Morbidity and Mortality in the United States," in Abortion and Sterilization, Hodgson, Jane, M.D., ed. (1983), pp. 158-159.

36. Boston Women's Health Book Collective, The New Our Bodies, Ourselves (1986), p. 303.

37. Where D & Es are available, they may cost more than induction abortions. Ibid., pp. 302-304.

38. Ibid.

39. Scott, Julia, "How Are Changing Values Affecting Reproductive Health Care for Adolescents, Women of Color, and Low-Income Women," presentation to the National Council on Foundations Meeting, Kansas City, Missouri (April 1986), pp. 1-2.

40. Rapp, Rayna, "XYLO: A True Story," in Arditti, Test-Tube Women, p. 314.

41. American Academy of Pediatrics, and American College of Obstetricians and Gynecologists, Guidelines for Perinatal Care (1983), p. 209.

42. Title 11 of the Public Health Service Act (the Genetic Diseases Act) provided for direct federal support until the Reagan administration altered the regulations. The state governments now receive these funds through the categorical block grants (Maternal and Child Health Program). Disbursement of funds is discretionary within a range of services. Spokesperson for the Alan Guttmacher Institute, telephone conversation, January 16, 1988.

43. Alan Guttmacher Institute, The Financing of Maternal Care in the U.S. (December 1987). Eligibility for free or assisted care through the MCHP is income-related with considerable variations in thresholds by state.

44. "State Fiscal Year Expenditures for Prenatal Genetics Screening Reported to Be Only $8.6 Million," Family 10:1 (February 1981).

45. Thirteen states and the District of Columbia provide funding for abortion through Medicaid. North Carolina also provides funding through the State Abortion Fund, which has different requirements than Medicaid. S. K. Henshaw, Deputy Director of Research, Alan Guttmacher Institute, telephone conversation, January 7, 1988.

46. The Census Bureau reports that in 1983, 14 percent of whites, 21.8 percent of blacks, and 29.1 percent of Hispanics were not covered by health insurance of any kind. "15 Percent of Americans Found to Lack Health Insurance," New York Times, Feb. 18, 1985, p. A13.

47. There is only a four-week period during pregnancy when physicians recommend that amniocentesis be performed for genetic screening. See above, note 41. Women eligible for, but not enrolled in, Medicaid or the MCHP can easily pass the critical period enmeshed in the bureaucratic process of application.

48. Personal communication from Lorna Wilkerson, former genetics counselor at Harlem Hospital, New York City, June, 1986.

49. See, e.g., Kassam, M.J.G., Fernhoff, A., Brantley, K., Carroll, L., Zacharias, J., Klein, L., Priest, J., and Elsas, L., "Acceptance of Amniocentesis by Low-Income Patients in an Urban Hospital," American Journal of Obstetrics and Gynecology 138:1 (1980), pp. 11-15.

50. Ibid.

51. Wilkerson, op.cit.

52. Anne McDonald, Head of Genetics Counseling Unit at the Hospital of the University of Pennsylvania, telephone conversation, August, 1986.

53. Wilkerson, op.cit.

54. U.S. Department of Health and Human Services, Task Force on Black and Minority Health, Vol. VI, Low Birth Rate and Infant Mortality (January 1986), pp. 16, 89, 132-133; see also "Prenatal Care: Medicaid Recipients and Uninsured Women Obtain Insufficient Care," GAO Publication (September 30, 1987), Doc. GAO HRD 87137.

55. See Kassam et al., op. cit., pp. 13-14.

56. Ibid., p. 13.

57. See generally, Rapp, Rayna, "Moral Pioneers: Women, Men and Fetuses on the Frontier of Reproductive Technology," in Embryos, Ethics and Women's Rights: Exploring the New Reproductive Technologies, Baruch, E. H., D'Adamo, A. F., Jr., and Seager, J., eds. (1988), pp. 109-124.

58. Ibid., p. 117.

59. Anne McDonald, op. cit., reported, for example, that women who are not fluent in English and their counselors find the process overly difficult and time consuming.

60. Asian-Pacific Health News, Asian Health Project, Los Angeles, (Spring 1985), p. 3.

61. Ibid.

62. See Vargas, Maria G., and Wilkerson, Lorna, "Defining our Cultures and Bridging the Gap," in Strategies in Genetic Counseling: Religious, Cultural and Ethnic Influences on the Counseling Process, Magyari, P., Paul, N., and Bieseker, B., eds., March of Dimes/Birth Defects Foundation, "Birth Defects" Original Articles Series, Vol. 23, No. 6 (1987), pp. 167-187.

63. Farfel, Mark R., and Holtzman, Neil A., "Education, Consent and Counseling in Sickle Cell Screening Programs: Report of a Survey," American Journal of Public Health 74:4 (April 1984), pp. 373-375.

64. Ibid.

65. Sue, Donald W., Counseling the Culturally Different In Theory and Practice (1981), pp. 10-11.

66. Vargas and Wilkerson, op. cit.

67. Hubbard, Ruth, and Henifin, Mary Sue, "Genetic Screening of Prospective Parents and of Workers," in Biomedical Ethics Reviews, Humber, James M., and Almeder, Robert T., eds. (1984), pp. 92, 100.

68. Ibid.

69. Centers for Disease Control, "Recommendations for Assisting in the Prevention of Perinatal Transmission of Human T-Lymphotropic Virus Type III Lymphadenopathy-Associated Virus and Acquired Immune Deficiency Syndrome," Morbidity and Mortality Weekly Report 34:48 (Dec. 1985), pp. 721, 724.

70. Hubbard and Henifin, op. cit., p. 75.

71. George C. Cunningham, former Chief of Child and Maternal Health, California Department of Health Services, telephone conversation, January 25, 1987.

72. Ibid.; see also, World Health Organization Collaborating Center in Perinatal Care, Health Services Research in Maternal and Child Care, "Unintended Pregnancy and Infant Mortality/Morbidity," in Closing the Gap: The Burden of Unnecessary Illness, Amler, Robert W., and Dull, Bruce, eds. (1987).

73. The neonatal mortality rate for blacks was 13.4 per 1000 births; that for whites was 7.1 per 1000 births. Report of Secretary's Task Force on Black and Minority Health, Vol. VI, p. 55.

74. The overall infant mortality rate for Native Americans in 1978-80 was 14.6 per 1000 births. Ibid., p. 61. It should be noted that it is difficult to develop accurate statistics because Native Americans do not always identify themselves as such in official documents. U.S. Indian Health Service, Indian Health Service Chart Book Series, No. 421-166-4393 (June 1984), "Introduction."

75. Report of Secretary's Task Force on Black and Minority Health, Vol. VI, p. 60.

76. Ibid. at pp. 173-174, 12.

77. Report of Secretary's Task Force on Black and Minority Health, Vol. I, pp. 180-181.

78. Commission for Racial Justice, United Church of Christ, Toxic Wastes in the United States, 1987.

79. See discussion in the following section of this essay.

80. Kolder, V., Gallagher, J., and Parsons, M. T., "Court-ordered Obstetrical Interventions," 316 New England Journal of Medicine (May 7, 1987), pp. 1192-1196. Massachusetts reports that black and Hispanic women had the highest incidence of cesarean sections in the state. Massachusetts Department of Public Health, Cesarean Sections in Massachusetts (1982). There has been a similar finding in Baltimore. Gibbons, L. K., "Analysis of the Risk in C-Section in Baltimore," Doctoral Dissertation, School of Hygiene and Public Health, Johns Hopkins University, Baltimore, 1976. The 1972 National Natality survey found the same for all women of color. Black,

P. J., "Type of Delivery Associated with Social and Demographic Maternal Health, Infant Health, and Health Insurance Factors: Findings from the 1972 U.S. National Natality Survey," paper presented at the American Statistical Association Meeting, Chicago, Ill., August, 1977.

81. Janet Mitchell, M.D., former Director of Maternal and Fetal Medicine, Boston City Hospital, telephone conversation, 1986.

82. "Depo-Provera and the Indian Women," New York Times, Aug. 17, 1987, p. A16.

83. Corea, Gena, The Hidden Malpractice: How American Medicine Mistreats Women (1985), pp. 200-203; Scully, Diana, Men Who Control Women's Health; The Miseducation of Obstetrician-Gynecologists (1980), pp. 120-140 (see discussion of women as "teaching material").

84. One indication of the general health problems people of color face is that black workers are almost one and one-half times more likely than white workers to be severely disabled by job-related injuries and illnesses. Chicago Reporter 10:3 (March 1981), p. 2.

85. U.S. Commission on Civil Rights, Health Insurance Coverage and Employment Opportunities for Minorities and Women, Clearinghouse Publication 72 (Sept. 1982).

86. As Leith Mullings points out, even in two-parent households, the black woman often bears greater responsibility for providing sustenance to her family than the average white woman. Op. cit., p. 125.

87. Navarro, Vicente, Medicine Under Capitalism (1976), p. 38. Within hospitals, women of color work primarily in food services, laundry, and housekeeping, and as nurse's aides.

88. Massachusetts Coalition for Occupational Health and the Boston Women's Health Book Collective, Our Jobs, Our Health (1983), pp. 15, 40-42.

89. Schnoor, Teresa A., and Stellman, Jeanne, unpublished study cited in Women's Occupational Health Resource Center News, No.4 (Sept. 1982), p. 1. Breast cancer may be a particular risk for black women.

90. Brown, Joanna, and Scheir, Ronnie, "Workplace May Be Hazardous to Health of Blue Collar Minorities," The Chicago Reporter 10:3 (March 1981), p. 1.

91. "Women's Occupational Health Resource Center Fact Sheet," (March 1980), p. 1.

92. "The Jobs Puerto Ricans Hold in New York City," Monthly Labor Review 98:10 (Oct. 1975), pp. 12-16.

93. Abeles, Schwartz, Haukel, and Silverblatt, Inc., The Chinatown Garment Industry Study, A Report Submitted to Local 23-28, International Ladies Garment Workers Union (June 1983).

94. Mullings, op. cit., p. 129.

95. Ibid.

96. Ibid.

97. Westcott, Diane, "Blacks in the 1970s: Did They Scale the Job Ladder," Monthly Labor Review 105:6 (June 1982).

98. Blair, A., DeConfle, P., and Grassman, D., "Causes of Death Among Laundry and Dry Cleaning Workers," American Journal of Public Health 69:5 (May 1979), pp. 508-511.

99. Women's Occupational Health Resource Center News 3:5 (Sept./Oct. 1981), p. 4.

100. Kutz, F. W., Yobs, A. R., and Strassman, S. C., "Stratification of Organochlorine Insect Residues in Human Adipose Tissue," Journal of Occupational Medicine 19:9 (1977), pp. 619-622; Davis, Morris, "The Impact of Workplace Health and Safety on Black Workers: Assessment and Prognosis," Labor Law Journal 31:12 (Dec. 1980), p. 724.

101. Naeye, Richard, and Peters, Ellen, "Working During Pregnancy: Effects on the Fetus," Pediatrics 69:6 (June 1982), pp. 724-727.

102. In 1977, women workers in the lead pigment department of an American Cyanamid plant in West Virginia were given the "choice" of being sterilized or moving to lower paying jobs. Several women "chose" to become sterilized and regretted their decision. Laws against sterilization did not protect these women, because theoretically they had chosen that option. Clarke, op. cit., p. 188. These issues are discussed further in Hubbard and Henifin, op. cit., pp. 107-111.

103. Bloom, Arthur D., ed., Guidelines for Studies of Human Populations Exposed to Mutagenic and Reproductive Hazards (1981), pp. 41-42; Manson, Jeanne M., and Simms, Ruth, "Influence of Environmental Agents in Male Reproductive Failure," in Work and the Health of Women, Hunt, Vilma, ed. (1979), pp. 155-179.

104. For a more detailed discussion of the above, see Hubbard and Henifin, op. cit., pp. 79-112.

105. See generally Joan Bertin's position paper, "Reproductive Hazards in the Workplace," in this book.

106. Ibid.

107. See generally, The Childbearing Rights Information Project, For Ourselves, Our Families and Our Future, The Struggle for Childbearing Rights (1982).

108. Ibid., p. 2.

109. Levin and Taub, op. cit., pp. 10A-27-28.

110. Alliance Against Women's Oppression, "Caught in the Crossfire: Minority Women and Reproductive Rights," (Jan. 1983), p. 5.

111. Walker v. Pierce, 560 F.2d 609 (4th Cir. 1977).

112. Clarke, "Subtle Forms of Sterilization Abuse," cited in n. 6 above, p. 197.

113. One study reported that 45 percent of the sterilized black women interviewed did not know the procedure was irreversible. Forty percent of the sample said they regretted having been sterilized. Clarke, op. cit., p. 195.

114. Olive Karen Stamm, Committee on Sterilization, National Women's Health Network, telephone conversation, January 17, 1987.

115. Clarke, op. cit., p. 201.

116. Martha Eliot Health Center, Reproductive Health Report, 1985.

117. Clarke, op. cit., p. 200.

118. Corea, Gena, The Mother Machine: Reproductive Technologies from Artificial Insemination to Artificial Wombs (1985), pp. 204-205.

119. Indeed, there is evidence that vigorous contraceptive counseling reduces sterilization rates and increases usage of other forms of birth control. Clarke, op. cit., pp. 196-197.

120. In 1981, a psychologist found that 65 percent of the Puerto Rican women in Hartford and 55 percent of all Hispanic women in Springfield, Massachusetts had been sterilized. Personal communication from Dr. Vickie Barres, Brookside Family Health Center, Jamaica Plain, Boston, April 1985.

121. U.S. Department of Health and Human Services, National Center for Health Statistics, Data from the National Survey of Family Growth, "Contraceptive Use, U.S., 1982," Series 23, No. 2 (Sept. 1986), p. 32.

122. Ibid., p. 11.

123. Centers for Disease Control, Surgical Sterilization Surveillance: Tubal Sterilization and Hysterectomy in Women Aged 15-44 from 1979-1980 (Sept. 1983).

124. Alliance Against Women's Oppression, "Caught in the Crossfire: Minority Women and Reproductive Rights," (Jan. 1983), p. 5.

125. Scully, op. cit., p. 223.

126. Clarke, op. cit., pp. 193-194.

127. Scully, op. cit., pp. 120-140.

128. Clarke, op. cit., p. 205, n. 12.

129. Ibid., p. 206, n. 25.

130. See Scully, op. cit., pp. 102-253 for a discussion of differential gynecological/obstetric practice based on race and socioeconomic status and, specifically, its relationship to the potential for surgical abuse; see also, Corea, The Mother Machine, pp. 160-162 for detailed discussion of specific causes of iatrogenic infertility.

131. Clarke, op. cit., pp. 198-199; Corea, The Mother Machine, pp. 146-147.

132. See section below on "Infertility."

133. Mosher, W.D., and Pratt, W.E., "Reproductive Impairments Among Married Couples: United States," National Center for Health Statistics, Vital and Health Statistics (Dec. 1982), p. 9 states, for example, that 23 percent of black couples had impaired fecundity, compared to 15 percent of white couples. Of married black women with zero parity, 35 percent had impaired fecundity, but among white married women with zero parity, 21 percent had impaired fecundity. Once again, available data refer only to "white" and "black" and, therefore, do not adequately reflect the range of "nonwhite" populations.

134. U.S. Department of Health and Human Services, Public Health Service, "Health Status of Minorities and Low-Income Groups," ASI 85-4118.55 (1985), p. 85.

135. Ibid., p. 65.

136. Childbearing Rights Information Project, op. cit., pp. 66-68; Alliance Against Women's Oppression, "Caught in the Crossfire: Minority Women and Reproductive Rights," p. 5.

137. Centers for Disease Control, Surgical Sterilization Surveillance: Tubal Sterilization and Hysterectomy in Women Aged 15-44 from 1979-1980 (Sept. 1983), p. 9.

138. Corea, The Mother Machine, p. 147.

139. Clarke, op. cit., p. 199.

140. Barlow, Susan, and Sullivan, Frank, Reproductive Hazards of Industrial Chemicals (1982); President's Council on Environmental Quality, "Chemical Hazards to Human Reproduction," (1981-337-130-8008), (January 1981).

141. Corea, The Mother Machine, p. 147.

142. "Health Status of Minorities and Low-Income Groups," as cited in note 134 above, p. 53.

143. Arditti et al., "Introduction," Test-Tube Women, as cited in note 6 above, p. 4.

144. Spokesperson for the American Fertility Society, Birmingham, Alabama, telephone conversation, January 20, 1988.

145. A national "success" rate is extremely difficult to determine because each provider defines "success" in different terms, uses different criteria for admission to its program, and so forth. A figure commonly quoted by providers is a 60 percent rate of live births after four attempts. But others have calculated an overall success rate of four to five percent. Corea, The Mother Machine, pp. 179-80. The American Fertility Society is developing a databank of in vitro programs, so that meaningful figures can be compiled. A preliminary report will be released in early 1988. American Fertility Society, op. cit.

146. Only four states (Massachusetts, Maryland, Hawaii, and Texas) mandate private insurance coverage of in vitro procedures. Insurance coverage of other fertility treatments can be erratic. For example, artificial insemination is sometimes exempted as an experimental technique, although it has been in use since the turn of the century. Ibid.

147. U.S. Department of Health and Human Services, Public Health Service, Promoting Health/Preventing Disease: Objectives for the Nation (1980), p. 11. Funding for in vitro fertilization is not available from any public source. American Fertility Society, op. cit.

148. Corea, The Mother Machine, p. 145.

149. U.S. Department of Commerce, Bureau of the Census, Statistical Abstract of the U.S. (1987), p. 38. In 1985, 32 percent of black women were never married, 11 percent were divorced, and 14 percent were widowed.

150. Quoted from a Norfolk, Virginia fertility clinic physician by Corea, The Mother Machine, p. 145. See also Somerville, Margaret, "Birth Technology, Parenting and Deviance," International Journal of Law and Psychiatry 5:2 (1983), pp. 123-153.

151. See above for examples of reproductive experimentation on women of color and poor women; see Corea, The Mother Machine, pp. 166-168 regarding lack of informed consent for in vitro fertilization procedures.

152. Ibid., pp. 169-172.

153. "Historical Continuity in the Afro-American Family in the African Diaspora," a special issue of The Research News, University of Michigan, Department of Black Studies (April/May 1982), p. 20.

154. Devera Foreman, Philadelphia Chapter-Association of Black Psychologists, telephone conversation, 1986.

155. Ibid.

156. See generally, Corea, The Mother Machine, pp. 213-249.

157. Developed by the Asian Pacific Planning Council Health Task Force (May 1985).

REPRODUCTIVE TECHNOLOGY AND DISABILITY

Adrienne Asch[1]

Disability, Family, and Community

Introduction

How do the new reproductive technologies affect people with disabilities? Answering this question, taking into account current and future generations of disabled women, men, and children requires putting a biological fact, impairment or disability, in a social context. In a different society than ours, the meaning of the new technologies for people with disabilities could resemble that for people without disabilities. In other words, any special implications for people with disabilities stem primarily, though not exclusively, from their position as the subjects of deep-rooted ambivalence on the part of the nondisabled population. Below, I will sketch out the social context for disability in our society and then discuss from the perspective of disability rights the six areas of reproductive concerns featured in other segments of this book: prenatal screening, time limits on abortion, fetus as patient, reproductive hazards in the workplace, alternative modes of reproduction, and interference with reproductive choice.

A point on terminology: I use the term "impairment" or "biological condition" to refer to the characteristics that reside exclusively in the physiological state of disability. When referring to people who have disabling conditions, I refrain from speaking of, for example, "hemophiliacs" or "the blind," not wishing to focus on an identity, but rather on the disabling conditions themselves; so I speak of "people with hemophilia" or "people who are blind." Except when referring to specific pieces of legislation using the term "handicapped," I avoid its use or restrict "handicap" to the social implications of a disabling condition. We must abandon the usage of deeply offensive words such as "deformed," "malformed," "damaged," "defective," or even "handicapped" when discussing fetuses, newborns, children, or adults who have disabilities. Although the disabled community has not settled upon one word for itself, several major activist organizations and publications concerned with improving the political and legal status of this population use the term "disability" in their titles: Disability Rights Education and Defense Fund, *Disability Rag,* World Institute on Disability, and Disabled in Action, to name only a few. Accordingly, I will refer to the millions of people who are the particular subject of this essay as people with disabilities, as people who have disabilities, and as disabled people. The last of these terms, although the shortest, unfortunately highlights disability rather than underscoring that it is only one of a person's characteristics.

I would like to preface this essay by emphasizing some central themes:

1. Two major strands run throughout my discussions below and originate in my twin commitments: first, to reproductive choice for all women, with and without disabilities, to bear or not to bear children or particular types of children; second, to my conviction that life with disability can be valuable and valued, and therefore, we must carefully consider the consequences of our disability-prevention activities.

2. The new reproductive technologies compel us to examine what children, parenting, life, health, family, and disability mean.

3. The new reproductive technologies will present increasing opportunities to make choices about the kinds of babies that will be born or not. Our decisions in this regard will depend on our notions of what sorts of lives are worthwhile and how much toleration we have for the variability of human experience.

4. My concern is with the social consequences of disability both for disabled women and for fetuses with impairments who may or may not be born, as a result of the new reproductive technologies.

The Social Context of Disability

Facts about Disabled People

Because the new reproductive technologies are in large part concerned with preventing or circumventing disability, it is important to have a sense of how prevalent disability is in our society. Numbering over 40 million, disabled people make up 15 percent of the population, a substantial portion. Studies using different methodologies indicate that between 9 and 17 percent of people of working age are disabled (Bowe, 1980; Dejong and Lifchez, 1983; Haber and McNeil, 1983). Five percent of women of reproductive age have disabilities that biologically or socially constrain their reproductive choices.

The Definition of Disability

At the most basic level of definition, our society links the biological fact of impairment to the social realm of role function. Rarely is disability defined or measured without reference to some task or social role commonly expected of Americans of a certain age or gender. The Rehabilitation Act of 1973 (as amended in 1978 and 1986) defines a handicapped individual as "any person who (i) has a physical or mental impairment which substantially limits one or more of such person's major life activities, (ii) has a record of such an impairment, or (iii) is regarded as having such an impairment" (Rehabilitation Act of 1973, 87 Stat. 355, Section 7 (7 B)). The Rehabilitation Act goes on to define disability in detail in a manner encompassing many conditions not commonly thought of as disabilities, but rather as chronic illnesses or health problems.

Disabilities both readily apparent and invisible can interfere with daily activities. Mobility, for example, can be affected not only by polio or amputation, but by arthritis, heart conditions, or respiratory or back problems. People with histories of institutionalization for mental illness or mental retardation may, in fact, not be hindered in any life task, but may carry records that haunt them and impede their access to education and employment because of the stigma they carry. People with cancer in remission, cosmetic disfigurements, or obesity, along with all other people who have disabilities, may find themselves "regarded" as impaired when they can perform in any social role. Thus, the social construction of disability, like that of gender, demonstrates that it is the attitudes and institutions of the nondisabled, even more than the biological characteristics of the disabled, that turn these characteristics into handicaps.

Disability and Age

Disability increases dramatically with age. Ten percent of individuals under 21 as compared to almost 50 percent of those over 65 have disabilities. Of the estimated 90,000 infants born annually with impairments likely to remain throughout their lives (March of Dimes, 1985; Zola, 1983), most have conditions that could not have been detected *in utero*. Prenatal diagnosis now and in the future will not eliminate or notably reduce the prevalence of disability. Many of the health conditions, speech impairments, or learning disabilities of the school-age population do not occur or are not detected until the school years. The vast majority of disability, moreover, does not occur until mid- or late adulthood. Most disability can be attributed to chronic conditions typically occurring in later life, accidents on or off the job, illness, stress, or toxins in the workplace and environment.

Disability, Ethnicity, and Class

Some congenital and adult-onset disabilities appear more frequently in particular ethnic groups or social classes. Among commonly discussed prenatally diagnosable conditions, for example, sickle cell anemia is found mainly in the black population, Tay-Sachs disease among Ashkenazi Jews, cystic fibrosis overwhelmingly among whites, and spina bifida among Northern Europeans and Latinos.

Most disabled men and women are not part of the middle class and not well placed to take advantage of the new reproductive technologies. Twenty-six percent of disabled people, as compared with 10 percent of the nondisabled of working age, live at or below the poverty level (U.S. Census Bureau, 1983). For some people, poverty causes disability, although for others, it is the consequences of disability that cast them into poverty. Low-income people, often black, Hispanic, Asian, or Native American, are more likely than others to acquire the impairments attributable to the poor nutrition, inadequate medical and prenatal care, hazardous jobs, high pollution levels, stress, violence, and drugs associated with inner-city life. Many disabled individuals, contending with unemployment or underemployment, are further impoverished by the lack of public services, which causes them to pay steeply for medical care not covered by health insurance and for unsubsidized special transportation and attendant care.

The Medical Model of Disability

In recent years, proponents of disability rights have been advocating the replacement of the long-dominant "medical model" of disability--the view that disability is exclusively a biological problem--with the "minority group model" of disability, which emphasizes the social components of having a disability. The conventional view in medical, rehabilitation, and social service literature and practice saw the poverty, low educational attainment, and mass unemployment of disabled people as a result of their health alone.

The tenets of the "medical model" of disability persist because disability differs from race and gender in one fundamental respect: Not all problems of disability are socially created and, thus, theoretically remediable. No matter how much broad and deep social change could ameliorate or eradicate many barriers to fulfillment encountered by today's disabled citizens, in no society would it be as easy or acceptable to have a disability as not to have one. The inability to move without mechanical aid, to see, to hear, or to learn is not inherently neutral. Disability itself limits some options. Listening to the radio for someone who is deaf, looking at paintings for someone who is blind, walking upstairs for someone who is quadriplegic, or reading abstract articles for someone who is intellectually disabled are precluded by impairment alone. Physical pain, the inconveniences and disruptions occasioned by medical treatments, or routines of medication, rest, restricted diet, and exercise programs are not desirable aspects of life. It is not irrational to hope that children and adults will live as long as possible without health problems or diminished human capacities.

The Social Construction of Disability

If some portion of the difficulty of disability stems from the biological limitations, the majority does not and is in fact socially constructed. In asserting that disability is socially constructed, I am making two claims: first, that even those characteristics we label as "disabling" are at least partly socially determined; second, that disability's all-too-frequent consequences of isolation, deprivation, powerlessness, dependence, and low social status are far from inevitable and within society's power to change.

Although some disabilities seem more inherently limiting than others, ascertaining their consequences is more bound up with social circumstance than at first appears. In her research on life in a late-nineteenth-century Martha's Vineyard village where many residents were deaf and all deaf and hearing inhabitants spoke sign language, Groce (1985) demonstrates how social circumstance shapes the perception and experience of disability. Deafness and sign language were such a part of the community that whether or not someone was deaf ceased to be a salient identifying characteristic or to inhibit work and social life. By contrast, in today's largely urban culture, conditioned to telephone, radio, and uncaptioned television, and where virtually no hearing person learns sign language, deafness is perceived as a seriously disabling condition.

Why has the social construction of disability been so negative? Burt (1979) and Hahn (1983a) argue that disability provokes both existential and aesthetic anxiety. Disability reminds people of limitations in functioning and autonomy and of what is ugly and imperfect, perhaps imperfectable, in the world and in the human condition.

In a 1987 *Harper's Magazine* forum on reproductive technology, Thomas H. Murray underscored just how much of what is perceived as disabling arises from social consensus when he noted:

> We make judgments all the time. We decide what is a disease and what is not a disease, what's a deformity and what's not a deformity. For example, society says: "If you have a harelip, that's a deformity, and it's enough of one to warrant trying to correct it. We'll even help you pay for it." That's a social consensus. Whereas if you want a tummy tuck because you don't like your paunch, we say we'll let you do it, but we sure as hell won't pay for it. We draw that line (p. 43).

As our society develops ever more refined prenatal diagnosis and medical technology, more and more characteristics may be considered unacceptable deviations and disabilities. Simultaneously, cultural requirements for what is considered desirable and attractive may become more rigid. By monitoring and evaluating our social attitudes toward disability, we have the option to encourage or halt such trends in intolerance of human variability.

Disabled People as a Minority Group

Like blacks and women, disabled people have come to see themselves as a minority group. They share a sense of differing from the majority in appearance or behavior in ways that are devalued. They are aware that, as Goffman (1963) points out, "normals" view the stigmatized, including the disabled, as "not quite human" (p. 5). The "minority group" model of disability emphasizes that all people with disabilities have contributions to make and must be accorded the full degree of human dignity and rights. Disabled people contribute to society not because of their impairments, as the right-to-life movement and some religious writers would argue, but because in addition to their impairments they have inherently valuable qualities no different from those of the nondisabled.

Disabled people are also coming to develop a sense of commonality because of their shared experience of socioeconomic and political oppression (Dworkin and Dworkin, 1976; Harris and Associates, 1986). Activists and scholars of disability have recognized that the nondisabled majority, by ignorance or design, has created environments, social institutions, and a host of practices that effectively excluded and segregated disabled people. They have discerned that obstacles to education, community and political participation, independent living, employment, and personal relationships resided not in the incapacities of individuals in wheelchairs to walk stairs, but in the existence of the stairs that kept them out of schools, public meetings, apartments, offices, or friends' homes. If people with Down syndrome were in residential institutions and sheltered shops, it was because their segregation resulted from the desires of educators, employers, and service professionals to keep them segregated, and not from their inherent incapacity to succeed in integrated schools, work, and community homes (Evans, 1983; Rothman and Rothman, 1984).

During the 1970s, grassroots groups of disabled people formed statewide and national membership organizations and coalitions, founded legal advocacy centers and newsletters, and worked along with parents of disabled children and nondisabled advocates inside and outside government to create and implement laws guaranteeing rights of access to education, employment, government services, and community life. At the end of the 1980s, although disabled people still have far fewer civil rights than those available to people fighting race or sex discrimination, their situation has definitely begun to improve (Asch, 1986; Bowe, 1980; Funk, 1987; Hahn, 1987; Scotch, 1984, 1988).

Creating an Inclusive Society

Bowe (1980), Gliedman and Roth (1980), and the National Council on the Handicapped (1986) all provide excellent blueprints for creating a society that would truly include, and not exclude, people with disabilities. Gliedman and Roth (1980) offer this powerful vision of the reforms necessary to create an inclusive society:

> Suppose that somewhere in the world an advanced industrial society genuinely respected the needs and the humanity of handicapped people. What would a visitor from this country make of the position of the disabled individual in American life? . . .
>
> To begin with, the traveler would take for granted that a market of millions of children and tens of millions of adults would not be ignored. He would assume that many industries catered to the special needs of the handicapped. Some of these needs would be purely medical . . . But many would not be medical. The visitor would expect to find industries producing everyday household and domestic appliances designed for the use of people with poor motor coordination. He would look for cheap automobiles that could be safely and easily driven by a paraplegic or a quadriplegic. He would anticipate a profusion of specialized and sometimes quite simple gadgets designed to enhance the control of a handicapped person over his physical world--special hand tools, office supplies, can openers, eating utensils, and the like. . . .
>
> As he examined our newspapers, magazines, journals, and books, as he watched our movies, television shows, and went to our theaters, he would look for many reports about handicap, many fictional characters who were handicapped, many cartoon figures on children's TV programs, and many characters in children's stories who were handicapped. He would expect constantly to come across advertisements aimed at handicapped people. He would expect to find many handicapped people appearing in advertisements not specifically aimed at them.
>
> The traveler would explore our factories, believing that handicapped people were employed in proportion to their vast numbers . . .

> He would walk the streets of our towns and cities. And everywhere he went he would expect to see multitudes of handicapped people going about their business, taking a holiday, passing an hour with able-bodied or handicapped friends, or simply being alone.
>
> He would take for granted that in many families one or both parents were disabled. He would assume that an elaborate network of services existed to help them in raising their children and that a parallel and partially overlapping service network existed to serve the needs of able-bodied parents who had handicapped children.
>
> He would explore our man-made environment, anticipating that provision was made for the handicapped in our cities and towns. He would look for ramps on curbs, ramps leading into every building. He would expect the tiniest minutiae of our dwellings to reflect the vast numbers of disabled people.
>
> Observing our elections, he would take for granted that the major parties wooed the disability vote just as assiduously as they pursued the labor vote or the black vote. He would assume that disabled individuals had their share of elected and appointive offices. He would expect to find that the role played by the disabled as a special interest group at the local and national levels was fully commensurate with their great numbers (pp. 13-15).

In today's society, much of the grief and burden posed by intimacy with a person who has a serious health problem or disability stems from lack of social support services or lack of social acceptance of disabled people as fully human and, therefore, naturally deserving of a place within family and society equal to that of nondisabled people. Social services, financial support, education, employment, and recreational opportunities will not improve for disabled children or adults until we genuinely believe that it is acceptable to be a person with a disability, and to claim a share of familial and societal resources at least as great as that claimed by those without disabilities. Arguing that fully integrating disabled people into national life poses insuperable problems of resource allocation only reinforces the view that disabled people are not fully human or fully part of society. If it were otherwise, we would automatically understand that we should evaluate existing resources and structures in terms of whether they assist all citizens, including those with impairments, rather than perpetuate the notion that disabled people should get only as much as is left over. Some of the necessary changes, such as creating barrier-free buildings and transportation, are initially costly. However, other changes, such as eliminating obstacles to employment for the millions of disabled people willing to work, require primarily changes in attitudes and institutional practices rather than expensive modifications. Committing monies to making people independent would enable society to reduce current expenditures on unconstructive "dependence" programs for the disabled.

Changing attitudes of the nondisabled public, the medical profession, and many disabled people themselves requires recognizing that the disabled can lead worthwhile and fulfilling lives. Incapacity in one area need not cause incapacity in other areas (Wright, 1983). Because most disability arises though traumatic injury or progressive chronic illness during adulthood, people who become disabled must determine how they can pursue their goals and relationships with changed health, energy, or appearance. They are likely to find that adaptation and reorientation may take months or years, ideally including contact with others who have made similar life adjustments and acceptance of necessary changes on the part of intimates and employers. Given such time, opportunity for contact with professional services and other disabled people, and opportunity to maintain old goals and relationships or find new ones, most people succeed in attaining or preserving a substantial degree of satisfaction with their lives, as reported by rehabilitation experts and studies of disabled people across the nation (DeLoach and Greer, 1981; Harris and Associates, 1986; Schultz and Decker, 1985).

Disability and the New Reproductive Technologies

Reproductive decisions starkly compel us to confront the profound connection between the political and the personal. Although many throughout society have started to accept individuals with disabilities as coworkers, fellow students, and casual acquaintances, people's attitudes toward having the disabled in their intimate circles of friends, lovers, mates, or children are changing far more slowly. As Fine and Asch (1988) and Stroman (1982) have pointed out, nondisabled people generally assume that expected and desired reciprocity will be absent from relationships with a disabled person. People fear that intimacy with a disabled person will bring great grief and burden, and little in return for their care. These comments apply both to men considering women with disabilities as potential partners and mothers of children, and to women and men deciding about raising children with disabilities (Asch and Fine, 1988; Sawisch, 1978; Siller et al., 1976).

Disabled Women as Mothers

Unlike any other group of women in the world, women with disabilities have not been expected to become mothers and have frequently been impeded from doing so (Hyler, 1985; LeMaistre, 1985; Shaul et al., 1985; Thurman, 1985). At least 5 percent of women of reproductive age are disabled in ways that limit their reproductive choices. Some women's disabling conditions present difficulties in using contraception methods or in conceiving and maintaining a pregnancy. The woman with circulatory problems, for example, may be limited in her options for birth control. However, she may face fewer obstacles to becoming a mother than does a woman whose disability of deafness or mental retardation imposes no medical restrictions upon her birth control or pregnancy choices, but considerable restriction on her social and economic options.

Far more than their nondisabled sisters, disabled women live alone, with partner absent, and in poverty (Bowe, 1984; Fine and Asch, 1985). Asch and Fine (1988), Asch and Sacks (1983), Hahn (1983b), Matthews (1983), and Safilios-Rothschild (1970) all note that intimate relationships are absent from the lives of many disabled women and men who are otherwise successful in work, community involvement, and friendship. Women with disabilities are less likely to be married and more likely to be separated or divorced than nondisabled women (Asch and Fine, 1988). Bonwich (1985) reports that most of the marriages or serious heterosexual relationships of the spinal cord-injured women she studied did not survive their becoming disabled. Women with disabilities are far less likely than nondisabled women or similarly disabled men to be employed. Across all racial groups, only about one in four women with disabilities participates in the labor force. Those who are employed are often trapped in low-wage, dead-end jobs (Bowe, 1984).

Many disabled women encounter substantial legal, medical, and familial resistance to their choice of motherhood. The *Braille Monitor* (1986) and Gold (1985) describe instances where blind parents' disabilities were given as a reason to deny custody or even unsupervised visitation during divorce. The *Braille Monitor* (1986) and the *Washington Post* (1988) describe instances in which children were taken from mothers disabled by blindness and cerebral palsy, respectively, because of the mothers' alleged inability to care for the children. Referring to Elizabeth Stern, the adoptive mother with multiple sclerosis in the celebrated *Baby M* case, Pollitt (1987) queried "why a disease serious enough to bar pregnancy was not also serious enough to consider as a possible bar to active mothering a few years down the road. If the Sterns' superior income could count as a factor in determining 'the best interests of the child,' why shouldn't Mary Beth Whitehead's superior health?" (p. 682). Well over three-quarters of Americans view a woman's physical disability as an acceptable reason for her having an abortion (Lamanna, 1984), and physicians, the media, state laws, and courts have thwarted women and men with physical or intellectual impairments who have wished to be parents.

As with so much else about the situation of disabled people, professionals and the public assume that, if children of parents with impairments incur difficulties, the disability is to blame, not the social or economic circumstances imposed on the family by the treatment of the disabled parent. It is asserted that children of physically disabled parents are unduly self-reliant, because they have had to help their parents with physical and household tasks and that, in addition to too early responsibility, the children "risk growing up with a sense of inferiority and intense self-consciousness about themselves and their families, and a burden of guilt when they find the responsibilities too heavy" (Walker, 1985, p. C1). People with mild or moderate retardation encounter even greater institutional opposition to their raising children (Macklin and Gaylin, 1981), despite evidence that, with proper social and family support, many such women and men prove adequate at supplying children with the needed physical and emotional nurturance (National Public Radio, 1988; Zetlin et al., 1985). Those opposing the parenting aspirations of disabled individuals have often facilely accepted false stereotypes of the disabled--assuming that incapacity in one area means incapacity in all--without thoughtfully examining what successful parenting may really require.

Having Disabled Children

For most nondisabled people, giving birth to a child with an impairment is rarely welcomed. The medical profession does not celebrate with the new mother or parents; the grandparents, other relatives, and friends do not delight in the arrival of a child with impairments; and the immediate question raised is how much the child with a disability will burden the woman, her mate if she has one, siblings, relatives, and society as a whole. White society may discriminate against black adults or children, but historically, black women and children could count on care, love, support, and a modicum of respect within their own families and communities. Not so for the woman with a disability or for most children with impairments. The lack of a natural communal or familial structure can be psychologically and socially devastating. Successfully raising a child with a disability requires discovering that disability in itself need not preclude a rewarding life for child and family.

Implicit in many of the reproductive technologies and explicit in some is the goal of preventing future disability. Sometimes this means that women and men with hereditary disabilities are presumed immoral if they choose to create children who themselves will also have the disability. Notwithstanding such pressure from society and health professionals, some disabled people--especially those in the tight-knit deaf community (*Insight*, 1986) or in the Little People of America--count on raising children and may prefer having a child like them in what they consider an essential way.

I will now turn to examining the six areas of reproductive issues identified by the Project on Reproductive Laws for the 1990s, beginning with prenatal screening, because it most strongly embodies the psychological and moral dilemmas for feminists also committed to the rights and welfare of disabled people.

Prenatal Screening

Two voices struggle within those committed to both feminism and disability rights. One voice says: "Women want the best for their children, and in no way can disability be seen as 'the best.' Any child will have enough difficulty, so let us spare what we might be able to prevent through prenatal screening. Let us recognize that we will cope with and love a child who incurs a disability, but let us avoid such disability if possible." The other voice says: "Every life has difficulties. Disability need not be insuperable. Do not disparage the lives of existing and future disabled people by trying to screen for and prevent the birth of babies with their characteristics."

This section of this essay emphasizes concerns and apprehensions about the meaning and possible misuse of genetic screening and prenatal diagnosis, thus giving more weight to the less common and more controversial latter voice. Although I support both efforts to prevent disability and the right of women to reproductive choice, I believe that genetic screening and prenatal diagnosis followed by abortion differ morally and psychologically from other methods of preventing impairments, and from all other abortions save those for sex selection. What differentiates ending pregnancy after learning of impairment from striving to avoid impairment before life has begun is this: At the point one ends such a pregnancy, one is indicating that one cannot accept and welcome the opportunity to nurture a life that will have a potential set of characteristics--impairments perceived as deficits and problems. What differentiates abortion after prenatal diagnosis (and abortion for sex selection) from all other abortions is that the abortion is a response to characteristics of the fetus and would-be child and not to the situation of the woman. Although in the discussions below I highlight concern about the proliferation of prenatal screening and diagnosis, I appreciate their value for individual women and families. I in no way claim that these practices should ever be legally prohibited or that they can never be justified.

The Capabilities of Prenatal Diagnosis

Over 100 chromosomal, structural, and metabolic disorders can now be diagnosed prenatally through the use of amniocentesis at about 16 weeks of gestation, chorionic villus sampling (CVS) at about 12 weeks of gestation, ultrasonography, amniography, fetoscopy, and the screening of alpha-fetoprotein levels (Cohen, 1986; March of Dimes, 1985). These include Down syndrome, spina bifida, sickle cell anemia, Tay-Sachs disease, and adult-onset Huntington's disease. The diagnostic capabilities are developing rapidly, so that, as of 1987, it was possible to detect hemophilia, cystic fibrosis, and some muscular dystrophy in the fetus. Scientists have begun an extensive national project of mapping all genetic markers, which will enable them to identify genetic predispositions to a wide range of conditions, including some forms of Alzheimer's Disease, manic depression, diabetes, and various cancers (National Public Radio, 1987). Some diagnoses reveal certainty of disability, but not the severity of the condition, as is true for diagnoses of Down syndrome and spina bifida, whereas other diagnoses reveal only predispositions to acquiring disability, but no information about how likely it is that the condition will occur or how severe the manifestations will be in any instance.

Medicalization of Pregnancy

One important feminist argument launched against the increasingly routine usage of genetic screening in prenatal care is that this practice contributes to the further medicalization of pregnancy (Arditti et al., 1984; Rothman, 1986). The use of new technological interventions such as genetic screening offers greater power to a medical profession that already exercises considerable control over pregnant women. Furthermore, it may represent an unnecessary complication of a woman's life, especially in the case of chorionic villus sampling, which enables prenatal diagnosis before the end of the first trimester of pregnancy. In some instances of disability diagnosed through CVS, women will undergo abortions of fetuses that would miscarry within weeks or months even without medical intervention. Thus, with CVS even more than with amniocentesis, women will live with the knowledge and possible psychological consequences of having ended pregnancies that would have ended anyway without medical intervention.

The Pressure to Abort after Diagnoses of Disability

Many proponents of prenatal diagnosis exert subtle and overt pressures for users of the technology to abort after diagnoses of disability. Much writing lauding the technology is simplistic and founded in ignorance, prejudice, or the medical model of disability (Annas and Elias, 1983; Bayles, 1984; Steinbock, 1986). Being disabled is portrayed as being permanently ill and in pain. Fletcher (1984) advocates prenatal diagnosis and fetal therapy, because he believes disability to be inherently destructive of productive life, saying: "People free from genetic disorders would finally be able to live full lives and contribute a great deal more to society" (p. 36). In this kind of reasoning, all problems are located in the presence of disability itself, rather than in the social conditions causing lack of productivity and defeatism. Such proponents of prenatal diagnosis argue that disability imposes steep socioeconomic and emotional costs on society, on children who are its "victims," and on families.

Costs to Society

In my view, it is neither moral nor practical to base abortion decisions on presumed costs to society. My moral argument flows from my conviction of the worthwhileness of life with disability, and my belief that a just society must appreciate and nurture the lives of all people with varying endowments in the natural lottery. Although some disablement entails costs in medical treatment, support services, and professional assistance, universal prenatal screening and selective abortion would eliminate a negligible portion of the economic costs of disability. Since most impairment does not occur until adulthood, prenatally diagnosable disability now and in the future will account for only a fraction of impairment and costs in medical care or special disability-related programs.

Costs to Children

Concerns about implications for children and their families warrant more consideration. What of the arguments that urge us to spare children physical pain, psychic anguish, and social isolation imposed by disability? Disability is likely to entail some of these for the foreseeable future. Some women may feel it intolerable to watch their children suffer hurts that could have been avoided by not letting pregnancies go to term. It is one thing to know that you yourself can accept pain for yourself in your own life and another to know that, by bringing life into the world, you guarantee pain to another that you will be able to do relatively little to ease. In order to imagine bringing a disabled child into the world when abortion is possible, one must be able to imagine saying to a child: "I wanted you enough and believed enough in who you could be that I felt you could have a life you would appreciate even with the difficulties your disability causes." Instead of thinking of avoiding the life itself, might we think about how the expected problems can be reduced or avoided? If parents and siblings can love and enjoy the disabled child; if neighbors, friends, relatives, child care centers, and schools routinely included disabled children; and if those producing television commercials and programs, children's books, and toys took disabled people into account, the child might not live with the anguish and isolation replete in the lives of disabled children of yesterday (Brightman, 1984; Browne et al., 1985; Orlansky and Heward, 1981; Roth, 1981).

Costs to Families

Many who are willing to concede that people with disabilities could have lives they themselves would enjoy nonetheless argue that the cost to families warrants promoting abortion of fetuses with impairments. Women, especially, are seen to carry the greatest load for the least return in caring for the disabled child. Raising such a child epitomizes what women have fought to change about their lives as mothers: unending care; little relief in the form of partner or societal assistance; constant juggling of all resources to give the child the best; uncertain recompense in terms of the relationship with the child; and little opportunity to maintain a life with adult interests and relationships apart from the child.

If we see the child with the disability only as a set of deficits and problems and never discover other enjoyable, pride-giving characteristics, families naturally would take such a grim view. Unfortunately for prospective parents, the wealth of fine writing by parents of disabled children (*see* Appendix A for examples) is known primarily by other such parents and specialists in disability, and not by geneticists, obstetricians, pediatricians, and mainstream or feminist prospective parents. Rarely do mainstream or feminist books about mothering provide glimpses into the lives of the mothers of children with disabilities (Bernard, 1974; Dally, 1983; Dowrick and Grundberg, 1980; Friedland and Kort, 1981; Thorne and Yalom, 1982; Zelizer, 1985). On the contrary, one well-known feminist commentator wrote with regard to decisions about motherhood:

> Suppose, as an easy starter, that an unborn child is genetically defective--has Down's Syndrome, let us say, or Tay-Sachs disease, or hemophilia, or some other defect which, though not necessarily genetic or lethal, is a serious handicap, costly to parents. The moral issue here is fairly clear, and there is in fact a consensus that such a pregnancy should be terminated (Bernard, 1974, p. 253).

Some feminists have claimed that prenatal diagnosis and selective abortion are mandated by a society that gives so little of its resources to disabled citizens, and leaves families and women to bear the entire burden alone (Rothman, 1986). They note that the burden is manifold: families with disabled members may endure hardships imposed by emotional distress, social isolation, and financial difficulties.

The argument that a woman's only realistic decision is to abort a fetus with a disability until society is more willing to include disabled people is at first powerful, persuasive, and mindful of the toll disability takes on families. It loses some force, however, when we consider that women of color bring children into the world even knowing that their children will grow up in a racist society, and may suffer economically, socially, emotionally, and psychologically as a result. Nonetheless, these women look forward to having children, and believe that as mothers they can provide love and inculcate strength with which their children can face the world. A major difference between the Native American, black, or Latina woman contemplating pregnancy and the nondisabled woman contemplating raising a disabled child is that, in the former case, the woman can see herself as like her future child and can imagine that the child will fulfill her aspirations for motherhood; in the latter case, she sees the child as neither like her nor as constituting the fulfillment of her parental aspirations.

I believe that women today who choose to abort after prenatal screening do so for a host of reasons, some of them economic, but most having to do with personal feelings and needs. Why would women contemplating motherhood be heartbroken or devastated by a child's disability? Is that heartbreak a response to the conviction that the child's life will be stunted, or is it largely that dreams of raising a child will not be fulfilled? We all, disabled and nondisabled, women and men, must honestly acknowledge what we value and individually seek in being parents (Hoffman and Hoffman, 1973). We must also acknowledge what frightens, repulses, or distresses us about disability in others, in ourselves, and in those close to us. Seeking to avoid the experience of raising disabled children is no crime or callous, selfish statement, as some may claim. It is an honest, understandable, if perhaps misinformed, response to the fears that a disabled child will not fulfill what most women seek in mothering--to give ourselves to a new being who starts out with the best we can give, and who will enrich us, gladden others, contribute to the world, and make us proud. Let us frame our thinking about prenatal diagnosis and selective abortion in a sincere discussion of what we long for in the experience of having children. Let us then ask how a child's disability will compromise that dream. Such discussion will help us to answer the question of whether it is disability inherently that pains or the consequences of disability that might be changed with genuine societal commitment to change them. If we believed that the world was a problem to the child and not the child a problem to the world, we might be better able to imagine how raising a child with a disability could give much the same gratifications as raising another child who did not start life with a disabling condition.

Cross-cultural Attitudes towards Disability: A Model for Acceptance of Disability

Studies in the last two decades have demonstrated differential responses to disability based on the values of different racial, ethnic, and economic groups (Darling, 1979; Rapp, 1988). Whites and middle-class people in general showed more discomfort with Down syndrome and retardation, whereas people of color and those of lower socioeconomic status expressed more fear of physical vulnerability.

The "Position Paper on Prenatal Diagnosis of Sickle Cell Anemia" authored by the National Association for Sickle Cell Disease gives a clue to how different cultures respond to impairment and offers a model of thoughtful presentation (National Association for Sickle Cell Disease). Sickle cell anemia affects one in every 500 blacks, making it somewhat more common than Down syndrome, which affects one in every 800 infants. Striking about the position paper is its matter-of-fact treatment of the topic. Its message: part of being black is knowing that a small percentage of individuals carry the gene for the trait and a smaller percentage have the disabling condition. The discussion neither exaggerates nor minimizes the consequences of the condition and yet embodies much of what one would value in a consideration of prenatally diagnosable disability. Suppose Down syndrome, cystic fibrosis, or spina bifida were depicted not as an incalculable, irreparable tragedy but as a fact of being human? Would we abort because of those conditions or seek to limit their adverse impact on life? We might continue to favor abortion, but a matter-of-fact attitude promises a greater chance for acceptance and social inclusion.

Drawing Lines

In a forum in *Harper's Magazine* (1987), the ethicist Nancy Dubler stated with reference to prenatal screening and selective abortion: " . . . I think that society has a shared perception on certain diseases. We can draw lines. We are human beings; we deal with difficult problems all the time" (p. 42). Even those most apprehensive about the anti-disability sentiment in prenatal diagnosis would probably tolerate or welcome abortions after diagnoses of Tay-Sachs disease and other degenerative conditions that cause extreme pain, loss of awareness of self and others, and death in early childhood. For others, atypical sex chromosomal conditions such as an extra Y chromosome represent the extreme of a characteristic for which women should not abort, one that may not really be an "impairment" at all. As genetic screening improves, moreover, we are able to learn about predispositions to such common conditions as diabetes, asthma, heart disease, or cancers. At some not-too-distant point in the future, we may confront pressures to abort for these conditions that we now take to be a part of what it means to be alive. On the other hand, Down syndrome, spina bifida, Huntington's disease, cystic fibrosis, sickle cell anemia, Turner's syndrome, short stature, and muscular dystrophy exemplify a spectrum of controversy. For some disabled and nondisabled people, aborting fetuses with those impairments means preventing the existence of people whose lives could be enriching to themselves and others despite difficulties. To most people, however, at least partly because of the lack of accurate information about and contact with disabled individuals, failure to abort is cruel to the potential child, destructive to the family, socially irresponsible, and possibly immoral.

The incomplete nature of prenatal screening's predictive powers makes drawing lines particularly difficult. Screening may reveal that a child will have cystic fibrosis, but the test does not indicate whether the child will live ten years or into adulthood, or whether the child will have constant pain and frequent hospitalization, or whether she or he will have a milder form of the disease with fewer disruptions of study, play, work, marriage, and childraising. Predisposition to cancer at fifty tells a parent nothing about the first fifty years of a would-be child's life, and nothing about what the cancer will be like when and if it occurs.

The fantasy--and fallacy--of childrearing is that parents can guarantee or create perfection for the child. We cannot do away with hardships for our children. In doing away with one hardship, we may nonetheless fail to prevent one that is worse. Prenatal diagnosis stimulates our fantasies of the perfect world and the perfect life, but also forces us to recognize that these are only fantasies. The fetus diagnosed with no disability may be born with an undetectable severe health problem or may acquire it at five, fifteen, or thirty. The one diagnosed with spina bifida may lead a life she loves, but her nondisabled sibling may, for reasons having nothing to do with a sister's disability, be angry, frustrated, and unfulfilled.

I believe that we can and should draw lines concerning the ethical use of selective abortion following prenatal screening. I would urge us to question whether we wish to abort for most disability that will not cause great physical pain or death in early childhood. Accepting that the life of a man with profound retardation can be worthwhile for him and his family increases the chance that we will strive to ensure such an outcome (Evans, 1983; Rothman and Rothman, 1984; Turnbull and Turnbull, 1984). Yet we still have to be prepared to respect the right of a particular pregnant woman in a particular situation to decide what she can accept for herself and for her child.

Genetic Counseling

Ideally, prospective parents should not come to genetic counseling uninformed about the potential of disabled children. Information about living with physical, sensory, emotional, and intellectual impairments should be part of high school curricula and prenatal education and care. However, until such time as the public is more knowledgeable about disability, genetic counselors bear the responsibility of providing a general orientation to disability.

Genetic counseling can be an undeniable boon to prospective parents, but there is evidence that genetic counselors need to educate themselves to be more nuanced and thorough in conveying information about disability. Recent investigations of the attitudes and practices of genetic counseling do not bode well for helping users of prenatal diagnostic techniques to gain any real insight into the meaning of disability in today's society. The counseling described in Applebaum and Firestein's (1983) *A Genetic Counseling Casebook* failed to include information about life with the disabling conditions under consideration. Although the cases therein dealt with a wide range of impairments, only in one case report involving Down syndrome did the counselor discuss services, schools, or the likely quality of life of the disabled child in the particular locale. Rothman (1986) reports that counselors offered only the barest medical information, when asked by clients, and that details about diagnosable conditions were given only *after* one of these conditions had been detected. Most results from amniocentesis are not conveyed until about 19 weeks of gestation. This is just 5 weeks short of the beginning of the third trimester of pregnancy, and in the third trimester, abortion is a far more complex procedure, available from fewer providers. Women have little time between learning of fetal disability and the end of the period in which they would usually decide whether to continue a pregnancy or abort. They would be hard pressed to obtain and absorb the range of information that would be optimal for genuinely informed choices in those remaining weeks.

Elias and Annas (1987) conclude that "although most counselors claim to use a nondirective approach, few deny that an element of counselor bias always exists" (p. 40). Rothman (1986) found that nearly all of the genetic counselors she interviewed would themselves abort for Down syndrome; half would abort for a disability that required use of a wheelchair such as some instances of spina bifida; half would abort for blindness; and one-third would abort for deafness. Despite their professional commitment to nondirective counseling, it is hard to believe that counselors with such attitudes do not communicate them to anxious clients by words chosen, questions asked, information given, or topics omitted.

I wish to argue that entitlement to the new technology obliges people to obtain detailed information about the disabilities that the screening detects. Exposure to this information should be as much a part of the counseling procedure as is the medical data that counselors now provide. This assumption should be built into the licensing process for all of the health care professionals involved in providing prenatal screening. No licensed professional should offer screening to a client who will not consent to receive information about disabilities.

Genetic counseling that leads to truly informed decision making about disability must include:

1. a detailed description of the specific biological or psychological limitations associated with particular disabilities and what they mean in terms of day-to-day functioning
2. information about services to benefit children with specific disabilities in a particular area, and about which of these a child is likely to need immediately
3. discussion of the laws governing education, entitlement to financial assistance for children and families, access to buildings and transportation, employment, and housing for disabled people
4. contact with a representative of parent groups
5. contact with a representative of a disability rights group and an independent living center
6. a visit with a child and family living with the diagnosed disability and
7. articles and books by parents of similarly affected children.

Counselors can convey much of this material at the time women go for tests and during the waiting period between test and result. Learning about equal opportunity laws for disabled people and government and philanthropic services for disabled children and their families need not await a diagnosis of a particular fetal impairment, but should be part of the initial screening and pretest workup. People could receive a few articles by parents of disabled children among the informational handouts they get. Contact with disability rights groups, an independent living center, and disabled children and their families could begin during the waiting period between taking the test and getting the diagnosis.

Only that information pertinent to the specific diagnosis must come after the woman receives test results. Although the time is short between receipt of results and the beginning of the third trimester of pregnancy, counselors should work with women closely and quickly to ensure that they begin gathering disability-specific knowledge, and have personal contact with children and families within days of the news. A balanced presentation should include exposure to women who chose not to bear or to rear disabled children, and to parents of one disabled child who have used the technology to avoid having another, as well as to mothers and fathers who chose to raise children with disabilities.

Information is only valuable in a context. The counselor should work with the woman who has received a diagnosis of fetal disability to examine how having a child with that particular condition will affect her life, the life of others to whom she is committed, and how, if at all, it will diminish her hopes for herself and her child. Women who learn that they are carrying a fetus with Down syndrome or spina bifida may decide that they need no information beyond the diagnosis before making an appointment for an abortion. The same professionals who counseled and tested these women should indicate that responsibly concluding the counseling process requires time for contact with such disabled children and their families as well as some time for reflection.

The Training of Genetic Counselors

The genetic counselor cannot convey this information if she herself (and the counselor usually is a woman) does not have it. Typically, the genetic counseling curriculum covers only the science of genetics, medical information about some genetically transmitted conditions, and techniques of counseling. The ideal genetic counseling training program should include courses on cultural diversity and the social meaning of disability. Appendix A is a sample course on the disabled person in contemporary society. Originally designed for liberal arts undergraduates, the course can be adapted for professionals by requiring more experiential work with disabled people and their families, and more involvement with service programs and advocacy groups. Without a grounding in how people with various disabilities live their lives and exposure to a social perspective on disability, counselors cannot knowledgeably and compassionately assist women and families in these decisions.

Access

All individuals who wish to undergo prenatal genetic screening should have financial access to it. Medicaid programs have begun to fund such access, and all private insurance arrangements should do so as well.

However, guaranteeing access must not be tied to coercion in any form. There must be no enforced prenatal diagnostic testing of any pregnant women, including those with AIDS. Providers of prenatal screening should not give their clients any test results that the clients do not wish to receive. Users of the technology should be permitted to declare which test results they wish to receive and which they wish withheld. Fetal gender is often withheld at the client's request. So, too, could information about disabilities be withheld if what is discovered is an impairment outside the set for which someone originally sought testing. Rothman (1986) calls attention to the dilemma that can be posed for parents if they learn of chromosomal characteristics that represent deviations from the norm, such as XYY in males, but whose disabling consequences have never been established. If women decide that they will abort only for particular disabling conditions, they may wish to assert their right not to know of any other conditions the tests disclose. However, since it is always possible that tests could reveal something entirely unexpected and likely to be of grave consequence to the fetus, informed refusals must be designed to allow some latitude to physician and counselor.

Nor should there be any coercive pressure to abort if certain conditions are detected. We must vigorously oppose all directive counseling by genetic counselors, physicians, hospital administrators, and insurance companies. Such concern is not alarmist. Stanworth (1988) cites a British survey that found that 75 percent of physicians "required women to agree to abortion of an affected fetus before they gave amniocentesis" (p. 31). We must oppose any efforts by insurance companies to deny women and their babies third-party payment, if the company believes medical problems should have been detected and avoided by abortion. Women who are emotionally prepared to raise children with disabilities must not be pressured out of doing so because the medical or counseling professions oppose it or because insurers view disabilities as preexisting conditions. If women truly have the right to choose, they must have the means to choose to raise whatever type of child they wish.

Wrongful Birth and Wrongful Life

The advent of screening, testing, and selective abortion has given rise to two types of lawsuits commonly known as "wrongful birth" and "wrongful life." In name and connotation, they go beyond medical malpractice suits. The former is brought by a woman on behalf of herself for the pain, anguish, and extraordinary expenses incurred in raising a child with a disability whom she bore because of lack of information or improper professional conduct of screening and testing. Suits for wrongful life are brought on behalf of the child born with a disability, with someone claiming that the child's life is itself a wrong deserving compensation and that, but for improper professional behavior toward the child's mother, the child would not exist.

Commitment to access and choice for women commits us to support those women who believe that they were denied testing or adequate counseling and, thus, bore a child with a disability whom they would have aborted had they known. We can support such suits only because they may be one of the few ways to compel the medical profession to inform and care for women responsibly during pregnancy. For this reason alone, feminists as well as advocates of disability rights can accept such actions.

Wrongful birth and especially wrongful life suits both respond to and perpetuate serious social problems. The suits are brought chiefly in order to obtain substantial financial settlements to compensate for the medical malpractice that gave rise to the child's birth, and to assist the woman and child with the expenses disability either imposes in reality or is thought to impose. Supporting women's rights to wrongful birth suits, however, must not lead to accepting a system of care and services to people with disabilities based on the origin of the condition and not on the needs the condition engenders. If the parent can claim that the disability results from another's negligence in providing information or carrying out medical procedures, she or he can obtain funds to buy equipment and services to ease the lives of the child and the family. The child whose parents did not know about the existence of such suits, chose not to test, or chose not to abort may have exactly the same condition and require the same financial and human assistance, yet be forced to do without it because the parents lack the resources a settlement provided. We would better address the needs of all disabled children by working for comprehensive health and family disability insurance and adequate disability-related public services, regardless of the origin of the disability.

There is reason for us to fear wrongful birth suits and to oppose suits for wrongful life: it is the message they send to the children themselves, disabled people, and society about the worth of lives with impairments (Steinbock, 1986). Although the child with profound retardation will not understand that the parent is suing for damages because of his or her birth, an adolescent with spina bifida, cystic fibrosis, or sickle cell anemia could easily be psychologically damaged by discovering that a parent took legal action because of his or her birth. It is sad but true that the unintended birth of a disabled child may injure a mother herself financially, socially, and psychologically. Thus, we can accept her need to take action on her own behalf in the form of a suit for malpractice or wrongful birth.

Concerns about the meaning of such suits for disabled children, disabled people, and society in general lead to unequivocal opposition to wrongful life suits. Justifications have been advanced for these suits on two grounds: first, to ensure financial compensation to the child for disability-related expenses that will occur in adulthood; and second, to compensate for the wrong of forcing someone to live with a disability that could have been avoided by not bringing the child into the world. Neither ground justifies suits claiming wrongful life, as opposed to medical malpractice. Adverse economic consequences should be alleviated by the general societal changes already discussed. We have to acknowledge that, even in a nondiscriminatory society, some disabilities will be so severe that independence, self-support, and productivity will be precluded; nonetheless, it is impossible to condone suits that claim that a particular child's life should not exist or is itself an injury deserving financial compensation. Courts awarding damages in wrongful life actions have argued that living with a disability is worse than the harm of nonexistence. However, this is an incorrect comparison if one is genuinely interested in assessing the damages disability imposes on a child or adult, which should be financially remunerated. It would be far better to assess only disability-related costs and provide compensation that would put the child and family in a situation more comparable to that of a child and family without disability. Claiming that life with disability is worse than no life at all offends self-respecting disabled people and represents the extreme of what is dangerous about testing, diagnosing, and suing.

Time Limits on Abortion

The question of whether or not states would be justified in restricting postviability abortions has significant relevance for the disabled. Women with disabilities may be especially vulnerable to situations that give rise to late abortions. It is crucial to protect their access to late abortions.

Access for Disabled Women

All of the circumstances that make teenagers figure prominently among those who have late abortions might also characterize the situation of the institutionalized disabled woman, especially one who is young or who has severe limitations in verbal or intellectual skills. Such a woman may take longer to recognize pregnancy, or, once aware of it, to get those who assist her to listen, to believe, and to help decide what to do. Even in circumstances where she might be able to continue with the pregnancy if she wished, learning about all of the programmatic options--services, financial support, attendant care, and status within a residence or an institution--could be quite time-consuming. Alternatively, discovering that her circumstances are compromised if she continues with a pregnancy could be delayed until the time after viability, causing her to need an abortion that could be difficult to obtain.

In the case of any woman, her economic or social arrangements may alter during pregnancy, causing her to decide that a once-wanted pregnancy is now out of the question. Women with disabilities even out of institutions run a greater-than-average risk of finding that their circumstances are seriously constrained by forces beyond their control: they may lose jobs, find that attendants will not remain after the birth of a child, or discover that relatives, friends, or service providers refuse assistance if a pregnancy is carried to term. Such constraints should be opposed by feminists and disability rights advocates, but meanwhile, the disabled woman needs the option of abortion or adoption just as does any other woman.

Certainly health changes during pregnancy can seriously affect the woman with a disability. Women with diabetes, cardiovascular disease, high blood pressure, kidney disease, suppression of the immune system, or breast, ovarian, or cervical cancer are at increased risk of health problems during pregnancy that might lead them to seek a late abortion. Women with these conditions may successfully carry pregnancies to term, but sometimes conditions that appear to pose no problem in early stages of pregnancy may become aggravated later and threaten the woman's life or health. Poor women, black women, and other women of color have higher rates of such conditions as diabetes, cardiovascular disease, high blood pressure, cervical cancer, and other disabilities that may lead women to need abortions late in pregnancy. Obviously, these and other disabled women need the ability to safeguard their health that perhaps only the option of late abortion can give them. We must make sure that no further cutbacks in family planning or abortion services erode such women's access to options that they need to protect their health.

Fetus as Patient

For advocates of women's rights and disability rights, questions raised about the status of the fetus-as-patient have two foci: the rights of a disabled pregnant woman and the welfare of an affected fetus. Disabled pregnant women are in a double bind. On one hand, they lack adequate access, to potentially valuable perinatal technology, and on the other hand, if they gain greater access, they risk possible exploitation in the name of medical experimentation. The quest for developing interventions to treat the fetus-as-patient is unfortunately propelled by negative stereotypes about disability. The unpleasant connotations of disability in the public mind glorify fetal therapy. Cases in which maternal and fetal needs come into conflict over the treatment of the fetus *in utero* represent a troublesome problem--promoting women's autonomy during pregnancy may at times run afoul of the worthy goal of doing what we can to prevent avoidable impairment in our children.

Access for Disabled Women

Although I believe that doctors and prospective parents have ethical obligations to the fetus before and after viability, the entire discussion about fetal intervention must be cast in the framework of respect for the pregnant woman's commitment to do the best she can for the being she has elected to carry to term. Most women will not willfully engage in activities known to harm their fetuses, and thus, common sense dictates that society should allocate considerable resources to public education and to high-quality prenatal care for all women rather than to the search for ever-more-sophisticated technological fetal interventions for some. Certainly, all measures should be taken to clarify to the pregnant woman what her acts or omissions can do. After that, it must be up to her. As long as fetuses reside within women, women must make the decisions about what happens to their own persons, bodies, and lives.

Women with disabilities will be in an extremely vulnerable position when it comes to their making medical decisions. Disability may deny them access to technology or may result in their being at an especially great risk of medical interference. Their disabilities may cause medical complications during late stages of the pregnancy, and the fetuses they carry may run increased risks over those carried by women without cardiovascular, diabetic, epileptic, or other conditions. They may need more extensive prenatal care, with all the technological armamentaria feminists tend to fear, as well as an unusually sensitive and supportive obstetrician or birthing center. Prejudice and poverty may preclude access to high-quality prenatal care, and thus, the children of such women, like the children of other low-income and minority women, run the risk of incurring needless disability.

Women with stigmatized conditions such as paraplegia, blindness, deafness, mental retardation, or histories of psychiatric illness face prejudices on the part of medical staff who may mistrust their powers of comprehension or decision making. Such mistrust could lead the health care providers either to deny the women technological interventions they would provide to others or to compel the women to undergo monitoring, medication regimens, hospital confinement, cesarean sections, or fetal surgery. Disabled women are at least as vulnerable as those without impairments to condescending, paternalistic treatment at the hands of the medical establishment, and thus, our efforts to educate women about technology and medical staff about the capacities of women as decision makers must take account of the particular concerns of disabled women. In instances of demonstrable decisional incapacity, medical professionals will continue to be able to rely on the courts for appropriate guidance. However, in general, we must support women's rights to full, clear information about the pros and cons of particular courses of action, and then endorse their decision-making rights.

Fetal Intervention to Prevent Disability

Those favoring medical intervention to protect the fetus argue that their mission is to prevent needless illness and impairment in children. We must remember, however, that the efficacy of medical interventions on behalf of the fetus, such as medication, surgery *in utero*, or cesarean delivery, is not proven. Most impairments are best treated after birth. Only three conditions--hydrocephalus, herniated diaphragm, and hydronephrosis--may be more responsive to surgical correction before birth than afterward (Blakeslee, 1986). In this arena, it is especially important to be wary of exaggerated claims of the value of high-tech interventions as crucial to the welfare of the fetus.

Proponents of intervention make the same claims for fetal therapy and for many cesareans that are made for prenatal diagnosis and selective abortion: children, families, and society would suffer irreparable harm from disability. John Fletcher (1984), for example, reports an estimated 1.5 billion dollars in direct costs and 1.4 billion dollars in foregone earnings as a result of congenital disability. He notes that parents of disabled children will need to reduce or curtail employment to care for them (a concern particularly for mothers), and that children with disabilities will grow up to be forever unemployed or underemployed. Here is the epitome of the medical model of disability: disability must be prevented, because disabled people cannot function within existing society. If we evaluate reproductive technologies within a medical model of disability, we will all suffer. Instead, we must seek to prevent disability while working to change the conditions that impose such dire consequences on children, families, and society.

Feminists rightfully fearful that the trend of seeing the fetus as a separate patient will erode women's autonomy during pregnancy have sometimes suggested a halt to all attempts at fetal surgery. Such a halt seems unwise, just as ending searches for other cures of disabling conditions would be unwise. Our task is to use technology for good, not to oppose its creation because we fear the motives of those who control it. We would be best advised to suggest that most high-tech fetal intervention could be avoided with better general prenatal care for all. Some fetuses, however, will benefit immensely from monitoring, cesarean birth, or other interventions. We should provide such therapies only in instances where the intervention cannot be done successfully or effectively after delivery of the newborn. We can support judicious use of the technology in those few instances where it will benefit the fetus as long as it is coupled with consideration for the needs and rights of the pregnant woman whose body will be invaded.

When Pregnant Women Disobey Medical Advice about the Welfare of the Fetus

Anyone who knows that her error in judgment created otherwise avoidable disability in her child will live with extreme distress for a long time. We can support a woman's right to make such decisions, while acknowledging and asking her to acknowledge that the consequences can be grave for a child she sought to have and for someone she loves. She will have to explain to herself, others in her world, and possibly the child that she caused impairment. In cases such as these, we have to do more than simply safeguard the pregnant woman's decision-making rights and protect her from criminal liability. Women who refuse suggested medication, fetal surgery, or cesarean sections must be given clear information about what the social, psychological, and institutional consequences of a particular disability can be.

I may deplore what some women will do, but I am not yet prepared to take away their rights of self-determination. Although it is true that another potential person is involved--a being in whom women have invested something and at least implicitly have accepted as belonging to them and to the world by accepting the responsibility of pregnancy--all that we can and should do is inform, encourage, and warn. People misuse their bodies daily. Even as parents of existing children, people undoubtedly act or fail to act in ways that harm themselves or their children. Pregnant women should be subject to advice and counsel, but only in extreme cases might I ever countenance governmental interference.

Reproductive Hazards in the Workplace

Creating safe workplaces is probably the least controversial area of the new reproductive technology topics under discussion. The goal of cleaning up workplaces to ameliorate workplace hazards that may cause impairment in adults or fetuses is a laudable one, but I would like to offer a few caveats.

We must demand protection against hazard-induced disability without exaggerating the disastrous consequences of disability or perpetuating stereotypes of disability. Feminists striving to broaden corporate understanding of the risks to all workers should consciously work with disability rights advocates to articulate an informative, persuasive, and nonstigmatizing message. In this way, people with disabilities can join in the battle against reproductive hazards in the workplace without it being prejudicial to them.

The effort to eliminate hazards in the workplace has to include attention to the needs of special segments of the disabled population. For those disabled women and men in competitive employment, there are not necessarily greater risks of fetal impairment than there are for any other worker. Many disabled people, however, work in noncompetitive employment, in their own homes, in institutions, or in sheltered workshops. None of these employment settings is likely to come under close government scrutiny for safety generally, yet disabled people need the same protection as others and need allies beyond the disability community to fight for it. Similarly, designers of equipment used by disabled people in and out of the labor force must be regulated to guarantee that they not ignore potential reproductive hazards of any new devices produced.

Exclusionary Medical Standards and Screenings

My last concern is that we not work only for reproductive safety and ignore workplace practices that bar already disabled people from the workplace. Those of us who wish to ensure that employers not screen out workers who may carry genetic disabilities or who may be at higher-than-average risk for having their genes altered by exposure to toxic substances must work with disability rights activists to prohibit all medical standards and screenings that needlessly bar disabled people from jobs that they can perform. Women used to find themselves prevented from working by laws designed to promote their safety. Disabled people still face the same benevolent paternalism that results in exclusion and poverty. Companies throughout the nation exclude disabled people from jobs by using medical standards that have nothing to do with job performance and pertain only to the worker's future risk of disability (and therefore future use of an employer health insurance policy).

As employers, the federal government, federal contractors, and federal grantees are prohibited from using discriminatory medical standards under Sections 501, 503, and 504 of the Rehabilitation Act, and some states such as New York and California have used anti-discrimination laws to curtail the use of prejudicial medical standards. Nonetheless, the fight to have only job-related medical screenings is far from won. Most campaigns to outlaw genetic screening of workers and to promote workplaces free of reproductive hazards have been undertaken without any awareness of how these reproductive issues fit into the broader framework of disability rights in the workplace. The issues of worksite genetic screening and reproductive hazards can be important for building coalitions among feminist, labor, and disability groups if we are willing to make them so. (Those wishing for more information should contact the Employment Law Center in San Francisco, the Disability Rights Education and Defense Fund in Berkeley, and the Compliance Investigation Unit of the New York State Division of Human Rights.)

Alternative Modes of Reproduction

From the arena of least controversy, I turn to the reproductive topic arousing the most debate among feminists and members of the public. Other sections of this book articulate a range of feminist perspectives on in vitro fertilization, egg and sperm donation, and various forms of surrogacy. Feminists in particular and society in general may decide to restrict, regulate, ban, or encourage some or all such practices based on a host of economic, moral, and social considerations quite apart from their implications for the lives of disabled people. Commitment to disability rights does not automatically lead to a particular position on the merits or drawbacks of these reproductive practices, but it does raise questions that might otherwise go undiscussed amid the tumult of conflicting positions and voices. As with the other topics, my concern here is both with implications for people with disabilities who may be freed or restricted by these practices and with implications for fetuses with impairments who may or may not be born, as a result of the technologies.

All forms of technologically assisted reproduction cause us to prize biological and genetic ties to children. So too do they increase our efforts to ensure that our offspring are free from genetic impairment. Feminists evaluating these technologies must grapple with concerns about enshrining the genetic and biological components of parenting and must squarely face the possibilities posed for genetic screening, genetic manipulation, and selection of gender and other characteristics. From the perspective of disability rights, four concerns are uppermost: Will women with disabilities have access to these forms of artificial reproduction? Are there any disabled women who will especially benefit from these new technologies? Will women or men likely to pass on disabilities be increasingly pressured to forego reproduction unless they employ means to avoid transmitting their genes? Should people with disabilities who will not raise children, but whose eggs or sperm will pass on genetic conditions be permitted to participate in artificial reproduction?

Benefits for Disabled Women

In the introductory sections of this essay, I discussed the biological and social obstacles to motherhood for disabled women. Some disabilities include inability to conceive, others create difficulties in maintaining a pregnancy, and still others limit social opportunities for forming relationships with men who would choose to share in raising children. Also, unrelated to an original disability, a disabled woman might discover that she is among the growing number of the nation's women who find themselves unable to become pregnant. For all of these reasons, disabled women may, in fact, be even more interested in technologically assisted reproduction than are their nondisabled counterparts.

Whether using artificial insemination or any of the newer technologies, disabled women may be driven to try the biological reproductive options simply because adoption administrators have been as reluctant as everyone else to accept disabled women as mothers. Some strides are being made, especially with foreign adoptions, but adopting for the disabled woman whether in a couple or alone is hazardous, and very likely to involve being subjected to demeaning, intrusive investigations, court battles, and attempts to force her to take a hard-to-place "undesirable" child. Social stereotypes are perpetuated, with adoption agencies insisting that a disabled parent, seen as less desirable, will be allowed to raise only a child considered less desirable. As a blind man said recently when discussing the possibility that he and his wife, also blind, might adopt: "I knew that as blind parents we weren't going to get the pick of who was out there. It makes me want to take my chances on the potluck of biology."

Interest in the technology will not lead to access for women with disabilities unless we remove the barriers of poverty and prejudice that so many disabled women confront. Many disabled women who would be potential users of the technology are unemployed and dependent upon government transfer payments from SSI (Supplemental Security Income) or SSDI (Social Security Disability Insurance) and government-funded medical care supplied through Medicaid and Medicare. Since infertility has not been considered a health problem, Medicaid and Medicare do not now fund infertility treatments. Certainly, disabled women who are not affluent, whatever their employment or health insurance status, will not have the resources to finance using a surrogate carrier. The relatively inexpensive and uncontroversial technique of artificial insemination will still be beyond the economic reach of many women who would gladly mother a child. Even the employed middle-class or affluent disabled woman in a stable heterosexual relationship is likely to be denied her chance to use these techniques by the hostility and prejudice of the medical profession. Collins (1983) reports on how few obstetricians feel themselves qualified to offer care to spinal cord-injured women who become pregnant.

In *New Conceptions*, Lori Andrews (1985) mentions that gestational surrogacy may provide new options for women with diabetes, hypertension, and heart conditions to be the genetic mothers of their offspring, but not their gestators. These are good examples of women who may be unable to undertake conventional childbearing because of biological obstacles, apart from the social barriers to their becoming mothers. They may welcome the new reproductive technologies. Similarly, although many women with orthopedic impairments have successfully carried pregnancies, others have been advised against pregnancy because of its likelihood of increased physical stress and health risks, as evidenced by Elizabeth Stern in the *Baby M* case. These women may have no more likelihood than anyone else of passing on a genetic condition, and they may have no problem in conceiving. Using surrogates to carry their children offers them the chance to be genetic mothers and to avoid the medical complications of gestational motherhood. Such an option might greatly enrich their lives. Whatever stance we take on in vitro fertilization, surrogacy, and other alternative modes of reproduction, realizing that we might distinguish among them and have different positions about each, we must insist that women with disabilities have equal access to these and other means of becoming mothers.

Use of Alternative Modes of Reproduction to Prevent Disability

Although most people seeking to reproduce with donated genetic material do so because of male or female inability to beget, conceive, or carry a child, about 5 percent of those using artificial insemination by donor (AID) do so to avoid transmitting genetic conditions to their offspring (Ontario Law Reform Commission,

1985). No doubt, some women also use donated eggs or embryos to avoid having a child who will have a disabling condition. Sometimes the prospective mother or father has the particular disability to be avoided, and at other times the prospective parent carries the gene but is not herself or himself affected. If, after extensive experience of having a disability or of living with someone who has it, a person decides that she or he does not want that experience for a child, it makes good sense to take steps to prevent it. However, as in the case of prenatal screening, prospective parents may be pressured by their intimates, doctors, and genetic counselors not to "inflict" a disability on offspring. It is very sad to see people forego becoming parents because they are told that they have no right to bring a disabled child into the world. For myriad mysterious reasons, the desire to produce our own biological children is extraordinarily powerful, and there is no reason to think it any less so for disabled people or for those at greater-than-usual risk of passing on impairment. We grant women autonomy during pregnancy, and that occasionally means carelessly or inadvertently bringing into being a child with a disability. We should certainly grant people the right to decide that having their own biological child is sufficiently important to them that they are willing to help a child live with any disability that may result.

Donor Screening

Another problematic area is the medical and genetic screening of donors of genetic material. Commentators on new reproductive practices and on the older AID all report that practitioners have no uniform standards or procedures for ascertaining that users of donated gametes receive material free of the genetic impairments they wish to avert. Many reports call for regulations requiring practitioners to obtain thorough medical and genetic histories of sperm or egg donors, to screen such donors, and to reject gametes found to contain serious impairment (Annas and Elias, 1983).

I would like to offer a radically different recommendation: that prospective users should be given the histories of and screening results for donors and should decide for themselves which, if any, gametes to reject. Calling for high standards of donor screening and of professional accountability to those who use donated material to create their offspring is not synonymous with automatic rejection of gametes found to contain genes for disabling conditions. Whatever standards a woman or couple would employ for themselves if they were using prenatal screening, they should apply to the information they receive about donated material that they might use for assisted reproduction. Should a donor who carries the gene for a correctable heartbeat irregularity or for a predisposition to asthma be rejected? All of the same uncertainties of drawing lines and predicting outcomes that characterize prenatal screening exist as well in the area of donor screening.

Many would-be donors will have no knowledge of having any genetic predisposition to impairment or of carrying a potentially disabling condition. Someone who knew that she or he was a carrier of a gene for a serious impairment would not be likely to donate. We need an additional safeguard for prospective users in the form of thorough genetic screening of donated reproductive material. Practitioners should urge first-time and repeat donors not to pass on knowingly any serious impairment, when they have no intention of helping the offspring they create to live with the consequences of a condition that may be physically painful and very likely stigmatizing. Ultimately, however, the appropriate method of protecting prospective parents is through information about donor history and not through donor rejection. We may want to develop guidelines indicating that no sperm or eggs should be used if there is a likelihood of genetic impairment above some specified percentage. However, since we all carry some genes with the potential for creating disability, just what that percentage realistically could or should be is not yet evident.

Interference with Reproductive Choice

Interference with reproductive choice for disabled women begins long before laws about access to reproductive services and technology. It starts in exclusion from sex education classes and in parental silence about sexuality and motherhood. It continues in the inaccessibility of affordable gynecological services; the lack of safe birth control for some disabled women, who have little choice left but sterilization; and the still-prevalent sterilizations of people with slight and severe degrees of mental retardation in and out of institutions (Macklin and Gaylin, 1981; U.S. Commission on Civil Rights, 1983). Authorities on the mentally retarded advance three reasons, some of them parallel to those advanced for not having disabled children, for considering sterilization, even if involuntary, for persons with some level of retardation: concern for the welfare of the resulting children; concern about the drain on the state's resources entailed by assisting parents and their children; and concern about curbing the sexual expression of retarded people unable to manage safe birth control, but desirous of and able to manage sexual intimacy (Macklin and Gaylin, 1981). As recently as 1983, 15 states had statutes permitting the compulsory sterilization of people with mental illness or mental retardation in or out of institutions, and 4 states authorized sterilizations for persons with epilepsy (U.S. Commission on Civil Rights, 1983).

Accepting or encouraging the desire of disabled women to be mothers requires us to evaluate what we expect of parents in general and of mothers in particular. If state welfare offices and family courts remove children from mothers who can soothe them with words and smiles but cannot themselves hold, cradle, or feed them, are these bureaucracies not equating mothering with physical acts rather than psychological ones? Is it fair for a welfare department to refuse a disabled woman additional attendant services to help her with her children, when the same agency appropriately provides additional money for food and clothing for nondisabled women with children? What about the woman whose mental retardation or mental illness does not preclude physical contact with her child and some level of emotional attention, but does suggest limited ability to make wise judgments about a child's physical safety or emotional growth? Arguably, there may be some people with or without diagnosable disabilities who are inadequate or unfit parents. However, there is no body of evidence to suggest that a parent's disability in itself harms children, and there is some evidence to suggest that raising children enriches the lives of disabled people as it enriches the lives of the nondisabled (Collins, 1986; Thurman, 1985).

Access for Disabled Women

Assuring disabled women reproductive choice, then, means working for birth control options for those women hindered in the use of the pill, diaphragms, or IUDs (Boston Women's Health Book Collective, 1984), so that disabled women are not compelled to risk unwanted pregnancies or sterilization in order to engage in heterosexual relationships. It also requires committing our resources not only to aid the infertile in general to have or to adopt children, but also to aid disabled women and men to have the economic means, legal sanction, and social and medical supports to participate in this valued area of life. We must oppose any efforts to deny disabled women access to the new reproductive technologies, just as we would oppose attempts to preclude usage of the technologies by single women or lesbians. As important as any other single reproductive issue for disabled women is the right to choose childbearing.

Conclusion

Feminists and advocates of disability rights share important beliefs and values. They prize rights to self-determination. They believe that biology is not destiny, and is far less important than social values and arrangements. Promoting reproductive freedom for all women requires recognizing and affirming our commitments to these shared views. It also requires clear support for the diversity among women. Endorsing options and choice for all women, believing that neither for women nor for people with disabilities must biology be controlling of life opportunities, and respecting the fact that different circumstances will cause women to make different decisions all should aid feminists and disability rights advocates to work together for policies on reproductive technology that we can all live with and struggle to attain.

Appendix A

Syllabus: The Disabled Person in American Society[2]

Primary Texts

Asch, A. and Fine, M. (Eds.). (1988). *Journal of Social Issues*, 44 (1). Special issue entitled *Moving Disability Beyond "Stigma."*

Gartner, A. and Joe, T. (Eds.). (1987). *Images of the Disabled, Disabling Images*. New York: Praeger.

Gliedman, J. and Roth, W. (1980). *The Unexpected Minority: Handicapped Children in America*. New York: Harcourt Brace Jovanovich.

International Center for the Disabled. (1986). *The ICD Survey of Disabled Americans, I*. New York: Louis Harris and Associates, Inc.

International Center for the Disabled. (1987). *The ICD Survey II: Employing Disabled Americans*. New York: Louis Harris and Associates, Inc.

Mappes, T. A. and Zembaty, J. S. (Eds.). (1986). *Biomedical Ethics*, second edition. New York: McGraw-Hill.

National Council on the Handicapped. (1986). *Toward Independence*. Washington, D.C.: U.S. Government Printing Office.

Unit 1: Defining the Problem:
Conceptual and Methodological Issues

1. Introduction/Defining Disability: Facts and Values
   - Mappes and Zembaty (1986). pp. 243-263. See Texts above.
   - Berscheid, E. and Walster, E. (1972, March). Beauty and the Beast. *Psychology Today*, 32-42.
   - Federal Register (1977). 45 CFR 84.3J. Vol. 42, No. 6.
   - Benjamin, M., Muyskens, J., and Saenger, P. (1984). Growth Hormones and Pressures to Treat. *Hastings Center Report*, 14 (2).

- National Council on the Handicapped (1986). pp. 3-6. See Texts above.
- 96th Congress, 1st Session, *Rehabilitation, Comprehensive Services, and Developmental Disabilities Legislation.* Washington, D.C.: U.S. Government Printing Office. Title I, Sections 7 and 13.
- *School Board of Nassau County vs. Arline*, 94 L.Ed.2d 307 (1987).

2. Disability Cross-culturally
   - Scheer, J. and Groce, N. (1988). Impairment as a Human Constant: Cross Cultural and Historical Perspectives on Variation. *Journal of Social Issues*, 44 (1), 23-37.
   - Groce, N. (1980). Everyone Here Spoke Sign Language. *Natural History*, 89 (6), 10-16.

3. Getting Data about People with Disabilities
   - *ICD* (1986). Introduction, pp. 1-9. See Texts above.
   - U.S. Department of Commerce, Bureau of the Census. (1983). *Labor Force Status and Other Characteristics of Persons with a Work Disability: 1982. Current Population Reports.* Washington, D.C.: U.S. Government Printing Office. pp. 1-8.
   - *Disability Rag* (1984, June). Entire issue.
   - *New York Times* (1986, December 23). Census Study Reports 1 in 5 Adults Suffers from Disability. p. B7.

4. Disability as Seen in Personal Narratives and Poetry
   - Asch, A. (1984). Personal Reflections. *American Psychologist*, 39, 551-552.
   - Baird, J. L. and Workman, D. S. (Eds.). (1986). *Toward Solomon's Mountain.* Philadelphia: Temple University Press.
   - Brightman, A. J. (Ed.). (1984). *Ordinary Moments: The Disabled Experience.* Baltimore: University Park Press. Introduction - 5, and "Impressions."

5. Disability as Seen in Fiction and Drama
   - Examples: Greenberg, *In This Sign.* Medoff, *Children of a Lesser God.* Williams, *The Glass Menagerie.*
   - Gartner and Joe (1987). Chapters 2, 3, and 4. See Texts above.

6. Psychology and Disability - Part 1: How Nondisabled Social Scientists Report Disabled People's Life Experiences
   - Bulman, R. and Wortman, C. (1977). Attribution of Blame and Coping in the "Real World": Severe Accident Victims React to Their Lot. *Journal of Personality and Social Psychology*, 35, 351-363.
   - Taylor, S. E., Wood, J. V., and Lichtman, R. R. (1983). It Could Be Worse: Selective Evaluation as a Response to Victimization. *Journal of Social Issues*, 39 (2), 19-40.
   - Frank, G. (1988). Beyond Stigma: Visibility and Self-Empowerment of Persons with Congenital Limb Deficiencies. *Journal of Social Issues*, 44 (1), 95-115.
   - Schneider, J. W. (1988). Disability as Moral Experience: Epilepsy and Self in Routine Relationships. *Journal of Social Issues*, 44 (1), 63-78.

   Supplemental book-length treatments:
   - Ablon, J. (1984). *Little People in America*. New York: Praeger.
   - Evans, D. (1983). *The Lives of Mentally Retarded People*. Boulder: Westview Press.
   - Higgins, P. C. (1980). *Outsiders in a Hearing World*. Beverly Hills: Sage.
   - Macgregor, F. C. (1979). *After Plastic Surgery: Adaptation and Adjustment*. Brooklyn: J. F. Bergin.
   - Schneider, J. W. and Conrad, P. (1983). *Having Epilepsy: The Experience and Control of Illness*. Philadelphia: Temple University Press.

7. Psychology and Disability - Part 2: Attitudes and Interactions
   - Makas, E. (1988). Positive Attitudes toward Disabled People: Disabled and Non-disabled Persons' Perspectives. *Journal of Social Issues*, 44 (1), 49-61.
   - Gliedman and Roth (1980). pp. 67-86, Appendix 3. See Texts above.
   - Asch, A. (1984). The Experience of Disability: A Challenge for Psychology. *American Psychologist*, 39, 529-536.
   - Fine, M. and Asch, A. (1988). Disability Beyond Stigma: Social Interaction, Discrimination, and Activism. *Journal of Social Issues*, 44 (1), 3-21.

8. From Deviance and Stigma to Minority Group Theory-Part 1
   - Goffman, E. (1963). *Stigma: Notes on the Management of Spoiled Identity.* Englewood Cliffs, N.J.: Prentice-Hall.
   - Katz, I. (1981). *Stigma: A Social-Psychological Analysis.* Hillsdale, N.J.: Erlbaum. Chapter 2 and pp. 102-124.
   - Bogdan, R. and Taylor, S. (1976). The Judged Not the Judges. *American Psychologist,* 31 (1), 47-52.
   - Mest, G. M. (1988). With a Little Help from Their Friends: The Use of Social Support Systems by Persons with Mental Retardation. *Journal of Social Issues,* 44 (1), 117-125.

9. From Deviance and Stigma to Minority Group Theory-Part 2
   - Stroman, D. F. (1982). *The Awakening Minorities.* Washington, D.C.: University Press of America. Chapter 1.
   - Hahn, H. (1988). The Politics of Human Differences: Disability and Discrimination. *Journal of Social Issues,* 44 (1), 39-47.
   - Gartner and Joe (1987). Introduction. See Texts above.

Unit 2: Disability over the Life Cycle

10. Growing Up Disabled
    - Gliedman and Roth (1980). Chapters 4-7. See Texts above.
    - Lussier, A. (1980). The Physical Handicap and the Body Ego. *International Journal of Psychoanalysis,* 61, 179-185.
    - Diamond, S. (1981). Growing Up with Parents of a Handicapped Child: A Handicapped Person's Perspective, in J. L. Paul, (Ed.), *Understanding and Working with Parents of Children with Specific Needs.* New York: Holt, Rinehart, and Winston, pp. 223-250.
    - Rousso, H. (1984, December). Fostering Healthy Self-Esteem. *The Exceptional Parent,* 9-14.

11. The Disabled Child and the Family
    - Gliedman and Roth (1980). Appendices 1 (pp. 305-363) and 5 (pp. 412-416). See Texts above.

- Darling, R. B. (1988). Parental Entrepreneurship: A Consumerist Response to Professional Dominance. *Journal of Social Issues,* 44 (1), 141-158.
- Massie, R. and Massie, S. (1984). *Journey,* second edition. New York: Knopf.
- Turnbull, H. R. and Turnbull, A. P. (1985). *Parents Speak Out: Then and Now.* Columbus, Ohio: Charles E. Merrill.

Supplemental reading:

- Darling, R. B. (1979). *Families against Society.* Beverly Hills: Sage. Chapters 1, 2, 7, and 8.

12. Education - Elementary and Secondary
    - Biklen, D. P. (1988). The Myth of Clinical Judgment. *Journal of Social Issues,* 44 (1), 127-140.
    - Gartner and Joe (1987). Chapter 6. See Texts above.
    - Federal Register (1977). 45 CFR 84.31-84.39. Section 504, Regulations.
    - *Board of Education of the Hendrick Hudson Central School District vs. Rowley.* 458 U.S. 176, 204 (1982).
    - National Council on the Handicapped (1986). pp. 47-49. See Texts above.

13. Education - Post-Secondary Education
    - Federal Register (1977). 45 CFR 84.41-84.49.
    - *Southeastern Community College vs. Davis,* 99 S.Ct. 2361 (1979).

14. Sexuality and Disability
    - Rousso, H. (1988). Daughters with Disabilities: Defective Women or Minority Women? in M. Fine and A. Asch, (Eds.), *Women with Disabilities: Essays in Psychology, Culture, and Politics.* Philadelphia: Temple University Press. pp. 139-171.
    - Bullard, D. G. and Knight, S. E. (Eds.). (1981). *Sexuality and Physical Disability.* St. Louis: C. V. Mosby. Introduction, pp. 1-2, Chapters 1-7, 9, 10, 16, and 21.
    - Kilmartin, M. (1984, May). Disability Doesn't Mean No Sex. *Ms.,* 114-118, 158-159.

15. The Rehabilitation System - Medical and Vocational
    - Wright, B. A. (1983). *Physical Disability - A Psychosocial*

Approach, second edition. New York: Harper and Row. Chapter 17.
- U.S. Department of Health, Education, and Welfare (1975). *Report of the Comprehensive Services Needs Study.* Contract No. 100-74-03-09. Washington, D.C.: U.S. Government Printing Office. pp. 39-55.
- 96th Congress, 1st Session. *Rehabilitation, Comprehensive Services, and Developmental Disabilities Legislation.* Title I, Sections 100-103, and 112.

16. Rehabilitation Part 2 - Critiques, Micro and Macro Levels
    - Scott, R. A. (1969). *The Making of Blind Men: A Study of Adult Socialization.* New York: Russell Sage Foundation. pp. 56-104.
    - Kemp, E. (1981, September 3). Aiding the Disabled: No Pity Please. *New York Times.*
    - Sheed, W. (1980, August 25). On Being Handicapped. *Newsweek*, p. 13.
    - Mappes and Zembaty (1986). pp. 601-626. See Texts above.
    - *Hastings Center Report*, 17 (4) (1987, August/September). Ethical and Policy Issues in Rehabilitation Medicine.

17. Employment
    - Kent, D. (1978). Close Encounters of a Different Kind. *Disabled USA*, 1 (5).
    - 96th Congress, 1st Session, *Rehabilitation, Comprehensive Services, and Developmental Disabilities Legislation.* Title VI, Sections 621-623.
    - Federal Register (1977). 45 CFR 84.11-84.19.
    - ICD (1986). pp. 22-27, 46-60, 69-81. See Texts above.
    - ICD (1987). Entire issue. See Texts above.
    - National Council on the Handicapped (1986). pp. 18-29. See Texts above.
    - Gartner and Joe (1987). Chapter 7. See Texts above.
    - *Nelson vs. Thornburgh*, 567 F. Supp. 369 (1983).
    - *Miller vs. Ravitch*, 60 N.Y.2d 527; 458 N.E. 2d 1235; 470 N.Y.S.2d 558 (1983).
    - *McDermott vs. Xerox Corp.*, 65 N.Y.2d 213; 480 N.E.2d 695; 491 N.Y.S.2d 106 (1985).

18. Living in the Community: Housing, Transportation, and Social Life

- ICD (1986). pp. 32-34, 62-69. See Texts above.
- National Council on the Handicapped (1986). pp. 32-46, 50-54. See Texts above.
- Crewe, N. M. and Zola, I. K. (Eds.). (1983). *Independent Living for Physically Disabled People.* San Francisco: Jossey-Bass. Chapters 9, 12, and 13.
- *Disability Rag* (1985, September). Entire issue.
- *Disability Rag* (1986, January). Entire issue.
- *City of Cleburne, Texas vs. Cleburne Living Center, Inc.*, 87 L.Ed.2d 313 (1985).

19. Marriage and Parenting
   - Bullard and Knight (1981). Chapters 11 and 12. See syllabus section 14 above.
   - Mappes and Zembaty (1986). pp. 306-326. See Texts above.
   - *Carney vs. Carney,* 157 Cal. Rptr. 393, 598 P.2d, 36 Sup. Ct. (1979).
   - Zola, I. K. (1982). And the Children Shall Lead Us, in I. K. Zola, (Ed.). *Ordinary Lives: Voices of Disability and Disease.* Cambridge: Applewood Books, 34-37.

20. Becoming Disabled in Midlife
   - Brickner, R. (1976). *My Second Twenty Years: An Unexpected Life.* New York: Basic Books.
   - Bonwich, E. (1985). Sex Role Attitudes and Role Reorganization in Spinal Cord Injured Women, in M. J. Deegan and N. A. Brooks, (Eds.). *Women and Disability: The Double Handicap.* New Brunswick, N.J.: Transaction Books, 56-67.
   - Burish, T. G. and Bradley, L. A. (Eds.). (1983). *Coping with Chronic Disease.* New York: Academic Press. Chapters 4 and 5.
   - Meyerowitz, B., Chaiken, S., and Clark, L. (1988). Sex Roles and Culture: Social and Personal Reactions to Breast Cancer, in M. Fine and A. Asch, (Eds.). *Women with Disabilities: Essays in Psychology, Culture, and Politics.* pp. 72-89.

21. Disability and Gender
   - Asch, A. and Fine, M. (1988). Introduction: Beyond Pedestals, in M. Fine and A. Asch, (Eds.). *Women with*

Disabilities: Essays in Psychology, Culture, and Politics. pp. 1-37.
- Russo, N. and Jansen, M. (1988). Women, Work, and Disabilities: Opportunities and Challenges, in M. Fine and A. Asch, (Eds.). *Women with Disabilities: Essays in Psychology, Culture, and Politics.* pp. 229-244.
- Gartner and Joe (1987). Chapters 2 and 3. See Texts above.

22. Disability and Later Life
    - Ainlay, S. (1988). Aging and New Vision Loss: Disruptions of the Here and Now. *Journal of Social Issues,* 44 (1), 79-94.
    - Hiatt, L. G. (1981). Aging and Disability, in N. S. McClusky and E. F. Borgatta, (Eds.). *Aging and Retirement.* pp. 133-152.
    - Veatch, R. M. (Ed.). (1979). *Life Span: Values and Life-extending Technology.* Beverly Hills: Sage. Chapters 4 and 9.
    - Simon, B. L. (1988). Never-married Old Women and Disability: A Majority Experience, in M. Fine and A. Asch, (Eds.). *Women with Disabilities: Essays in Psychology, Culture, and Politics.* pp. 215-225.

    Supplemental reading:
    - Becker, G. (1980). *Growing Old in Silence.* Berkeley: University of California Press.

Unit 3: Disability Policy, Politics, and Ethics

23. Disability Rights Movement
    - Gartner and Joe (1987). Chapters 1, 11, and Conclusion. See Texts above.
    - Scotch, R. K. (1988). Disability As the Basis for a Social Movement: Advocacy and the Politics of Definition. *Journal of Social Issues,* 44 (1), 159-172.
    - ICD (1986). pp. 109-116. See Texts above.

    Supplemental reading:
    - Asch, A. (1986). Will Populism Empower Disabled People? in H. C. Boyte and F. Reissman, (Eds.). *The New Populism and the Politics of Empowerment.* Philadelphia: Temple University Press. pp. 213-230.
    - Scotch, R. K. (1984). *From Good Will to Civil Rights:*

Transforming Federal Disability Policy. Philadephia: Temple University Press.

24. Disability Policy and Resource Allocation - Part 1
    - Brown, S. E. (1985). Infusing Blues, in S. E. Browne, D. Connors, and N. Stern, (Eds.). *With the Power of Each Breath.* San Francisco: Cleis Press. pp. 15-23.
    - Massie, R. and Massie, S. (1984). *Journey,* second edition. Chapters 18 and 25.
    - Mappes and Zembaty (1986). pp. 601-625. See Texts above.

25. Disability Policy and Resource Allocation - Part 2
    - Starr, R. (1982, January). Wheels of Misfortune: Sometimes Equality Just Costs Too Much. *Harper's*, 7-15.
    - ICD (1986). pp. 90-109. See Texts above.
    - Mappes and Zembaty (1986). pp. 563-585. See Texts above.

26. Defining the Quality of Life from the Disabled Perspective-Baby Doe
    - Gartner and Joe (1987). Chapter 10. See Texts above.
    - Hubbard, R. (1984, May). Caring for Baby Doe: The Moral Issue of Our Time. *Ms.*, 84-88, 165.
    - Mappes and Zembaty (1986). pp. 433-446. See Texts above.

27. Defining the Quality of Life from the Disabled Perspective-Genetic Screening and Prenatal Diagnosis
    - Rapp, R. (1984, March). "XYLO: A True Story," in R. Arditti et al. (Eds.). *Test-Tube Women.* Boston: Pandora Press. pp. 313-328.
    - Asch, A. (1986). Real Moral Dilemmas. *Christianity and Crisis, 46* (10), 237-240.
    - Mappes and Zembaty (1986). pp. 513-527. See Texts above.

28. Defining the Quality of Life from the Disabled Perspective-Right to Die: Should Disability Matter?
    - Gartner and Joe (1987). Chapter 5. See Texts above.
    - *Disability Rag* (1984, February). Entire issue.
    - Mappes and Zembaty (1986). pp. 359-365. See Texts above.

Notes

1. This essay reflects the views of the author and not those of the New Jersey Commission on Legal and Ethical Problems in the Delivery of Health Care.

A number of people helped to refine this essay. Among the participants in the Project on Reproductive Laws for the 1990s, I owe special debts to Sherrill Cohen, Irene Crowe, Mary Sue Henifin, Judy Norsigian, Rosalind Petchesky, and Nadine Taub. For their close reading and important suggestions, I would also like to thank: Ellen Baker, Douglas Biklen, Michelle Fine, Anne Finger, Alan Gartner, Judy Heumann, Mary Johnson, Bob Kraft, Betty Levin, Vivian Lindermayer, Abby Lippman, Marsha Saxton, Richard Scotch, William Weil, and Irving Kenneth Zola. I must say a special thank you to Ellen Baker and Alan Gartner, who took time to think through the topics as I struggled with them, and whose intellectual integrity and steadfast friendship have sustained me in this endeavor and much else besides.

2. A version of this syllabus appears in Quina, Kathryn, and Bronstein, Phyllis A., eds., Teaching a Psychology of People: Resources for Gender and Sociocultural Awareness (copyright 1988 by the American Psychological Association; adapted by permission of the publisher and author).

References

Andrews, L. B. (1985). New Conceptions: A Consumer's Guide to the Newest Infertility Treatments Including In Vitro Fertilization, Artificial Insemination, and Surrogate Motherhood.

Annas, G. J. and Elias, S. (1983). In Vitro Fertilization and Embryo Transfer: Medico-Legal Aspects of New Techniques to Create a Family. Family Law Quarterly, 17 (2) Summer, 199-223.

Applebaum, E. and Firestein, S. K. (1983). A Genetic Counseling Casebook.

Arditti, R., Klein, R. D., and Minden, S. (Eds.), (1984). Test-Tube Women: What Future for Motherhood?

Asch, A. (1986). Will Populism Empower Disabled People? in H. C. Boyte and F. Reissman, (Eds.), The New Populism: The Politics of Empowerment.

Asch, A. and Fine, M. (1988). Introduction: Beyond Pedestals, in M. Fine and A. Asch, (Eds.), Women with Disabilities: Essays in Psychology, Culture, and Politics, 1-37.

Asch, A. and Sacks, L. (1983). Lives Without, Lives Within: The Autobiographies of Blind Women and Men. Journal of Visual Impairment and Blindness, 77 (6), 242-247.

Bayles, M. D. (1984). Reproductive Ethics.

Bernard, J. (1974). The Future of Motherhood.

Blakeslee, S. (1986, October 7). Fetus Returned to Womb Following Surgery. New York Times, pp. C1 and C3.

Bonwich, E. (1985). Sex Role Attitudes and Role Reorganization in Spinal Cord Injured Women, in M. J. Deegan and N. A. Brooks, (Eds.), Women and Disability: The Double Handicap, 56-67.

Boston Women's Health Book Collective. (1984). The New Our Bodies, Ourselves.

Bowe, F. (1980). Rehabilitating America.

_______. (1984). Disabled Women in America. Washington, D.C.: President's Committee on Employment of the Handicapped.

Braille Monitor. (1986, August-September). Legalized Kidnapping: State Takes Child Away from Blind Mother, 432-435.

Brightman, A. J. (1984). Ordinary Moments: The Disabled Experience.

Browne, S., Connors, D., and Stern, N. (Eds.), (1985). With the Power of Each Breath.

Burt, R. A. (1979). Taking Care of Strangers.

Cohen, L. G. (1986). Selective Abortion and the Diagnosis of Fetal Damage: Issues and Concerns. Journal of the Association of Persons with Severe Handicaps, 1 (3), 188-195.

Collins, G. (1983, October 12). The Success Story of a Disabled Mother. New York Times, pp. Cl and C2.

_______. (1986, December 15). Children of Deaf Share Their Lives. New York Times.

Dally, A. (1983). Inventing Motherhood: The Consequences of an Ideal.

Darling, R. B. (1979). Families Against Society.

DeJong, G. and Lifchez, R. (1983). Physical Disability and Public Policy. Scientific American, 48, 240-249.

DeLoach, C. and Greer, B. G. (1981). Adjustment to Severe Physical Disability: A Metamorphosis.

Dowrick, S. and Grundberg, S. (Eds.), (1980). Why Children.

Dworkin, A. and Dworkin, R. (Eds.), (1976). The Minority Report.

Elias, S. and Annas, G. J. (Eds.), (1987). Reproductive Genetics and the Law.

Evans, D. (1983). The Lives of Mentally Retarded People.

Fine, M. and Asch, A. (1985). Disabled Women: Sexism Without the Pedestal, in M. J. Deegan and N. A. Brooks, (Eds.), Women and Disability: The Double Handicap, 3-21.

_______. (1988). Disability Beyond Stigma: Social Interaction, Discrimination, and Activism. Journal of Social Issues, 44 (1), 1-22.

Fletcher, J. (1984). Healing before Birth: An Ethical Dilemma. Technology Review, 87 (1), 27-36.

Friedland, R. and Kort, C. (Eds.), (1981). The Mothers' Book.

Funk, R. (1987). Disability Rights: From Caste to Class in the Context of Civil Rights, in A. Gartner and T. Joe, (Eds.), Images of the Disabled, Disabling Images, 7-30.

Gliedman, J. and Roth, W. (1980). The Unexpected Minority: Handicapped Children in America.

Goffman, E. (1963). Stigma: Notes on the Management of Spoiled Identity.

Gold, S. (1985, November). A Year of Accomplishment: Sharon Gold Reports to the Blind of California. Braille Monitor, 629-640.

Groce, N. E. (1985). Everyone Here Spoke Sign Language: Hereditary Deafness on Martha's Vineyard.

Haber, L. and McNeil, J. (1983). Methodological Questions in the Estimation of Disability Prevalence. (Available from Population Division, U.S. Census Bureau.)

Hahn, H. (1983a, March-April). Paternalism and Public Policy. Society, 36-46.

_______. (1983b, December). "The Good Parts": Interpersonal Relationships in the Autobiographies of Physically Disabled Persons. Wenner-Gren Foundation Working Papers in Anthropology, 1-38.

_______. (1987). Civil Rights for Disabled Americans: The Foundation of a Political Agenda, in A. Gartner and T. Joe, (Eds.,) Images of the Disabled, Disabling Images, 181-204.

Harper's Magazine. (1987, September).

Harris, Louis and Associates. (1986). The ICD Survey of Disabled Americans: Bringing Disabled Americans into the Mainstream. New York: International Center for the Disabled.

Hoffman, L. W. and Hoffman, M. L. (1973). The Value of Children to Parents, in J. T. Fawcett, (Ed.), Psychological Perspectives on Population.

Hyler, D. (1985). To Choose a Child, in S. Browne, D. Connors, and N. Stern, (Eds.), With the Power of Each Breath, 280-283.

Insight. (1986, November 24). Deaf Pride.

Lamanna, M. A. (1984). Social Science and Ethical Issues: The Policy Implications of Poll Data on Abortion, in S. Callahan and D. Callahan, (Eds.), Abortion: Understanding Differences, 1-24.

LeMaistre, J. (1985). Parenting, in S. Browne, D. Connors, and N. Stern, With the Power of Each Breath, 284-291.

Macklin, R. and Gaylin, W. (Eds.), (1981). Mental Retardation and Sterilization.

March of Dimes. (1985, November). Birth Defects: Tragedy and Hope.

Matthews, G. F. (1983). Voices from the Shadows: Women with Disabilities Speak Out.

National Association for Sickle Cell Disease. (No date). Position Paper on Prenatal Diagnosis of Sickle Cell Anemia.

National Council on the Handicapped. (1986). Toward Independence: An Assessment of Federal Laws and Programs Affecting Persons with Disabilities--with Legislative Recommendations. Washington, D.C.: U.S. Government Printing Office stock number 052-003-01022-2.

National Public Radio. (1987, December 29). Morning Edition.

________. (1988, January 16). Weekend Edition.

Ontario Law Reform Commission. (1985). Report on Human Artificial Reproduction and Related Matters.

Orlansky, M. D. and Heward, W. L. (1981). Voices: Interviews with Handicapped People.

Pollitt, K. (1987, May 23). The Strange Case of Baby M. The Nation, 681-688.

Rapp, R. (1988). Moral Pioneers: Women, Men, and Fetuses on the Frontier of Reproductive Technology, in E. H. Baruch, A. D'Adamo, Jr., and J. Seager, (Eds.), Embryos, Ethics, and Women's Rights.

Rehabilitation Act of 1973, Pub. L. No. 93-112, 87 Stat. 355 (1973).

Roth, W. (1981). The Handicapped Speak.

Rothman, B. K. (1986). The Tentative Pregnancy: Prenatal Diagnosis and the Future of Motherhood.

Rothman, D. and Rothman, S. (1984). The Willowbrook Wars: A Decade of Struggle for Social Justice.

Safilios-Rothschild, C. (1970). The Sociology and Social Psychology of Disability and Rehabilitation.

Sawisch, L. P. (1978). Expressed Willingness to Parent Handicapped Children (unpublished doctoral dissertation submitted to Department of Psychology, Michigan State University).

Schulz, R. and Decker, S. (1985). Long-term Adjustment to Physical Disability: The Role of Social Support, Perceived Control, and Self-Blame. Journal of Personality and Social Psychology, 48 (5), 1162-1172.

Scotch, R. K. (1984). From Good Will to Civil Rights: Transforming Federal Disability Policy, 159-172.

________. (1988). Disability as the Basis for a Social Movement. Journal of Social Issues, 44 (1).

Shaul, S., Dowling, P., and Laden, B. F. (1985). Like Other Women: Perspectives of Mothers with Physical Disabilities, in M. J. Deegan and N. A. Brooks, (Eds.), Women and Disability: The Double Handicap, 133-142.

Siller, J., Ferguson, L., Vann, D. H., and Holland, B. (1976). Structure of Attitudes toward the Physically Disabled. New York: New York University School of Education.

Stanworth, M. (1988). Reproductive Technologies and the Deconstruction of Motherhood, in M. Stanworth, (Ed.), Reproductive Technologies: Gender, Motherhood, and Medicine, 10-35.

Steinbock, B. (1986). The Logical Case for "Wrongful Life." Hastings Center Report, 16 (2), 15-20.

Stroman, D. F. (1982). The Awakening Minorities: The Physically Handicapped.

Thorne, B. with Yalom, M. (1982). Rethinking the Family: Some Feminist Questions.

Thurman, S. K. (Ed.), (1985). Children of Handicapped Parents.

Turnbull, H. R. and Turnbull, A. (Eds.), (1985). Parents Speak Out: Then and Now.

U.S. Census Bureau. (1983). Labor Force Status and Other Characteristics of Persons with a Work Disability: 1982. Current Population Reports, Series P-23, No. 127.

U.S. Commission on Civil Rights. (1983). Accommodating the Spectrum of Individual Abilities.

Walker, L. A. (1985, June 20). Children of the Disabled Face Special Pressures. New York Times, pp. C1 and C6.

Washington Post. (1988, January 18). Second Child to Be Taken from Disabled Mom. p. 2.

Wright, B. (1983). Physical Disability: A Psycho-Social Approach.

Zelizer, V. A. (1985). Pricing the Priceless Child: The Changing Social Value of Children.

Zetlin, A. G., Weisner, T. S., and Gallimore, R. (1985). Diversity, Shared Functioning, and the Role of Benefactors: A Study of Parenting by Retarded Persons, in S. K. Thurman, (Ed.), Children of Handicapped Parents.

Zola, I. K. (1983). Individual Choice and Health Policy: A Socio-political Scenario for the 1980's, in I. K. Zola, (Ed.), Socio-Medical Inquiries, 285-298.

IV. REPRODUCTIVE POLICY POSITION PAPERS AND COMMENTARIES FOR THE 1990s

1. Prenatal and Neonatal Technology, Abortion, and Birth

TIME LIMITS ON ABORTION

Nan D. Hunter

Position Paper

Fourteen years ago, the Supreme Court found unconstitutional every state law then in existence that prohibited abortions. Yet, despite the dramatic immediate effect of *Roe v. Wade*, 410 U.S. 113 (1973), the Court relied on what was very much a compromise approach. The Court did not hold that a woman's rights to autonomy and privacy yielded her sole control over the abortion decision. Rather, it adopted the by now well-known trimester system, under which abortion was more or less susceptible to state interference depending on when in the pregnancy it was performed. After the fetus attained viability, that point in the pregnancy when it could be sustained outside the uterus, the state acquired a compelling interest in preserving fetal life. Even that compelling interest must yield, the Court ruled, when an abortion was needed to protect the woman's own life or health. But absent those exigencies, abortions after viability could be restricted or even banned.

Since *Roe v. Wade*, an intensified focus on the fetus as an entity, both in political terms and within the medical profession, has characterized much of the policy discussion about abortion. Indeed, this shifting of the frame of public discussion about abortion from the woman to the fetus may constitute the greatest success of the right-to-life movement, which so far has failed in its major goals of overturning *Roe v. Wade* or enacting a constitutional amendment to ban abortion. The heightened attention being accorded post-viability abortions stems more from these contested political and ideological agendas than from major scientific advances or any increase in the number of these procedures being performed.

Post-viability abortions are extremely rare. Since abortion became legal nationwide, women have been able to make their abortion decisions earlier and earlier in pregnancy.[1] Today, 99 percent of all abortions are performed in the first 20 weeks of pregnancy; after 24 weeks, the period in which viability becomes possible, only 0.01 percent are performed.[2]

Despite their infrequency, however, such abortions present troubling ethical dilemmas. A pregnancy which has progressed for 26 weeks can result in a neonate with a significant chance of survival as well as in an abortion. Such a situation can be extremely traumatic for the woman who is seeking a late abortion and for the medical staffs who are called on to save neonates as well as perform abortions.

Moreover, theoretical issues associated with post-viability abortions present, in perhaps its sharpest form, the inevitable, irreducible tension of pregnancy: is the pregnant woman to be treated by the law as one person, or two? Was the Supreme Court in *Roe* correct in its interpretation that her privacy-based right to control her body had to be formulated differently from the norm because she was "not isolated in her privacy"? 410 U.S. at 159. Or should pregnancy be thought of, as historian Linda Gordon has suggested, as "an ongoing activity [of the woman's body], concretized"?[3]

The most common factual contexts of such abortions only deepen the ethical difficulties. Although these very late abortions are few, a disproportionate number of them occur in two situations. First, teenagers tend to have later abortions much more frequently than adult women, because of the greater impediments they face in obtaining the procedure. Moreover, the youngest teenagers tend to have the latest abortions. Second, abortions performed because of fetal anomalies, although constituting only a tiny fraction of all abortions, cluster in number after 20 weeks of pregnancy because amniocentesis results are generally not available until then. In both these situations--the young pregnant teenager and the woman who has learned that the fetus she is carrying may be severely impaired--the harm to the woman of forcing her to continue the pregnancy is particularly acute.

The Supreme Court attempted to resolve this tension between the woman's right to decide to have an abortion and the claim for continuation until term of substantially developed fetal life by drawing a line at viability. Viability, however, is inexact at best; it varies with the individual pregnancy and is by definition double speculation--an estimate of the point in time during the pregnancy when a certain probability (of extracorporeal survivability) has been reached. Its uncertainty is heightened by the possibility that technological advances may shift the medical estimates about when survivability is possible.

The question presented for legislatures is what time limits, if any, the state should impose on the woman's right to choose abortion. The majority of states currently abstain from imposing limits. In those states that deny women the right to have an abortion past a certain point in time (unless necessary for life or health), the law would impose on each woman months of forced pregnancy followed by forced labor and forced childbirth. Some have argued that the best resolution of this tension is to allow the pregnancy to end, but to compel the use of those abortion techniques that are most likely to result in preserving the survivability of the fetus.

This position paper favors an approach that combines both an abstentionist and an affirmative role for the state. Compelling the use of a fetal-preserving technique is rejected primarily because it results in coerced parenthood, an awesome infringement of autonomy. Additionally, compulsory use of such techniques poses risks of a variety of physical and psychological harms to both the woman and, if it survives, the child. Rather, we urge that the state not criminalize abortion or restrict the method selected, but instead use its considerable potential for intervention to address directly the underlying problems that tend to result in very late abortions.

Ultimately, the ascertainment of rights necessitated by this issue is more about pregnancy than abortion. The question of who should decide how and when pregnancy should cease and what should result probably has more impact on childbearing than on preventing of motherhood. There are many more cesarean section deliveries performed each year than post-viability abortions.[4] Increasingly intrusive fetal therapies are becoming more common in even normal pregnancies.[5] In all of these contexts, medical experts are proclaiming the fetus to be an independent patient.

Patienthood, like parenthood, implies a relational identity. For the fetus to be so denominated is not an inevitable, technologically determined, scientifically neutral development. New technologies do exist: we can now see the fetus, however hazily, by use of ultrasonography, and physicians can surgically alter and mend component parts. But whether we think of what we see or what is repaired as an entity with enforceable rights or as an especially delicate function of a person's reproductive process is an ideological construct.

Resonating just below the surface of this debate is the social acceptability of treating the pregnant woman as a maternal environment. Ancient cultural notions reappear in high-tech garb. Women long have been expected to be maternal and self-sacrificing; a woman's deviation from the nurturant role, or even failure to aspire to it, cuts against the grain of deep-seated cultural expectations. Laws that enshrine principles of protection for the fetus can arise in part from the sense that fetal primacy is for women the "natural" priority. Equally "natural," however, is the premise that women possess no lesser capacity to be moral decision makers, and that their decisions about late abortion are entitled to the same respect we accord to other difficult decisions of conscience.

This position paper first presents the factual background of why late abortions occur and describes how the relevant medical procedures are performed. Next, the current state of the law is outlined. The third section examines the deficiencies of the viability concept and of the primary ways in which states regulate the procedures used in post-viability abortions. The last section analyzes the doctrinal bases for constitutional protection of procreative liberty and concludes that social policy on late abortions should accord decision-making responsibility to each individual woman, while seeking ways to allow those decisions to be made earlier and earlier in pregnancy.

1. Factual Background

The beginning point for considering the issues raised by post-viability abortions is an examination of why these abortions occur. No one has ever studied the reasons for post-viability abortions *per se*, doubtless because there are so few of them. Most of the studies that are available examine delay factors for the category of all second-trimester abortions, ranging from 13 to 24 weeks. The few that have focused on abortions performed after 20 weeks are suggestive more than conclusive, but do nonetheless identify some important factors.

Teenagers obtaining abortions, especially young teenagers, are more likely to obtain late abortions than are older women. According to the most recent data available, teenagers obtain 43 percent of abortions performed after 20 weeks of gestation (compared to 28 percent of all abortions).[6] Moreover, the youngest teenagers are considerably more likely than older women to obtain the latest abortions. Abortions after 20 weeks comprise 3.7 percent of all abortions performed for girls 14 and younger, contrasted to the national average of 1 percent of all abortions.[7]

There are no published studies that pinpoint the reasons why teenagers tend to have a disproportionate share of the very latest abortions, but it is reasonable to conclude that the factors associated with youth that correlate with delay in getting second-trimester abortions generally apply also to the very latest abortions. Not surprisingly, those factors include low income, failure to recognize the symptoms of pregnancy, self-denial of the possibility of pregnancy, irregular menses, ambivalence about the abortion decision, and impeded access.[8] The barriers to access for young women include special restrictions such as parental consent and notification laws.[9] Teenagers also suffer more from the scarcity of providers outside metropolitan areas because traveling even relatively short distances is much harder for women who do not own or have independent access to a car and must explain absences from home or school.[10] Cumulatively, these factors lead to substantial delay.

A second reason for very late abortions is the detection of fetal anomalies. A diagnosis based on amniocentesis testing is generally not available, using current technology, until approximately the 20th week of pregnancy. A woman may well want time to think through her options and decide how to proceed. Because so few doctors will perform abortions at that stage of pregnancy, simply finding a doctor, scheduling the procedures, and traveling can easily consume a week or more. The Centers for Disease Control estimate that from 1500 to 3750 abortions each year result from detection of fetal anomalies.[11] Although these represent a tiny fraction of the total number of abortions, they cluster in the post-20-week period, and therefore could account for up to 35 percent of all abortions performed after 20 weeks.[12]

A third major reason for very late abortions arises when there is a sudden reversal of the woman's health. The onset or worsening of such diseases as pre-eclampsia, diabetes, cardiovascular disease, cervical cancer, ovarian cancer, breast cancer, high blood pressure, kidney disease, suppression of the immune system,[13] and certain respiratory, urinary, and neuromuscular disorders late in pregnancy may present a health risk sufficient to require termination. Some maternal conditions carry the likelihood of fatal disabilities for the child; others may create sufficient uncertainty about the pregnant woman's own survival that the decision whether to bring a child into being is one she feels compelled to rethink.[14] Thus, sudden complications in the woman's own health may lead not only to a need for the pregnancy to end sooner than planned, but also to a reconsideration (even if the pregnancy was initially wanted) of whether the woman wants to attempt to save the fetus.

These medical problems may correlate with youth or income. Teenagers are 92 percent more likely to have anemia and 15 percent more likely to have toxemia during pregnancy than are older women.[15] In New York City, 18 percent of pregnant women receive no prenatal care or care only in the last trimester; for teenagers, the figure is 34.4 percent.[16] Advanced pregnancy may be the first opportunity for diagnosis of health problems. Psychological distress may also reach a level where treatment, which may include abortion, is indicated.[17]

Some of the factors that contribute to the incidence of late abortions are avoidable by public policy mechanisms, and some are not. Chorionic villi analyses are being developed that allow earlier detection of some of the fetal anomalies now diagnosable only by amniocentesis. These developments may alleviate some delay, but they do not screen for all conditions. No medical advances will be able to eliminate the possibility of unforeseen life- or health-threatening complications arising during pregnancy. But many of the factors associated with teenagers' propensity for late abortions could be significantly altered by public policy initiatives.

Sexual shame and sexual ignorance are persistent themes in American culture. They reinforce each other and contribute to a social atmosphere in which young people are blocked from acquiring basic knowledge about sexuality and birth control. One result is the highest teenage pregnancy rate in the developed world.[18] A concerted effort to provide access to information and services for young people, especially those in low-income groups or in nonmetropolitan areas, would be a major step toward a decrease among teenagers in unwanted pregnancies, in the total number of abortions, and therefore in late abortions stemming from social factors.

Because some post-viability abortions will always occur, however, it is important to ensure that hospitals are available in which they can be performed. One byproduct of the increased public concern over late abortions has been a trend among hospitals to restrict the performance of abortion to the first 20 weeks of pregnancy.[19] Publicly funded hospital systems should be required to offer the full range of treatment options to women who must make these different decisions.

The concrete aspects of that decision-making process are that, if a pregnancy after 20 weeks is to be ended, there are three primary methods of abortion. Most common is saline instillation, which accounts for 58 percent of all such abortions.[20] After removal of as much amniotic fluid as possible, an equal amount of hypertonic saline solution is injected into the amniotic sac. The saline, which is virtually always fatal to the fetus, triggers uterine contractions, and the fetus is expelled. The second most widely used technique is dilatation and evacuation (D & E), which is used for 24 percent of such abortions.[21] D & E involves surgical removal of fetal tissue parts, followed by suction to ensure a complete abortion. Fetal survival is not possible. The last generally used method is prostaglandin instillation, used in 8 percent of these cases.[22] Like saline, the prostaglandin injection triggers contractions and fetal expulsion; unlike saline, it does not destroy fetal tissue and so allows for the possibility of live birth.

Live births are rare events, even with prostaglandin abortions. No national statistics are kept as to the number. In New York State in 1982, there were 18 live births following more than 160,000 abortions.[23] If that same ratio were applied to the total number of abortions nationwide, one would expect 150 live births to occur.[24] This figure is probably extremely misleading, however, because the percentage of New York State abortions that occurs past 20 weeks is nearly twice the national average.[25] An estimate that eliminates the New York differential would be closer to 75.

A live birth does not mean that a viable infant is being born.[26] It typically signifies only momentary movement before death: "If a 150 gram (five-ounce) fetus without a heartbeat or respiration twitches after being expelled in the course of a saline instillation abortion, a live birth should be reported."[27] Death may occur so quickly after expulsion of the fetus that the woman for whom a prostaglandin method has been used will deliver the fetus and see or feel it move and then die in the bed next to her before a nurse reaches her to attend the delivery. Even the possibility of such an experience has caused women to avoid prostaglandins when they have a choice of method.[28]

2. Current Law

> If the State is interested in protecting fetal life after viability, it may go so far as to proscribe abortion during

> that period, except when it is necessary to protect the life or health of the mother. *Roe v. Wade*, 410 U.S. 163-4.

Under the federal constitution,[29] a state may assert a compelling interest in the potential life of the fetus beginning at viability. "This is so because the fetus then presumably has the capability of meaningful life outside the mother's womb." *Roe v. Wade*, 410 U.S. at 163. The deceptively simple language of this rationale embodies major ambiguities (what level of capability is required? what is "meaningful life" in this context?) which the Court is still addressing.

In the cases since *Roe* that have addressed viability, the Court has relied repeatedly on the physician's exercise of professional judgment to resolve these questions. In *Planned Parenthood Assn. of Missouri v. Danforth*, 428 U.S. 52 (1976), the Court described viability as "a point purposefully left flexible for professional determination and dependent upon developing medical skill and technical ability." 428 U.S. at 61. It reiterated that viability is "essentially [ ] a medical concept . . . The time when viability is achieved may vary with each pregnancy, and the determination of whether a particular fetus is viable is, and must be, a matter for the judgment of the responsible attending physician." 428 U.S. at 63.

Colautti v. Franklin, 439 U.S. 379 (1979) is the decision that most stresses the themes both of relying on physician judgment and of protecting doctors from criminal liability for having made good-faith decisions that a prosecutor might later second-guess. In *Colautti*, a Pennsylvania abortion law that criminalized post-viability abortions (except for life or health reasons) was stricken as unconstitutionally vague for failing to specify that the doctor could rely on his or her subjective assessments, in good faith, rather than comply with an objective standard based on the medical community at large or a panel of experts, and for failing to include a criminal scienter requirement.[30] The *Colautti* decision begins by underscoring "the central role of the physician, both in consulting with the woman about whether or not to have an abortion, and in determining how any abortion [is] to be carried out." 439 U.S. at 387. The Court then renders its definition of viability, considerably elaborated from the *Roe v. Wade* original:

> Viability is reached when, in the judgment of the attending physician on the particular facts of the case before

> him, there is a reasonable likelihood of the fetus' sustained survival outside the womb, with or without artificial support. Because this point may differ with each pregnancy, neither the legislature nor the courts may proclaim one of the elements entering into the ascertainment of viability--be it weeks of gestation or fetal weight or any other single factor--as the determinant of [viability]. 439 U.S. at 388-9.

The Court in *Colautti* also ruled on a provision that would have mandated, for post-viability abortions, that

> "the abortion technique employed shall be that which would provide the best opportunity for the fetus to be aborted alive so long as a different technique would not be necessary in order to preserve the life of the mother." Statute quoted at 439 U.S. at 397.

This requirement was also stricken as too vague, again because it failed to specify that a doctor could be safe in making a good-faith selection of an abortion method based on a woman's overall health needs (even if the method chosen was not necessarily indispensible to the preservation of her life or health) and because it lacked a scienter requirement. In so ruling, the Court relied on physician testimony that described the contraindications for various methods and the difficulty in predicting which methods would satisfy the statute. "The choice of an appropriate abortion technique, as the record in this case so amply demonstrates," the Court wrote, "is a complex medical judgment about which experts can--and do--disagree." 439 U.S. at 401.

The Court decided *Colautti* on vagueness grounds, and explicitly stopped short of ruling on whether the privacy principle grounding the abortion right permitted "a 'trade-off' between the woman's health and additional percentage points of fetal survival" in the selection of a method of abortion. 439 U.S. at 400-401. The Court finally reached and resolved this issue in its most recent abortion decision. In *Thornburgh v. American College of Obstetricians and Gynecologists*, 54 U.S.L.W. 4618 (June 11, 1986), the Court ruled that a law requiring doctors to use the technique providing the "best opportunity" for live birth when they performed post-viability abortions was unconstitutional because it permitted an exception to that requirement only when the fetal-favoring technique presented "a significantly greater" medical risk to the woman's life or health. *Id.* at 4624. Compelling the woman to assume *any* increased medical risk in an attempt to save the fetus was found impermissible.

Four justices dissented from the ruling. Two, Justices Rehnquist and White, stated explicitly that the state's compelling interest in fetal life after the point of viability does justify the imposition of a measurably greater medical risk on the woman. 54 U.S.L.W. at 4634-5. Justice O'Connor accepted the state's argument that the word "significantly" meant only that the risk must be "real and identifiable." 54 U.S.L.W. at 4641. She expressed no opinion on the point at which an increased risk to the woman would render such a provision an "undue burden" on the abortion right. Chief Justice Burger's dissenting opinion did not address this issue.

The regulatory scheme for post-viability abortions allowed by the Court thus includes the power to prohibit all medically non-necessary abortions after viability; if such an abortion *is* necessary, however, the state may not force onto the woman a technique for it which endangers her health more than another available technique.[31] Implicit in the Court's analysis (but not explicitly ruled on) is the proposition that fetal-favoring techniques *can* be required if there is no proof that they create more risk to the woman's health than fetal-destructive techniques.[32]

Obviously a key component of this scheme is the meaning and scope of "health." The Court has repeatedly stressed that the "health" needs of the woman are to be interpreted broadly. It is "critical," the Court has said repeatedly, that a physician's "best clinical judgment" "'may be exercised in the light of all factors--physical, emotional, psychological, familial and the woman's age--relevant to the well-being of the patient.'" *Colautti v. Franklin*, 439 U.S. at 388, quoting *Doe v. Bolton*, 410 U.S. at 192. Every member of the Court (prior to Justice Scalia's appointment) had endorsed in some context the principle that what constitutes health is to be broadly defined, and within the physician's discretion. *Roe v. Wade*, 410 U.S. at 207-8 (Burger, C.J., concurring); *City of Akron v. Akron Center for Reproductive Health*, 462 U.S. at 467 (O'Connor, White, and Rehnquist, JJ., dissenting).

The extent of discretion invested in the physician to determine "health" protects the doctor from politically motivated prosecution and enables the woman to receive medical advice tailored to her needs rather than to the goal of bringing about live birth at all costs. But it also transfers the sticky moral and ethical questions surrounding post-viability abortion to the physician, in the guise of "professional judgment." The teenager who has denied to herself and others for months that she is pregnant may or may not qualify as having a "health" justification for an abortion, depending on her doctor's judgment. It will likely be precisely the "emotional, psychological, [and] familial" concerns of the woman that will raise the thorniest problems--the fear of having to endure a live birth only to watch the fetus die, the decision whether to bring into being a severely disabled child, the impact of motherhood on an impoverished teenager. The definition of "health" in this doctrinal structure masks issues of fundamental personal autonomy, and delegates to physicians (whether they want them or not) thinly disguised moral as well as medical judgments.[33]

Despite a relatively heavy dose of public attention for this subissue of the abortion debate, however, legislative regulation of post-viability abortions is still more the exception than the rule. Implicitly declining the Supreme Court's invitation, 37 states and the District of Columbia have enacted no law specifically restricting post-viability abortions. Louisiana uniquely mandates that the method chosen shall be the one "most likely to preserve the life and health of the mother. If the physician has a choice of methods or techniques consistent with this priority, he shall preserve the life and health of the unborn child."[34] The remaining 12 states utilize some form of a requirement that the method used be the one most likely to preserve the life and health of the fetus[35] or that the standard of care be the same as for the fetus intended to be delivered.[36] Three of these states use a reasonableness or availability test for the standard of care.[37]

The states also vary in their formulation of how the life and health interests of the fetus are to be weighed against those of the woman in selecting a method or making other decisions. Florida's law emphatically states that "the woman's life and health shall constitute an overriding and superior consideration."[38] Five states permit an exception to the requirement of a method allowing the chance of live birth when it would increase the risk to the woman.[39] The laws of three states,[40] which would have required risk of "serious" harm or "significantly greater" risk before another method could be used, have been rendered unenforceable by the Supreme Court's ruling in *Thornburgh*. The law of three other states is silent on this question.[41]

3. Deficiencies of Drawing the Line at Viability and Regulating Method Choice

Although the Supreme Court has continued to invoke the viability line, it has been one of the most frequently criticized aspects of the abortion cases. Because viability is the line that demarcates presumptively legal from presumptively illegal procedures, it shoulders an especially heavy weight in the structure of the Court's abortion jurisprudence. Many reasons, aside from the criticism already advanced, suggest that it is a weak link.

Viability is an intrinsically vague and indeterminate concept. It is defined as the point in a given pregnancy when the fetus is capable of sustained life outside the body of the mother, with or without artificial aid. There is no standardized definition of when it will occur or what elements constitute its diagnosis. It is generally expressed in terms of estimated gestational age, e.g., likely to occur by the 26th week of pregnancy. But estimates of gestational age, which is the best gauge of viability, are simply professional guesses, based on the woman's recollection of the dates of her last menstrual period, clinical examination of the uterine size, and ultrasound imaging of the fetus.[42] Such estimates are routinely off by one to two weeks during this stage of pregnancy, an error which can be critical in this context.

Even if one has accurately determined gestational age, there are changing estimates about what level of probability is associated with ages 23 to 27 weeks. Estimates of likelihood of survival vary in the reports from different medical centers.[43] Any medical problems of the particular pregnancy or of the woman may affect survivability, as would the presence--or absence--of sophisticated neonatal care units.

Nor is there consensus about the degree of likelihood that constitutes viability. Physicians' opinions about what chance of survival marks the threshold of viability range from 5 percent and up.[44] Thus, the very *definition* of viability, in operational terms, is unsettled. Whatever the Court may have wished, viability cannot be thought of as a bright line; given the range of variables, it is hardly a line at all.

The definitional problem is compounded by the possibility, although never considered by any court, that the development of lavage and reimplantation procedures may create viability at the very beginning of pregnancy. New methods of reproduction allow the removal of fertilized ova and their transference to another woman for gestation.[45] If such tiny embryos were to be considered viable because of the possibility of reimplantation, the same severe restrictions on very late abortions theoretically could apply to women ending at least some very early pregnancies.

In recent years, much literature has appeared that argues that the point of viability--wherever it may be now--is inexorably marching backward through pregnancy.[46] It is true that within the range of possible viability from 24 to 28 weeks,[47] the probabilities have shifted. That is, the estimated chances of survival have increased at each point within that range.

Nonetheless, as of today at least, the range of 24 to 28 weeks has not significantly expanded. The absence of lung surfactant prior to that stage of embryological development so far has created a barrier to any quantum leap breakthrough in technology that would lower viability to substantially before 24 weeks.[48] Such a development is not impossible of course. But there is no reason to assume that it is either inevitable or imminent.[49] Thus future shifts of viability into the middle of the second trimester or earlier, which would implicate many more abortions, are unknown and at present unknowable.

The major deficiencies of the attempts to regulate selection of method are the adverse impacts on health and on autonomy. A criminal law that mandates a fetal-favoring method absent harm to the woman inevitably will tip the scales of the physician's judgment. A doctor will necessarily wonder whether his or her measurement of the consequences of any given method choice on the woman's health will be sufficient to protect against later prosecution, if challenged. Different risks are known to be associated with each method.[50] For a doctor who is skilled in D & E, for example, that procedure might very well be considered the safest, most comfortable, and least traumatic one for the patient. But any such doctor would be in a quandary as to how much less the risk to the woman must be before the D & E would not be criminal.

The physician's dilemma is compounded by the fact that, because so few abortions after 24 weeks are performed, comparative morbidity data that segregate that time period are virtually nonexistent. So he or she is left with no reference point to assist in the assessment of risk and no defense if a local prosecutor later wants to second-guess that judgment call.

Further, the concept that has been accepted by courts of requiring a fetal-favoring method when the risks are precisely equal presumes that real cases conform to abstract equivalency. Any assessment involves comparison of different kinds of risks, as well as different probabilities of their occurrence. Exact equivalency, in this context, is a fiction.

These judgments, moreover, are precisely the kind that, because of the very gravity of their impact and the subjectivity of their meaning, are made in other kinds of cases by the patient.[51] Abortion patients, too, should be allowed the final weighing and assuming of whatever risks are involved. After having been fully advised of the range and nature of the possible complications associated with the various methods, the woman undergoing an abortion "is the one to decide which of the possible consequences [s]he wants to risk." *Dunham v. Wright*, 423 F.2d 940, 945 (3d Cir. 1970). "The weighing of these risks against the individual subjective fears and hopes of the patient is not an expert skill. Such evaluation and decision is a nonmedical judgment reserved to the patient alone." *Cobbs v. Grant*, 502 P.2d 1, 10 (Cal. 1972). This is, if anything, especially applicable to women seeking late abortions because the medical risks of death or harm to the woman during the procedure (by any method) increase the later in pregnancy it is performed.[52]

Ultimately, considerations of this magnitude will determine not just the preference for a particular method but the underlying decision itself. Even if her health condition "entitles" her to a post-viability abortion, a woman whose only option is labor and delivery with the possibility of a live birth in bed may find it less horrific to finish the pregnancy. In such a situation, a law that on its face restricts only method choice operates in effect to compel pregnancy and coerce parenthood.

4. Doctrinal Bases for Public Policy on Late Abortions

Although unsatisfactory and perhaps unstable, the viability compromise of *Roe v. Wade* has withstood attack in the courts and continues to establish at least the upper limit for how restrictive a state's approach can be. But it need not be the only model. Legislatures that face the issue of time limits on abortion--whether because of medical advances altering the survivability calculus, or shifting Supreme Court majorities reanalyzing *Roe v. Wade*, or simply a desire to revamp the health code--can adopt approaches that are different from, so long as they are not contradictory to, the trimester and viability framework outlined by the Court.

The remainder of this position paper examines the principles of law underlying the policy tension inherent in late abortions. Its goal is to identify the least restrictive alternative that reconciles the woman's interest in self-determination and the society's interest in tolerance with the unease many feel about what they perceive to be disregard toward fetal life.

A. Autonomy, Conscience, and Tolerance

The privacy doctrine relied on by courts in cases involving reproduction is based on principles of both bodily integrity and personal autonomy. Of the two, the bodily integrity arguments are older, often being viewed as the bedrock of limited government itself.

> In the history of the common law, there is perhaps no right which is older than a person's right to be free from unwarranted personal contact . . . It is this interest in the physical security of the person and integrity of the body upon which the modern tort of battery is premised . . . [T]his interest . . . is referred to in the . . . Magna Carta and Blackstone identified "the right of personal security" as one of the three elements of "liberty" . . . Our own constitutional history contains many references to the importance of the "inviolability of the person" . . . More specifically, a respect for bodily integrity, "as the major locus of separation between the individual and world," underlies the specific constitutional guarantees of the Fourth Amendment [and] the Eighth Amendment as well as the due process clauses of the Fifth and Fourteenth Amendments. *Davis v. Hubbard*, 506 F. Supp. 915, 930-1 (D. Ohio, 1980) (citations omitted).

Literal physical self-sovereignty concepts underlie much of American law, whether criminal law, tort law, or personal liberty claims under the Fourteenth Amendment. The Supreme Court has recognized that a coerced physical intrusion, as the compelled completion of a pregnancy would surely be, involves risks to both the medical and "dignitary" interests of an individual.

In 1985, for example, the Court ruled unconstitutional the unconsented-to surgery performed on a criminal defendant in order to recover a bullet, which was evidence of the crime. In that case, the Court found a severe intrusion on the defendant's dignitary privacy interest, premised largely on the loss of control necessitated by the forcible administration of anesthesia:

> When conducted with the consent of the patient, surgery requiring general anesthesia is not necessarily demeaning or intrusive. In such a case, the surgeon is carrying out the patient's own will concerning the patient's body and the patient's right to privacy is therefore preserved. In this case, however, . . . the Commonwealth proposes to take control of respondent's body, to "drug this citizen--not yet convicted of a criminal offense--with narcotics and barbiturates into a state of unconsciousness" and then to search beneath his skin for evidence of a crime. *Winston v. Lee*, 470 U.S. 753, 765 (1985).

The physical consequences of enduring a pregnancy one does not want because the state orders that it must be brought to term are, on their own, chilling. Even more profound, however, are the infringements on one's most personal sense of dignity and self. In deciding the abortion cases, courts have placed primary reliance on decisional rather than physical integrity.

> Th[e] right of personal privacy includes the interest in independence in making certain kinds of important decisions . . . The decision whether to bear or beget a child is at the very heart of this cluster of constitutionally protected choices . . . This is understandable, for in a field that by definition concerns the most intimate of human activities and relationships, decisions whether to accomplish or to prevent conception are among the most private and sensitive. *Carey v. Population Services International*, 431 U.S. 678, 685 (1977).

A constitutional shelter for independence in making reproductive decisions has obvious analogs in a range of other contexts also. The freedom to profess a belief or a nonbelief in a supreme being or an equivalently central life philosophy, *Welsh v. United States*, 398 U.S. 333 (1970) and *United States v. Seeger*, 380 U.S. 163 (1965); the freedom to pursue one's profession or vocation, *Yick Wo v. Hopkins*, 118 U.S. 356 (1886); the freedom to associate with others for the assertion of political beliefs or validation of cultural identity, *NAACP v. Alabama ex rel Patterson*, 357 U.S. 449 (1958); and the freedom to marry, *Loving v. Virginia*, 388 U.S. 1 (1967) and to live in family groupings, *Moore v. City of East Cleveland*, 431 U.S. 494 (1977)--all could be included under the rubric of autonomy.

At least some aspects of all these decisions go beyond the purely consequential sequelae of the particular choice and touch essential indicia of personhood. They implicate an individual's most deeply held precepts and most fundamental sense of identity. For most people, they express core life values that define the self.

For no decision is this more true than the decision whether to become a parent.[53] What it means to be a parent is more than physical risk taking, radical alteration of daily life, the assumption of financial obligation, or even the impact of lost options such as not finishing a degree. To bear or beget a child is to bring into being another human life, to project further into the future a genetic continuity, and to create a physiological kinship bond regardless of by whom the child is raised. Each of these considerations is profound.

Forcing parenthood upon the unwilling individual is an awesome degradation of the self. The impact of knowing that one has a child continues long after even a successful adoption.[54] For many birth parents (and adopted children), the knowledge that a parent-child relationship exists creates an issue that may never be fully put to rest and a sense of responsibility that always feels undischarged.

Altering the selection of abortion technique does nothing to alleviate this dimension of the problem. Even if one accepts as possible the hypothesis that requiring the use of method X would not result in any increment of greater risk by any measurement to the woman, so that she could be relieved of the physical bonds of pregnancy, a successful attempt to produce a live birth would leave her with psychological bonds perhaps as great or greater. The closest analogy for a man would be to have his sperm used--without his consent or contrary to his instructions--to inseminate a pregnancy that is then carried to term and produces a child, converting him against his will into a father. No risk to his physical well-being would be involved. But the harm to his right of self-determination would be real indeed.

In addition to the individual's interest in autonomy as to procreation, one must also weigh the society's interest in nonimposition of particular ethical choices or values. It is not surprising that, for many persons, deciding how to resolve one of the typically anguishing situations in which late abortions happen will implicate religious and other central beliefs. Whatever else may be said of the debate over who should have the decision-making authority over late abortions, there is clearly no consensus.[55] As with other questions of religious and quasi-religious belief in a pluralistic society, the very lack of consensus constitutes a strong claim for allowing the exercise of individual conscience.

B. Interest in Fetal Life

A protective interest in fetal life, as illustrated by the debate in *Thornburgh* between Justices Stevens and White, has two primary dimensions. The first is a desire for the continuation of an already well developed particular human potentiality, which is given less weight than the well-being of a living person, but significant weight nonetheless. The second is a concern that pregnancy termination bespeaks a general denigration of human life, regardless of when it occurs. 54 U.S.L.W. at 4626-7, 4630-1, 4634-5.

The first interest involves the issue of whether some responsibility or legal duty is owed each individual fetus that attains a possibility of extracorporeal survival and, if so, the nature of it. Justice Stevens ascribes identity and sentience to late-term pregnancies and distinguishes them on that basis from less developed embryos. 54 U.S.L.W. at 4626-7. The very specificity of the concern makes it appropriate that the resolution of it should be particularized. The current *Roe v. Wade* framework allows for a somewhat individualized approach, through its necessary-for-life-or-health escape clause from laws prohibiting post-viability abortion. Justice Stevens' expression of concern for developed fetal life appears in the context of an opinion upholding the principles of *Roe v. Wade*.

As part of an individualized approach, considerations of likely impairment, for example, would be relevant. The certainty of pain and severe impairment or the probability of early death[56] could mean that Justice Stevens' interest would be in some cases better served by an abortion that does not result in a live birth. At the least, such a situation would cause one to hesitate bringing the pregnancy to term as a gesture of aid to the potential child. Other situations return us to the original tension: a child's life, if the child is born with less severe disabilities or to a very young and poor mother, might be hard and uncomfortable but preferable to nonlife. Still, given a particularized approach and given the greater weight accorded to the living rather than the potential person, the gravity of harm to the woman of forced parenthood could lead to the same result of permitting a procedure in which the fetus is not preserved.

Justice White's perspective would admit of no abortions except those necessary to save the woman's life. *Roe v. Wade,* 410 U.S. at 173 (White and Rehnquist, JJ., dissenting). It is not the features of each human situation, but the revealed general principle about valuing fetal life which is determinative in this approach. For some, of course, this concern will mean an unyielding anti-abortion position. For many others, however, support for the principle of choice is coupled with an unease when no ethical claim is recognized for the value of fetal life.

Potential human life can be valued in many ways. To compel procreation is more punitive than respectful of the generative process. Policies such as comprehensive prenatal care and support services for young parents enlist state aid in efforts of far greater scope than any law concerned with the very small number of late abortions. Such social policies do not resolve an argument framed in terms of "abortion is murder," or the "fetus is a person." But they do express and concretize a state policy of providing care and material support for the potentiality of human life.

Conclusion

The question of how the law should treat post-viability abortions requires a balancing of harms, interests, and other options. The burden to a woman of continuing a pregnancy against her will is stupefying. Even with the pregnancy terminated, compelling parenthood by imposing alternative means (selection of a fetal-favoring abortion method or use of an artificial uterus) to bring a child into being can be devastating to conscience and identity, as well as emotional health. Moreover, the line upon which the *Roe v. Wade* structure rests--viability--is so riddled with uncertainty and potential for future meaninglessness that its chief asset, if this is one, is that it signifies the refusal to assign full weight to either the woman's right or the interest in fetal life, and the insistence on compromise of what is unquestionably a difficult contradiction.

The state has a range of possible policy schemes that it can employ in seeking to avert abortions in the last third (or even the second half) of pregnancy. A few exceptional emergency situations that necessitate abortion will continue to occur no matter what the state does. But, because no one asserts that women would, if given the option, *prefer* to make their abortion decisions later rather than earlier (and the data conclusively show that the opposite is true), it is reasonable to believe that choice-enhancing rather than choice-constricting policies would actually work. The most humane way to prevent late abortions is to alter the social conditions that are the preventable cause of them. This resolution of the problem produces a far better outcome than does reliance on the faulty structure of *Roe v. Wade*.

Proposed Public Policy Measures to Reduce Late Abortions

1. Comprehensive education in the schools concerning sexuality and contraception.
2. Free, confidential contraceptive and abortion service programs, widely advertised, geographically dispersed throughout each state, that include the development of school-based health clinics offering contraceptive care and counseling. Ideally, programs offering family planning services to low-income teenagers should be components of comprehensive social services programs which include vocational training and counseling and educational incentives.
3. Comprehensive prenatal care, available free or at low cost to low-income pregnant women.
4. Restoration of Medicaid funding for abortions.
5. Repeal of laws requiring parental consent or notification for teenagers seeking abortions.
6. Free or low-cost hospital-based genetic counseling programs for pregnant women and their partners. Such programs should include scientifically accurate information about the biological limitations of the specific disability at issue in any given case, as well as information as to the experiences of families who have coped with those disabilities.
7. Medical research for improving methods for the detection of fetal anomalies, so that testing if desired can be made available as early as possible during pregnancy.

Proposed Statute

As was seen in the foregoing memorandum, the majority of states currently do not regulate post-viability abortions. The following language is therefore proposed both as a model law and as a statement of the legal principle in favor of full reproductive choice, which can be invoked to maintain the absence of restrictions.

> The state shall not compel any woman to complete or to terminate a pregnancy, nor shall the state restrict the use of medically appropriate methods of abortion.

Notes and References

1. In 1973, the year Roe v. Wade was decided, 63 percent of all abortions occurred within the first 10 weeks of pregnancy. Today, 88 percent of all abortions occur in the first 10 weeks. Grimes, Second Trimester Abortions in the United States, 16 Family Planning Perspectives 260 (1984).

2. Henshaw, Binkin, Blaine, and Smith, A Portrait of American Women Who Obtain Abortions, 17 Family Planning Perspectives 90, 91 (1985). One study suggests that even the 0.01 percent figure may be too high. Staff from the Centers for Disease Control investigated all abortions reported as occurring during the third trimester in the state of Georgia from 1979 and 1980. Of the 78 with adequate data, only three had been classified correctly. Two of those involved anencephalic fetuses. Correction of the misclassified reports would reduce the rate of post-24-week abortions in Georgia to 0.004 percent. Spitz, Lee, Grimes, Schoenbucher, and Lavorie, Third Trimester Induced Abortion in Georgia, 1979 and 1980, 73 American Journal of Public Health 594 (May 1983).

3. Presentation to the working group of the Project on Reproductive Laws for the 1990s, March 18, 1986.

4. In 1977, 455,000 cesarean sections were performed in the United States, or approximately 13 percent of all deliveries. Marieskind, An Evaluation of Caesarean Section in the United States, Dept. Health, Education and Welfare (1979), Table 1 at 13.

5. See generally references cited in Janet Gallagher's position paper, "Fetus as Patient," in this book.

6. U.S. Centers for Disease Control (CDC), Abortion Surveillance 1981 (Nov. 1985), Table 14 at 37.

7. Id.

8. Alan Guttmacher Institute (AGI), 3 Public Policy Issues in Brief 1, 3 (1983); Grimes, Second Trimester Abortions in the United States at 261-2; Burr and Schulz, Delayed Abortion in an Area of Easy Accessibility, 244 Journal of the American Medical Association 44 (1980).

9. See Donovan, Judging Teenagers: How Minors Fare When They Seek Court-Authorized Abortions, 15 Family Planning Perspectives 259 (November/December 1983) and ACLU Reproductive Freedom Project, Parental Notice Laws: Their Catastrophic Impact on Teenagers' Right to Abortion (1986).

10. In 78 percent of all U.S. counties, where 28 percent of all women live, there was no identified abortion provider at all in 1980; nearly 90 percent of nonmetropolitan counties had no abortion services. AGI, Issues in Brief, at 3. Travel across county lines alone is associated with an average delay of 5 days. Henshaw and O'Reilly, Characteristics of Abortion Patients in the United States, 1979 and 1980, 15 Family Planning Perspectives 5, 154 (January/February 1983).

11. Grimes, Second Trimester Abortions in the United States, at 261. This number represents only 2-5 percent of all amniocentesis tests performed in a year. The overwhelming majority of amniocentesis tests are negative for anomalies, and have the effect of reassuring the prospective parents.

12. CDC reported 10,783 abortions performed past 20 weeks gestation in 1981. Abortion Surveillance 1981, Table 11 at 35.

13. The immune system weakens during pregnancy, with the ratio of T-helper to T-suppressor cells reaching its lowest point in the third trimester. This is one reason for the increasing concern that pregnant women may have especially strong reasons to learn whether they have been infected with the human immunodeficiency virus (HIV), the virus believed to cause AIDS. Another reason is that HIV infection can be transmitted in utero or at the time of birth. To date, approximately 350 children in the United States have been diagnosed as having perinatally acquired AIDS; more than 200 have died. An unknown number are infected with the virus. CDC, Recommendations for Assisting in the Prevention of Perinatal Transmission of HTLV-III and LAV and AIDS, 34 Mortality and Morbidity Weekly Report (1985).

14. See generally, S. L. Romney, et al., Gynecology and Obstetrics, (2nd ed. 1981) at 703, 705-708, 710, 712-714, 718, 722, 724, 726, 729, 732, 739, 756, 762, 764, 776, 778, 783-784, 793, 795. See also Williams, Obstetrics at 477. One of the leading causes of maternal mortality nationwide, hypertensive states of pregnancy (toxemia), does not appear until the late second trimester. Termination of the pregnancy is recommended for severe cases, and these are usually post-viability. Id. at 528, 543.

15. A. Radosh, New York City Office of Adolescent Pregnancy and Parenting Services, unpublished manuscript, 1986.

16. New York City Department of Health, Summary of Vital Statistics (1986), at 16, 17.

17. See, e.g., McRae v. Califano, 491 F. Supp. 630, 675-6 (E.D.N.Y. 1980) rev'd on other grounds sub nom. Harris v. McRae, 448 U.S. 297 (1980).

18. Jones, Forrest, Goldman, Henshaw, Lincoln, Rosoff, Westoff, and Wolf, Teenage Pregnancy in Developed Countries: Determinants and Policy Implications, 17 Family Planning Perspectives 53 (March/April 1985).

19. New York Times, "When Abortions Become Live Births." Feb. 15, 1984. The pool of service providers available even in the second trimester is small. Only 20 percent will perform abortions after 14 weeks. AGI, Issues in Brief at 3.

20. Grimes, Second Trimester Abortions in the United States at 262.

21. Id.

22. Id. The remaining abortions are performed by methods using a mixture of techniques, such as instillation of prostaglandin combined with urea (also a fetal-destructive substance), or injection with urea, followed by a D & E.

23. New York Times, "When Abortions Become Live Births." Feb. 15, 1984.

24. Abortion Surveillance 1981 reports 1.3 million total abortions.

25. Abortion Surveillance 1981, Table 11 at 35.

26. A live birth is defined for official statistical purposes in 45 states as a "product of conception, irrespective of the duration of the pregnancy which . . . breathes or shows any other evidence of life such as beating of the heart . . . or definite movement of voluntary muscle." National Office of Vital Statistics DHEW, International Recommendations in Definitions of Live Birth and Fetal Death, Washington, D.C. (1950), cited in Grimes at 263.

27. Grimes, Second Trimester Abortion in the United States at 263.

28. In a clinical study to compare D & E with prostaglandin instillation for second-trimester abortions, three of the women who had been randomly assigned to the prostaglandin group dropped out of the study because of their fear of a long, painful labor culminated by abortion in bed. Grimes, Hulka, and McCutchen, Midtrimester Abortion by Dilatation and Evacuation Versus Intra-Amniotic Instillation of Prostaglandin F_{2a}: A Randomized Clinical Trial, 137 American Journal of Obstetrics and Gynecology 785, 789 (1980).

29. No decision has addressed the question of whether state constitutions might be the basis for a right to abortion without the viability limit. Some state courts do interpret the privacy right underlying abortion provided by their state constitutions more expansively than the United States Supreme Court has interpreted the federal constitution, at least in the context of public funding. See, e.g., Right to Choose v. Byrne, 91 N.J. 287, 450 A.2d 925 (1982); Committee to Defend Reproductive Rights v. Myers, 29 Cal. 3d 252, 625 P.2d 779 (1981); Moe v. Secretary of Administration and Finance, 417 N.E. 2d 387 (Mass. 1981).

30. Scienter is the mental state necessary for criminal prosecution; typically it incorporates intentional, knowing, reckless, or negligent acts.

31. Justice White, in his Thornburgh dissent, decries this result as illogical because it allows the state to forbid entirely nonnecessary abortions past the point of viability, even if they are safer than childbirth, but prevents the state from imposing any quantum of enhanced risk on the woman in method selection for necessary abortions, even though the state's interest in the fetus is then at its height. 56 U.S.L.W. at 4635. In a sense he is correct. His response is to allow the state to implement more fully its disfavor of these (or any) abortions. Another

response would be to conclude that increased medical risk, which is impermissible, is no greater detriment to the liberty interest at stake than compulsory pregnancy and childbirth, which therefore should be impermissible. Thus one should adopt the doctrinal principles advocated herein.

32. Lower courts have explicitly adopted this position. See Wynn v. Scott, 449 F.Supp. 1302, 1321 (N.D. Ill. 1978).

33. The medicalization, and thus social control, of decisions which would otherwise be made by individuals is a common theme in contemporary sociology. See, e.g., Irving Zola, "Medicine as an Institution of Social Control" in Conrad and Kern (eds.), The Sociology of Health and Illness (1981).

34. La. Rev. Stat. § 1299.35.4.

35. Ariz. Code § 36-2301.01(B); Ark. Code §41-2564; Mass. Ann. Laws ch. 112, §12(O); Mo. Rev. Stat. §188.030(2); Okla. Stat. Ann. Title 63 §1-732(D); Pa. Code §3210(b) [declared unconstitutional in Thornburgh]; Wis. Code 1985.

36. Fl. Stat. Ann. §390.001(5); Ill. Rev. Stat. Ch. 38, Par. 81-26(1)(1983) [declared unconstitutional on various grounds in Charles v. Daly, 749 F.2d 452 (7th Cir. 1984), appeal dismissed sub nom. Diamond v. Charles, 56 U.S.L.W. 4418 (April 29, 1986)]; Pa. Code § 3210(b).

37. Iowa Code Ann. §707.10; Ky. Rev. Stat. §311.780; Minn. Stat. Ann. §145.412 (3)(3).

38. Ch. 390.001(5).

39. Arizona; Arkansas; Illinois; Missouri; Wisconsin.

40. Massachusetts; Oklahoma; Pennsylvania.

41. Iowa; Kentucky; Minnesota.

42. With an ultrasound image, the size of certain fetal structures (usually the cranium and the crown-rump length) can be directly measured. These measurements are then compared to standardized charts (e.g., a certain cranial diameter correlates with a certain number of weeks of development). Although the measurements themselves are precise, the age estimates are based on averages; an equivalent process would be guessing the age of a teenager based on height--although 5' 6" might be the average height of caucasian males at the age of 16, the particular 5' 6" boy might be a short 18 year old or tall 14 year old. See Rhoden, Trimesters and Technology: Revamping Roe v. Wade, 95 Yale Law Journal 639, 659 (1986).

43. See Rhoden, Trimesters and Technology: Revamping Roe v. Wade, 95 Yale Law Journal at 660.

44. Colautti v. Franklin, 439 U.S. at 396 n. 15.

45. See generally Lori Andrews' position paper, "Alternative Modes of Reproduction," in this book.

46. See Justice O'Connor's dissent in City of Akron v. Akron Center for Reproductive Health, 462 U.S. at 457.

47. Roe v. Wade, 410 at 160.

48. The author of one of the articles cited by Justice O'Connor in her dissent in Akron stated:

> I know of no current research which would lead me to believe that a fetus of less than 22 weeks gestation can be or soon will be sustained outside the uterus . . . Below 24 weeks gestation, the fetus' lungs are simply not adequately developed to sustain oxygenation even with ventilator support.

Kopelman, letter to Nan D. Hunter.

49. Another author cited by Justice O'Connor stated:

> . . . [i]t is highly likely that the lower limit beyond which human gestation is simply incapable of survival has already been virtually reached in its entirety, at least insofar as we intend such survival to be possible without the intervention of some form of artificial extrauterine

> environment that could be provided by the successful creation of an artificial placenta.

Stern, letter to Nan D. Hunter.

50. For prostaglandin instillations, for example, contraindications include hypertension, asthma, glaucoma, and epilepsy. W. Hern, Abortion Practice (1984) at 125. Gastrointestinal side effects, sometimes severe and prolonged, are common in prostaglandin abortions; almost 50 percent of the patients experience nausea, vomiting, and/or diarrhea. Kerenyi, "Intraamniotic Techniques" in Abortion and Sterilization: Medical and Social Aspects (J. Hodgson, ed. 1981) at 368. Hypertension is also a contraindication for saline injections, as are sickle cell disease, anemia, heart disease, kidney disease, and blood coagulopathy. Kerenyi, "Hypertonic Saline Instillation" in G.S. Berger, et al., Second Trimester Abortions: Perspectives After a Decade's Experience (1981) at 179. The most frequent complications are hemorrhage and infection. A disadvantage of both prostaglandin and saline procedures is the length of time patients are incapacitated. The overall injection to abortion interval is approximately 24 hours for saline (somewhat shorter if combined with oxytocin), and 16 hours for prostaglandin. Binkin, Schulz, Grimes, and Cates, Urea-Prostaglandin Versus Hypertonic Saline For Instillation Abortion, 146 American Journal of Obstetrics and Gynecology 947, 949 (1983). Actual labor, which carries its own set of risks, may last 4-12 hours. Rooks and Cates, Emotional Impact of D & E vs. Instillation, 9 Family Planning Perspectives 276 (1977). The literature contains no contraindications based on the patient's health condition for the use of D & E. As a surgical procedure, however, it always carries the risk of cervical laceration and even perforation. Other complications include excessive blood loss, hemorrhage, and pelvic infection. Peterson, et al., Second Trimester Abortion by Dilatation and Evacuation: An Analysis of 11,747 Cases, 62 American Journal of Obstetrics and Gynecology 185 (1983). In assessing the advisability of using D & E, medical experts believe that the primary factor is the individual physician's level of skill and training. Cates and Grimes, "Morbidity and Mortality of Abortion in the United States" in Abortion and Sterilization at 156-158.

51. Perhaps the strongest support for the patient's right to decide is found in the line of cases authorizing the patient or a substituted party to order the withdrawal of life-sustaining technology. Tune v. Walter Reed Army Medical Center, 601 F. Supp. 1452 (D.D.C. 1985); Bouvia v. Superior Court of California, 225 Cal. Rptr. 297 (Ct. App. 1986); Bartling v. Superior Court, 163 Cal. App. 3d 193, 209 Cal. Rptr. 220 (1984); Superintendent of Belchertown v. Saikewicz, 370 N.E.2d 417 (Mass. 1977); Matter of Quinlan, 70 N.J. 10, 355 A.2d 647 (1976).

52. Grimes, Second Trimester Abortions in the United States at 262-3.

53. "Conscience means moral awareness, and liberty of conscience means the exercise of moral awareness. Abortion presents a matter for individual moral decision, in a matter of ultimate concern respecting bringing a life into the world." Harris v. McRae, clergy testimony.

54. The following passage was contained in a letter written for the National Abortion Rights Action League "Silent No More" campaign, by a woman who was unable to obtain an abortion in 1951:

> I remember--and believe me, this is something one never forgets--my devastated father weeping over the news, the subterfuge and strategy involved in spiriting me out of sight, the lonely wait in a strange city, and the birth alone amongst strangers, completely devoid of the joy which should rightly accompany this event, the pressure to relinquish my daughter for adoption, then the time of agonizing readjustment, which phased into acceptance but left a never-totally assuaged sense of loss.

55. See Harrison, Our Right to Choose: Toward a New Ethic of Abortion (1983); Fost, Chudwin, and Wikler, The Limited Moral Significance of Fetal Viability, 10 Hastings Center Report, December 1980, pp. 10-13; and Macklin, Personhood in Bioethics Literature, 61 Milbank Memorial Fund Quarterly 35 (Winter 1983).

56. A child born with Tay-Sachs disease for example, with a life expectancy of four years, was found to suffer

> . . . from mental retardation, susceptibility to other diseases, convulsions, sluggishness, apathy, failure to fix objects with her eyes, inability to take an interest in her surroundings, loss of motor reactions, inability to sit up or hold her head up, loss of weight, muscle atrophy, blindness, pseudobulper palsy, inability to feed orally, decerebrate rigidity, and gross physical deformity.

Curlender v. Bio Science Laboratories, 106 Cal. App. 3d 811, 165 Cal. Rptr. 477, 480-1 (1980).

PRENATAL SCREENING

Mary Sue Henifin, Ruth Hubbard, and Judy Norsigian

Position Paper

The Supreme Court has long viewed a wide range of personal and family activities as individual liberties and privacy interests protected by the Constitution.[1] These include the right to decide whether to procreate and the right to rear and raise our children free from government intrusion absent a compelling state interest:[2]

> If the right of privacy means anything, it is the right of the individual, married or single, to be free of unwarranted governmental intrusion into matters so fundamentally affecting a person as the decision whether to bear or beget a child.[3]

Another aspect of individual liberty and privacy is the right to be free from coerced physical intrusion, including coerced medical treatment. Both the constitutional right of privacy and the common law right to bodily integrity protect the right to accept or refuse medical treatment. It is a "well-recognized common law right of self-determination that '[e]very human being of adult years and sound mind has a right to determine what shall be done with [her] own body. . . .'"[4] However, the Supreme Court has not yet decided whether procreative choice and bodily integrity protect the right to use or refuse new reproductive technologies.

State legislators and lower courts have already taken actions that limit the right to choose or refuse medical treatment during pregnancy. Ultimately, it is likely that an increasingly conservative Supreme Court will decide whether the Constitution protects choice in the use of new reproductive technologies. But court pronouncements have a limited role in shaping reproductive alternatives. Legal protections for reproductive rights are circumscribed by poverty and other social problems.[5] Procreative choice presumes access to medical care and other fundamental necessities, such as food and housing. It also presumes absence of oppression because of inborn conditions, and access to social supports for a child's particular needs such as those associated with a disability. Thus, the principle of procreative choice must be measured against the circumstances of prospective parents and the conditions they face when making reproductive decisions.

This position paper addresses issues pertaining to a set of new reproductive technologies--prenatal screening tests, which permit physicians to diagnose disabilities in the embryo and fetus. At present, there are no preventive therapies for most conditions that can be diagnosed before birth, and none at all for chromosomal abnormalities and other genetic conditions. New prenatal therapies for certain structural and metabolic fetal disabilities are being researched, but such treatments are still experimental. Usually, a prenatal diagnosis of a disability merely enables a pregnant woman to decide whether to carry the fetus to term or to terminate the pregnancy.

The Vatican, in its recent pronouncement urging legislative action on new reproductive technologies, condemned prenatal screening if women "request such a diagnosis with the deliberate intention of having an abortion should the results confirm the existence of a malformation or abnormality."[6] In contrast, some physicians and legal commentators have suggested that a pregnant woman should be liable for "prenatal abuse" if she knows she is at risk of carrying a fetus with a disability and refuses prenatal testing, or if she knows her fetus has a disability and refuses an abortion or treatment.[7]

It is our position that standard prenatal tests and counseling should be available to all women irrespective of ability to pay. However, no test should be mandatory, and no prospective parents or their child should be denied medical services or insurance coverage because they have previously refused a test. Nor should parents incur legal liability for refusing to be tested. However, we recognize that, even with these safeguards, the decision to accept or refuse a test is not a free individual choice because it is contingent on a number of economic and social factors.

Some physicians and legal commentators place pregnant women in a special category where their individual liberty and privacy interests would be denied by permitting physicians or the state to make decisions with regard to prenatal screening, fetal therapy, or abortion. Without going that far, to the extent that health care providers overtly or subtly pressure a woman who is carrying a fetus with a disability to have an abortion, they are interfering with her procreative choice. Furthermore, when they believe that screening the fetus is more important than honoring the woman's refusal to be screened, they can pressure her into renouncing her right to make health care decisions. Both instances constitute infringement of a woman's right to bodily integrity because the tests are physically intrusive and because the decision to abort involves the woman's body. We believe that a woman's right to bodily integrity encompasses the fetus that she carries, and thus, her decisions must be respected over third-party attempts to intervene on behalf of the fetus. Any other approach dilutes her constitutionally protected right to privacy, including freedom from compelled physical invasion. Perhaps more importantly, it denies that pregnant women are in the best position to make decisions about their own health and the interests of the fetuses they are carrying. Experts are frequently mistaken in their judgments about what is in the best interest of the pregnant woman and/or her fetus, as has been documented by women and their children who have suffered the adverse effects of unnecessary forceps and cesarean deliveries, X-ray pelvimetry, DES, thalidomide, and medicated labor.[8]

* * *

We begin this position paper with a description of prenatal screening programs and some of their problems, including the stigmas attached to persons at risk of having children with a disability, the fear during the early stages of pregnancy that the fetus may later be diagnosed as having a disability, and the problems arising from uncertainty about how to interpret test results. Next we document legal pressures on women to be tested and problems with informed consent for prenatal testing. Finally, we discuss how lack of prenatal care for poor women and discrimination against persons with disabilities profoundly limit reproductive choices.

The New Medical Technologies of Prenatal Diagnosis and Screening

At present, the incidence of inborn diseases is on the order of 3 to 4 percent, depending on how one defines disease.[9] It is difficult for most people to evaluate what it would mean to *them* to have a child with a disease or disability, because most people have limited experience with disabilities and regard the possibility of having a child with a disability as a "nightmare" to be prevented at all cost.[10] Therefore, parents may hope for prenatal assurances that their child will be healthy. Although science cannot offer such blanket assurances, since most health problems cannot be anticipated, there is considerable pressure among professionals and the public to foresee increasing numbers of disabilities and diseases. As it becomes possible to screen for a growing number of genetically transmitted health problems, it is important that prospective parents be offered a range of options and full information and counseling about them, so that they can evaluate various alternatives.

Since scientists first developed prenatal tests 20 years ago to detect fetal sex and Down syndrome, more than 4000 genetic traits have been cataloged, and prenatal tests now exist for over 300 of them with new tests being developed all the time.[11] Most of these tests are used to detect rare diseases, not to screen routinely for common difficulties, and can be useful only when there is a family history of a particular disease.[12] The usual procedure is to perform relatively simple biochemical tests on both partners to determine whether they are recessive carriers of the disease in question. If they are, then for a number of such diseases, methods have recently been developed to test fetal cells obtained by amniocentesis to determine whether the fetus will exhibit the disease. When this is the case, there is the option to terminate the pregnancy. Even if the pregnant woman decides to continue the pregnancy, early diagnosis gives the family time to prepare for the birth of a child with special needs. Screening of prospective parents can become problematic when a particular racial, age, or occupational group has a higher incidence of a particular genetic condition and therefore is singled out for testing. Associating Down syndrome with the pregnancies of older women, Tay-Sachs disease with Jews, and sickle cell anemia with blacks are examples of this phenomenon.[13]

Sickle Cell Testing and Stigmatization

The story of sickle cell testing in the United States illustrates some of these problems. Approximately one in 500 Afro-American babies is born with sickle cell anemia (SCA), and about one in 10 is a carrier. Carriers do not exhibit any symptoms of the disease. Several blood tests have been available since the 1960s to determine whether one has the disease or is a symptom-free carrier. Recently, a test has been developed to determine by means of amniocentesis whether a fetus has SCA.[14] This gives a woman whose child would have the disease the option to abort. However, because of the disproportionately large number of black women living in poverty, many of them do not have access to amniocentesis and/or abortion. Even if they have access to these procedures, the choice whether to abort is ambiguous because the symptoms of SCA range from mild to severe, and prenatal diagnosis does not predict how seriously the child will be affected.

In the early 1970s when screening programs for carriers of sickle cell trait proliferated--some of them compulsory on entering school or before marriage, and some carried out at the expense of the screenees--stories of discrimination against healthy carriers began to appear. By autumn 1972, at least one airline stewardess had been grounded. Other people reportedly were denied employment, and some insurance companies refused carriers of the sickle cell trait health or life insurance or inflated their premiums, even though there was "no evidence that trait carriers have a higher risk of disease or a shorter than normal life-span."[15] One physician pointed out that "the requirement of testing before entering school erroneously implies that some steps need to be taken by parents of six-year-old children with respect to the possession of the trait," when in fact there is nothing to do.[16] As Dorothy Wilkinson emphasizes:

> There are negative political implications to proliferating laws targeted solely at sickle cell with little or no legislation or enforcement of existing laws dealing with the myriad of socioeconomic problems confronting black Americans. These neglected areas are also related to their health and welfare.[17]

There is a further point that worries us. How are those who *have* sickle cell anemia to feel in the face of widespread publicity and legislation aimed explicitly at preventing others like them from being born? Surely, this is an extreme form of stigmatization.

Because of vocal criticisms by black leaders and many other people, large-scale screening for sickle cell trait was discontinued by the mid-seventies. However, New York State currently tests all newborns for sickle cell anemia to permit prompt treatment for those with SCA, in order to reduce increased risk of mortality from pneumonias. Several states have proposed similar testing programs, but these tests also detect those infants who are carriers and therefore are not expected to show symptoms. Unless comprehensive legislation is passed to protect individuals with genetic disease or trait carriers from discrimination, the welfare of these children, who are labeled at birth, may be compromised.

Down Syndrome and the Tentative Pregnancy

One of the most common uses of prenatal screening is to detect Down syndrome, which occurs in approximately one in 1000 newborns. The incidence of Down syndrome is not evenly distributed, but increases progressively among babies born to women above 30 years of age--from 1.54 per 1000 among infants born to women in the age range 30 to 34, to 2.63 per 1000 among those born to women 35 to 39, to 14.3 per 1000 for women 40 to 44, and up to 34.2 per 1000 for women 45 or older.[18]

An issue that has not had adequate attention in the wake of publicity over Down syndrome is the problem of stigmatizing older women as less fit for childbearing. At a time when increasing numbers of women must decide how to integrate having children and jobs, poorly supported assertions about the risks of postponed childbearing are a disservice.[19] There are no adequate studies of the reasons why the incidence of Down syndrome increases among children of older women. The role that fathers and their ages plays has barely been explored, and little attention has been paid to the health and environmental histories of the parents of affected children.[20] The assumption that the problem is "old mothers," and the resulting injunction that older women be screened, obscure important questions, which, if answered, might help decrease the incidence of the biological causes of Down syndrome.

Amniocentesis, the usual technique for obtaining fetal cells for diagnosis, has two technical limitations: it cannot be done until about 13 to 15 weeks into the pregnancy, and the procedure involves small, but measurable, risks of fetal injury, infection, and spontaneous abortion. Results usually are not available until well into the second trimester, so that if the woman decides to terminate the pregnancy, the abortion involves more risky and emotionally traumatic procedures than those done earlier.

Researchers report that, though parents are relieved when they get a negative answer, many pregnant women experience considerable tension and anxiety prior to amniocentesis and during the waiting period before they get the test results. The waiting period can last as long as four weeks.[21] This is the stage of pregnancy when most women begin to feel their fetuses move. With the possibility of abortion hanging over them, some women try not to become emotionally involved with their pregnancy until after they know they will carry it to term. Barbara Katz Rothman calls this "the tentative pregnancy" and eloquently documents the experiences of women waiting for the test results.[22] Furthermore, the decision to have an abortion following amniocentesis can be stressful and provoke depression in both parents.[23] To abort a wanted pregnancy because the fetus has a disability can be much more traumatic than to have an abortion for other reasons.

The "tentative pregnancy" will be shortened as a new technique, chorionic villus sampling (CVS), becomes more widely used. In this test, cells from the hairlike projections of the membrane that surrounds the embryo are sampled and cultured at about the 9th to 11th week of pregnancy. Results are available within days or even hours compared to the 2 to 4 weeks it sometimes takes to get results of amniocentesis. But even when the technique is perfected, the results may still not tell parents what they need or want to know to make an informed choice, so that they may have to undergo amniocentesis in addition. Because it is done early in the pregnancy, CVS detects abnormalities of many fetuses that would abort spontaneously as the pregnancy progresses. Therefore, it forces many prospective parents to make needless, difficult decisions about whether to terminate the pregnancy.

Procedures may become available by which scientists can test cells of fetal origin that enter the bloodstream of the pregnant woman. This would mean that blood samples collected at routine checkups during pregnancy could be used to diagnose genetic disorders of the fetus. This procedure would, of course, raise all of the ethical and legal problems associated with the procedures that are now in use, and perhaps intensify them because, although sophisticated technically, such blood tests would be relatively simple and not invasive for the pregnant woman.

At present, tests are being developed for use on early embryos before they become implanted in a woman's uterus. The embryos can be obtained by in vitro fertilization or by flushing them out of the uterus before they attach. The latter procedure is less invasive as far as the woman is concerned. At such an early stage of development, it is possible to remove one or two cells from the embryo and clone and test them, while the remaining embryo is frozen and stored. If the cells pass "genetic inspection," the embryo can be thawed and introduced through the cervix into the uterus of the woman who provided the egg, or the uterus of another woman who is prepared to gestate and give birth to the fetus. It may also be possible in the future to carry out genetic manipulation of the embryo, using gene insertion or deletion techniques. As far as scientists can tell, the fact that cells have been removed for testing does not affect the later development of the embryo, nor do freezing and thawing. At present, this still experimental procedure is intended for prospective parents who have a family history that makes them concerned about giving birth to a child with a life-threatening disease, but who cannot countenance the possibility of terminating a pregnancy after CVS or amniocentesis. Needless to say, it could also be used for a wide range of screening procedures and even become accepted practice for all pregnancies, if people become sufficiently preoccupied with preventing babies with all manner of potential health problems from being gestated and born.[24]

A major problem with screening tests is that the information they provide is frequently ambiguous. A fetal diagnosis of Down syndrome is just one example. The test cannot predict what extent of disability the child will experience, and the extent can range from mild to severe.

Neural Tube Defects and the Problem of Diagnostic Ambiguity

Current testing for neural tube defects (NTDs) provides another example of the difficulties of dealing with ambiguous results.[25] NTDs are among the most common birth defects in the United States with an incidence of 1 per 1000 births. NTDs result from the incomplete closure of the spinal cord and brain. Spina bifida is a NTD in which the spinal cord is exposed or improperly formed. The degree of disability associated with spina bifida depends on where and how the spinal cord is affected and may be quite mild. A total and fatal form of the defect is anencephaly, in which the brain or skull is missing or incomplete.

In 1972, researchers first proposed using alpha-fetoprotein (AFP) levels in a pregnant woman's blood serum to test for NTDs. Both the amniotic fluid and the blood serum of pregnant women show higher levels of AFP if the fetus has a NTD. The problem with testing maternal blood for high AFP levels is that it is a very imprecise technique. Levels of AFP change during pregnancy, so that a minor mistake in assessing the stage of pregnancy can result in an AFP level that appears to be too high or low for the assumed gestational age. Different laboratories also use different standards for assessing the AFP level. For every 1000 women screened, 50 will show an abnormally high reading, although only one of those women is likely to be carrying an affected fetus. All 50 require further tests, which may need to include ultrasonography or amniocentesis in order to detect that one woman.

Although only one out of 50 women who test positive on the first test is actually carrying a fetus with a NTD, some women have become discouraged with the lengthy testing process and the doubts it has raised about the health of their fetuses, and have terminated their pregnancies after the initial screening. At present, there are not enough genetic counselors to counsel women individually, or enough laboratories capable of accurately performing AFP tests and the followup tests that will be necessary if AFP screening is implemented on as large a scale as has been proposed. Despite these shortcomings, in 1983 the Food and Drug Administration approved, over objections by the American College of Obstetricians and Gynecologists (ACOG), the American College of Pediatrics, and a number of consumer groups, diagnostic kits developed by drug companies for use by private physicians and laboratories to detect the levels of AFP in a pregnant woman's blood. The groups opposing the kits feared that they would be used without adequate standards and supervision.

Although ACOG initially objected to the tests, its Department of Professional Liability informed physicians in 1985 that it is "imperative that every prenatal patient be advised of the availability of this test" to put the physician in "the best possible defense position" in terms of malpractice suits.[26] We agree with ACOG's initial assessment of AFP testing, in which it stated:

> maternal serum AFP screening should be implemented only when it can be performed within a coordinated system of care that contains all the requisite resources and facilities to provide safeguards essential for ensuring

> prompt, accurate diagnoses and appropriate follow-through services.[27]

Such a coordinated system of care does not exist for many pregnant women, nor is it likely to be funded when even the most basic prenatal services are not a public spending priority. Nevertheless, some states have proposed, and California has implemented, mandatory offering of AFP screening to all pregnant women. In California, every pregnant woman is required to sign a consent or refusal form for the test, stating that she has been offered it, understands its implications, and has made an informed choice. Given the diverse educational, national, cultural, and linguistic backgrounds of current residents of California, we cannot help but wonder how much this benefits pregnant clients.

Diseases of Ambiguous Origin

New genetic technologies are being developed to provide so-called probes that can identify specific small pieces of genetic material (DNA) that appear to be located close to the genes for genetic diseases such as cystic fibrosis or Huntington's disease, even though the genes themselves have not been identified. By collecting small tissue samples from different members of a family, some of whom have the disease and some of whom do not, scientists try to find what they call "markers" that appear only in the individuals who have the disease. In this way, it is becoming possible to identify those people in the family who will develop the disease before they exhibit its symptoms. The same test can be used to diagnose potential disease carriers within the family *in utero*, by testing fetal cells collected by means of CVS or amniocentesis. This will make it possible to diagnose many more genetic diseases prenatally.

The same diagnostic techniques are being expanded to attempt to identify markers for traits whose patterns of familial transmission are complex and ambiguous. Relatively few traits, and very few diseases, have simple, readily identifiable patterns of inheritance, because in general our biology is intricately interwoven with the ways we live: the food we eat, the climate in which we live, the pollutants and beneficial environmental influences to which we are exposed, and so on. The expression of even a clearly genetic disease, such as sickle cell anemia, depends on the kind of life a person who has it lives. For most traits that are said to run in families, the patterns of transmission are far more complicated than for sickle cell anemia, and the relative contributions of genetics and environment are far more difficult to pinpoint. However, scientists have begun to look for markers and to develop diagnostic probes for so-called familial predispositions to develop health problems such as high blood cholesterol levels (hypercholesterolemia), colon cancer, diabetes, bipolar manic depression, Alzheimer's disease, and a host of other diseases of complex and ambiguous origins, whose expression depends on many factors.

As prenatal tests are developed for such diseases that often do not become apparent until the child has grown up (or, indeed, grown old), prospective parents will be forced to make increasingly difficult decisions.[28] Who can be held responsible to act during pregnancy on predictions of a disease that may affect that future child in her or his twenties or even later in life? On the other hand, if someone becomes incapacitated with a disease, the tendency for which could have been diagnosed before birth, is his or her suffering the parents' (or mother's) "fault," if they (or she) refused an available test?

From Prenatal Screening to Prenatal Therapy

Through prenatal diagnosis, the fetus has become a new category of patient for whom experimental therapies are being developed.[29] A pregnant woman may feel compelled to accept every opportunity physicians offer to diagnose and "treat" her fetus despite the risks involved. In a recent review of advances in pediatric surgery, researchers noted that treatment is becoming available for a number of fetal impairments.[30] Detection of a fetal disability may now suggest a change in the timing of delivery, a change in the mode of delivery, or even prenatal treatment. For example, ultrasonography can be used to diagnose fetal urinary tract obstruction or intestinal blockage, leading the obstetrician to recommend an early delivery. A cesarean section may be suggested after a diagnosis of a large hydrocephalus ("water on the brain"). Other conditions can be treated by injecting medications and nutrients directly into the amniotic fluid. Most drastically, fetal surgery is recommended for several anatomic disorders. Researchers reviewing these treatments note that they are still experimental, but describe treatment of more than 80 fetuses and present a model for deciding when surgery on the fetus should be attempted. The authors recognize that the "therapeutic procedures involve significant risks for both fetus and mother, raising difficult ethical questions about risks versus benefits and about the rights of fetus and mother."[31]

They conclude that "[a]t this very early stage, fetal intervention should be pursued only in centers committed to research and development as well as (and prior to) responsible clinical application."[32]

In listing the minimum requirements for fetal intervention, the authors include the cooperative efforts of an obstetrician, a sonographer, a surgeon, a perinatologist, a bioethicist, and uninvolved professional colleagues for monitoring such innovative therapy. Where in all this is the pregnant woman? We believe that only she can decide whether a procedure should be performed on her body (as all of these procedures must be), particularly when such procedures involve significant risks to her own well-being, as all of these do. We question the propriety of performing experimental procedures on people, and challenge the perspective of a surgeon who reports that the "urinary tract was successfully decompressed" and only notes later that the infant died within a day of delivery.[33] Another surgeon reports that, of 95 fetuses who have been given shunts to drain excessive fluid from around their brains, 85 have survived, but of these survivors, two-thirds have serious brain damage.[34]

It is unethical to pursue fetal treatment where the commitment to research and development overrides the responsibility to the pregnant woman and the future child. Any assessment of the potential of such procedures must include a careful discussion of how to evaluate the risks to the pregnant woman and her fetus, and how to give her the available information, including both its diagnostic as well as prognostic uncertainties, in a way that enables her to make an informed decision.

Legal Pressures on Women to Consent to Prenatal Screening

Presently, women do not face governmental intervention when making a decision of whether or not to undergo prenatal screening. But some legal commentators suggest that a pregnant woman who is at risk of delivering a baby with a disability and refuses screening should be liable for "prenatal abuse." Others advocate that state courts allow children born with disabilities to sue their mothers for actions taken during pregnancy that may have contributed to their conditions. Both of these types of legal liability would severely limit the degree of autonomy experienced by women during pregnancy to make decisions about medical treatment and procreation.

Prenatal Abuse

The tort specialist Margery Shaw, a physician and jurist, has argued that a woman who decides to carry her fetus to term:

> incurs a "conditional prospective liability" for negligent acts toward her fetus if it should be born alive. These acts could be considered negligent fetal abuse resulting in an injured child. A decision to carry a genetically defective fetus to term would be an example. Abuse of alcohol or drugs during pregnancy . . . [w]ithholding of necessary prenatal care, improper nutrition, exposure to mutagens and teratogens, or even exposure to the mother's defective intrauterine environment caused by her genotype . . . could all result in an injured infant who might claim that his [sic] right to be born physically and mentally sound had been invaded.[35]

Shaw believes that it is unreasonable to carry a fetus to term knowing that it will suffer from a severe genetic disability and that such behavior should be recognized by courts as negligent, as though there were a universal set of values by which one can judge the appropriateness of a woman's concern for her fetus. Shaw encourages courts and legislatures to take action to ensure that fetuses are not "injured" by the "negligent acts" of others, including the pregnant women whose bodies sustain them. One tactic she advocates is to pass laws requiring those observing "potential and actual" fetal abuse to report it to the state, triggering court authority to compel "parents and prospective parents" to enter alcohol and drug rehabilitation programs, and, in the extreme, "to take 'custody' of the fetus."[36]

Shaw is not alone in her advocacy of prenatal liability. John Robertson, a specialist on procreation and the law, asserts that a "mother" who chooses not to abort her fetus has:

> a legal and moral duty to bring the child into the world as healthy as is reasonably possible. She has a duty to avoid actions or omissions that will damage the fetus and child, just as she has a duty to protect the child's welfare once it is born. . . . Once the mother decides not to terminate the pregnancy, the viable fetus acquires rights to have the mother conduct her life in ways that will not injure it.[37]

Robertson predicts that, because the trend in tort law is to eliminate parental immunity, suits by children against mothers for prenatal or even preconception injuries will be recognized in the future. He advocates the constitutionality of statutes that would forbid women the use of alcohol or tobacco during pregnancy and statutes imposing liability on women for exposing themselves to teratogenic substances.[38] He concludes that a woman's obligation to a fetus whom she decides to carry to term may require her to avoid work, recreation, and medical care choices that could be hazardous to the fetus. According to Robertson, a woman must also preserve her health for the fetus' sake, and accept established therapies for the fetus. Finally, he states that the woman's obligation to the fetus requires that she undergo prenatal screening where there is reason to believe that this screening may identify congenital defects correctable by available therapies.[39]

Permitting legal action to enforce the above policies would severely limit a prospective parent's, and in particular a prospective mother's, right to make decisions concerning her or his own body when doctors or lawyers assert that these decisions could have an impact upon a fetus. In the extreme case, all of a person's actions could come under public scrutiny if she or he planned to have children at some future time.

Two strands of legal precedent have already been used to limit a potential mother's right to refuse medical interventions prior to and during pregnancy. Courts have required pregnant women to undergo forced delivery by cesarean section for the health of their fetuses without their consent. Janet Gallagher discusses these cases in her position paper "Fetus as Patient" in this book. Related to these cases are ones where courts have considered whether or not a woman's actions during pregnancy constituted child abuse under state laws.

Courts have invoked laws designed to prevent child abuse and neglect in order to exercise temporary jurisdiction over the fetus of a drug-addicted woman and, in the process, over the woman herself.[40] For example, California's criminal child neglect statute, which requires a parent "to furnish necessary . . . medical attendance," was used to prosecute Pamela Rae Stewart for taking amphetamines, having sexual intercourse, and failing to stay off her feet and seek medical treatment for hemorrhage during her pregnancy.[41] Although the case against Ms. Stewart was dismissed on the grounds that the California statute was not intended to apply to prenatal acts and omissions, legislation was promptly introduced to extend the statute to such circumstances. We expect that an even greater number of prosecutions will be attempted against women as new reproductive technologies permit more types of prenatal diagnosis and treatment.

"Wrongful Birth," "Wrongful Life," and Maternal Liability

Even more alarming is the possibility that the approval by state courts of "wrongful life" causes of action may influence courts to permit the state or a child to sue a mother for what she did, or failed to do, during her pregnancy.

"Wrongful Life" Cases

"Wrongful life" and "wrongful birth" are the unfortunate labels courts have given to cases based on the claim that a woman would have terminated or avoided pregnancy except for medical negligence. "Wrongful birth" cases are brought by parents against health care providers. "Wrongful life" cases are also brought against health care providers, but they are based on the child's claim that he or she is entitled to damages for the very fact of being born. "Wrongful life" cases are usually attempted because the time in which the parent could sue has run out, whereas the child is able to sue until the age of majority. Because of the lack of a national health care program in the United States, obtaining a settlement from a malpractice insurer is often the only way to pay for the care for a child with a disability. Nevertheless, disability rights activists and others are alarmed by these decisions, because legal recognition of the principle of "wrongful life" demeans the lives of people with disabilities.

The majority of state courts allow claims for "wrongful birth" to go forward, although they reject claims for "wrongful life." However, recently the highest courts of New Jersey, California, and Washington have allowed wrongful life cases to proceed.[42] Although they refuse to award general damages for "wrongful life" that would cover all of the living expenses of a disabled child, these courts have awarded special damages for the extra expenses associated with the child's disabilities. Other state courts may follow these innovators.

The first case in which a state's highest court affirmed such damages was decided in California in 1982.[43] Joy Turpin was born completely deaf because of a hereditary disability. The time in which her parents could sue had run out, but as a minor she could still bring a court action. She sued a specialist in the diagnosis of speech and hearing conditions and the hospital employing him, because that specialist had failed to diagnose a similar genetic disability in Joy's sister before Joy was conceived. Joy's parents stated that they would not have conceived a second child had they known that their first daughter's deafness was hereditary. The court denied general damages for "depriv[ation] of the fundamental right of a child to be born as a whole, functional human being without total deafness," but granted damages for the "extraordinary expenses for specialized teaching, training and hearing equipment."

In deciding Joy Turpin's claim, the California court first overruled a lower court decision that had granted both general and special damages to another child who suffered from Tay-Sachs disease. In that case, a medical laboratory had been negligent in the conduct of its tests, and therefore had failed to warn the parents that their future children were at risk for Tay-Sachs disease.[44]

What impact will "wrongful life" cases have? Marjorie Shultz, in an innovative analysis of the legal doctrine of informed consent, has described these cases as directing attention to the wrong issues.[45] She argues that the real injury is denial of patient choice, that is whether to become parents, not the birth of a child with a disability. However, most courts have been unwilling to extend the doctrine of informed consent to find an injury where there has been no direct physical harm to the patient, and this has led to such novel and problematic doctrines as "wrongful birth" and "wrongful life."

Suing a Parent for Fetal Injury

Women now face the possibility that the trend in state courts to permit children to bring "wrongful life" cases against negligent health care providers will be extended to permit suits against mothers for actions taken during pregnancy. Such cases encourage increased public scrutiny of pregnant women's behavior. The motivation for such suits would be to get the parent's insurance plan to provide funds for expenses associated with the child's disability. But permitting such cases to proceed could have a chilling effect on women's right to make choices during pregnancy.

A lower California court foresaw such an extension of parental liability:

> [If] parents made a conscious choice to proceed with pregnancy with full knowledge that a seriously impaired infant would be born, . . . we see no sound public policy which should protect those parents from being answerable for the pain, suffering and misery which they have brought upon their offspring.[46]

California's legislature responded after the state's highest court recognized wrongful life suits by drafting legislation prohibiting such suits.[47] Other state legislatures, at the urging of abortion foes, have forbidden both wrongful birth and wrongful life suits.[48] Although we believe these statutes are overbroad because they prevent recovery for medical negligence and denial of patient choice, the California statute does not go far enough. It protects parents from claims "that a child should not have been conceived or, if conceived, should not have been allowed to have been born alive."[49] But it would not prevent a suit brought on behalf of a child for injuries allegedly suffered because of parental behavior. For example, under the California statute, a child with a genetic impairment could sue her father because he "permitted" his sperm to be exposed to ionizing radiation at work, thereby increasing the risk of genetic impairments in his offspring, or a child might be allowed to sue her mother for refusing to have her fetus treated *in utero* or because she used drugs that can harm a fetus.

An appeals court in Michigan has already allowed one such case to proceed. In *Grodin v. Grodin*, a son sued his mother for her consumption during pregnancy of the antibiotic tetracycline, which caused his teeth to develop mottled brown discolorations.[50] The Michigan court held that intrafamily tort immunity would not bar the suit if the mother's behavior was unreasonable. It stated that the mother would bear the "same liability for injurious, negligent conduct as would a third person. . . . The focal question is whether the decision reached by a woman in a particular case was a reasonable exercise of parental discretion." The case was remanded to the trial court to determine the reasonableness of the mother's behavior.

Generally, court cases like *Grodin* are filed to permit recovery against the parent's liability insurer for the child's medical care.[51] But *Grodin* has serious implications beyond insurance recovery. The court's opinion implicitly suggests that a pregnant women does not have the same freedom to make decisions about her health care as all other competent adults. Her behavior may be measured by a court against a hypothetical "reasonable parent" standard. Although the *Grodin* case was brought after the birth of the child, the same legal standard might be applied to obtain an injunction to control a woman's behavior during pregnancy.

We believe such a precedent is dangerous. The vast majority of women strive to do what is best for their future children, and any inability to do so most often arises because they lack quality, affordable health services and needed social supports such as decent housing, jobs, and child care. But what of those much rarer cases where, without intervention, the fetus is sure to be harmed, and where the intervention is available, affordable, and poses little or no risk to the pregnant woman?

We believe that even in this situation, prospective parents, with support from their communities, are better equipped than judges to evaluate the moral and ethical issues that arise in their lives. Nan Hunter, in her position paper "Time Limits on Abortion" in this book, describes legal precedent underlying this position and articulates a legal theory supporting autonomy over one's body, pregnant or not. She points to court decisions that recognize the right of competent adults to refuse medical diagnosis, treatment, and state-compelled physical invasion to obtain criminal evidence. These cases are grounded on constitutionally protected liberty and privacy interests affording bodily integrity and respect for individual conscience.

When the state interferes in women's reproductive choices, more than individual liberty is at stake. Such interference constitutes an unfortunate tradition of disadvantaging women as a class on the basis of our most distinctive biological characteristics:

> By subjecting women's decisions and actions during pregnancy to judicial review, the state simultaneously questions women's abilities and seizes women's rights to make decisions essential to their very personhood. The rationale behind using fetal rights laws to control the actions of women during pregnancy is strikingly similar to that used in the past to exclude women from the paid labor force and to confine them to the "private" sphere. Fetal rights could be used to restrict pregnant women's autonomy in both their personal and professional lives in decisions ranging from nutrition to employment, in ways far surpassing any regulation of the actions of competent adult men. The state would thus define women in terms of their childbearing capacity, valuing the reproductive difference between women and men in such a way as to

render it impossible for women to participate as full members of society.[52]

A woman who is forced to undergo medical diagnosis or treatment on behalf of her fetus is reduced to the role of "fetal container."[53] The state already exerts a powerful influence over individual choice through publicly funded and regulated screening programs and prenatal services; it must not also be able to use its police power to dictate the choices of competent pregnant women, or of potential parents of either sex, whether they involve genetic screening, prenatal diagnosis, fetal surgery, childbirth, or abortion. Only such freedom from state interference affords women all of the rights of competent adults, and permits reproductive choice for both women and men.

Informed Consent

The *Grodin* decision, described above, does not discuss whether the mother had been told by the prescribing physician that consumption of tetracycline during pregnancy could result in the discoloration of her child's teeth. Yet the ability to make decisions about health care during pregnancy requires both access to care, and information about the risks and limitations of medical advice, tests, and treatment. The right to information is protected by the legal doctrine of informed consent.

Two different standards are used by courts in deciding informed consent cases. Some state courts apply a "prudent patient" standard, looking to what information a reasonable patient would need to have in order to make an informed decision.[54] Other states retain the older, more difficult-to-prove standard that requires deviation from the accepted medical practice of a reasonable physician in similar circumstances.[55] As Marjorie Shultz explains in her review of this doctrine, informed consent has grown out of legal rules protecting individuals from unconsented-to physical contact and medical negligence. Patient autonomy "has never been recognized as a legally protectable interest."[56]

Many physicians acknowledge that, if they want to, they can present all the legally required information about a procedure and its risks in such a way as to ensure a patient's consent.[57] It takes commitment and precious time for the health care provider truly to seek the decision of the patient rather than coerce consent. More than one counseling session may be needed to afford the patient an opportunity to consult with family and friends, and mull over information before arriving at a decision. Even a commitment to seeking informed consent is not enough when differences in cultural and ethnic values come into play, because a health care provider may truly believe that she or he is "informing" the patient when no meaningful information is being communicated. Only when a physician or counselor is educated to be sensitive to a wide range of cultural and ethnic values and to the power dynamics and other intricacies of the counseling interaction are patients in a position to exercise informed consent. In the area of screening and prenatal diagnosis, informed consent is even more difficult to achieve. In this situation, it is essential for counselors to inform without communicating their own biases about the duties of parenthood, the worth of disabled individuals, and what they perceive to be "society's best interests," in order that prospective parents will be able to make their own informed decisions.

Informed consent is a misnomer in the sense that what should be sought is informed consent *or* refusal. A prospective parent has as much right to seek information that would lead to refusal of genetic screening or prenatal diagnosis as to seek information that would elicit consent. Prospective parents who consent to any screening or prenatal diagnosis procedures should also have the option on the informed consent form to note what specific information revealed by the procedures they want to receive and what they do not want to receive. For example, if they only want to know about Down syndrome, but not about other chromosome configurations of unknown significance (such as XYY), then they would not be told about them. The same would be true if they did not wish to know the fetus' sex. Such a restriction could lead to thoughtful discussions between genetic counselors and prospective parents on the use and relevance of certain kinds of information.

The Limits of Reproductive Choice: Poverty and Disability Discrimination

Most discussions of reproductive choice as it applies to prenatal screening and abortion revolve around recognition of a constitutionally protected right to make health care decisions during pregnancy and provisions for informed consent. But these concerns do not adequately protect the individual autonomy of the people who do not have access to any medical care or to other even more fundamental necessities, such as food and housing.

A woman who lacks access to prenatal care has no opportunity to exercise those aspects of "reproductive choice" that require the assistance of a medical practitioner. Similarly, a pregnant woman's reproductive choice is burdened if she lives on welfare in a third-floor walk-up apartment and is told that the fetus she carries has a neural tube defect. Her future child, who may be unable to walk, will have no constitutionally protected right to a ground floor apartment or an apartment in a building with an elevator. Only a society that guarantees equal rights and provides adequate economic and social supports, as well as medical care, for people with disabilities will afford prospective parents true freedom of choice in this area.

Finally, a pregnant woman who is addicted to drugs or alcohol does not have freedom to "choose" to improve her health and decrease the risks to her fetus when she cannot enter a residential drug treatment program because every bed is full (as is often the case), or because to do so she would have to give up custody of her children, if only temporarily. Unfortunately, this country does not recognize a constitutionally mandated "affirmative funding obligation" for medical care or for other basic societal prerequisites for health.[58]

Oppression of Persons with Disabilities and Reproductive Choice

The emphasis on reproductive choice as it applies to prenatal screening may satisfy potential parents who want every possible assurance that their babies will be "perfectly" healthy. Enthusiasm for screening among these couples has led some health care providers to conclude that the growth in screening is "consumer-driven," but this fails to acknowledge the pressures exerted by providers who genuinely fear malpractice suits if they do not urge the use of every new diagnostic test.[59] In reality, the reassurance parents feel when a fetus does not have a disability detectable by screening is misleading because most disabilities are not inborn, but acquired later in life. No matter how widespread prenatal screening becomes, people will continue to have unforeseeable disabilities, because most disabilities result from accidents and other mishaps in childhood and later life.[60] Prenatal screening will not greatly reduce the need for support services for individuals who have disabilities, nor will it reduce the need for laws to protect and improve the civil rights of persons with disabilities.[61] In fact, the widespread publicity about prenatal screening and implications that parents now can avoid bearing a child who has a disability contribute to the further stigmatization of people with disabilities.[62]

A pregnant woman is less than free to choose to forego prenatal screening or to carry to term a fetus with a diagnosed disability, if she knows that her child will face discrimination and oppression because of that disability. Parents of children with disabilities report that strangers feel free to ask, "How can you bring a child like that into the world."[63] To improve the lives of people who have disabilities, comprehensive laws must be enacted that prohibit discrimination on the basis of disability, modeled after state and federal laws forbidding discrimination on the basis of race and sex. Only in a society that values people irrespective of their disabilities, and that protects their civil rights and provides support services for people who need them will prospective parents have a genuine choice with regard to prenatal screening.

Poverty and Infant Morbidity

Most infant morbidity and mortality in the United States occur because babies are born premature or too small for their gestational age.[64] This happens, in large part, because the mothers are poor or very young (or both) and get no prenatal care of any kind, hence for social reasons and not because of inherited problems.[65] For example, a recent study found a sixfold increase in the risk of low birth weight associated with low income.[66] The disproportionate numbers of blacks living in poverty is correlated with nearly doubled rates of low birth weight and infant mortality.[67] Each year, a quarter of a million babies are born dangerously under weight, and one in ten of these infants dies before the first birthday.[68] Genetic screening and prenatal diagnosis do not touch these problems; indeed, they may make them worse by diverting money and time to genetic testing, particularly now when government funds for maternal and child health care and nutrition are being cut.[69]

While recognizing that disability discrimination and lack of access to prenatal care burden reproductive choice, we affirm the right of individual women to make choices about prenatal screening free from coercion. Neither health care providers nor the state should have the right to compel or withhold testing or treatment on behalf of the fetus for any reason. This perspective protects women's bodily integrity and permits fruitful discussions about what resources, including jobs, housing, health care, and food, need to be made available to ensure the health of adults and their children.

Solutions

The following proposals are intended to ensure and safeguard individual decision making at a time when there is growing momentum to limit the right of prospective parents, especially women, to refuse medical interventions on behalf of a fetus. The recommended policies would enhance noncoercive choice in the areas of genetic counseling, screening, and prenatal diagnosis by providing access to health care and support services. The model statute would prevent courts from controlling a potential parent's behavior to protect a fetus.

Public Policy Recommendations

1. A national health program should be implemented that offers access to health care regardless of income, and mandates allocation of resources, including social and medical resources, for basic prenatal care and support services for people with inborn or acquired disabilities.

2. State and federal laws, modeled after Title VII's prohibition against discrimination on the basis of race and sex, should be enacted to prohibit discrimination in employment, insurance, housing, public accommodations, and other areas based on medical status, disability, or hereditary traits.

3. Genetic counselors must be educated about the ways in which people from different backgrounds view the prospect of bearing a child with a disability. Training programs for counselors should require courses on differences in cultural and ethnic perceptions of disability and on disability rights as standard parts of the curriculum. Field training should include placement with an agency providing disability services or with a disability rights advocacy group, so that counselors have a better understanding of the experiences of people with a variety of disabilities and of their families.

4. Health and disability rights advocates, parents, and physicians jointly should develop fact sheets that describe what is known about screening of fetuses and provide contacts to disability rights groups, which would make available information about the range of experiences associated with different disabilities.

5. Short courses on the uses and abuses of prenatal screening should be offered by educational institutions, cable television, and other media.

6. Medicaid regulations should require the state or federal government to cover all standard screening that is available to individuals who are able to pay for these services out of pocket or through insurance policies. Information about prenatal testing and where to obtain it should be readily available.

7. As part of the informed consent or informed refusal process, it should be made explicit and stated on appropriate consent and refusal forms that refusal of tests will not lead to termination of medical care or to denial of state services.

8. An absolute and legally binding guarantee of confidentiality should protect information obtained from screening and prenatal diagnosis. The information should not be released to anyone without the informed consent of the screened person or her/his legal guardian.

9. Women who have undergone prenatal screening or who plan to do so should be encouraged to start or join a self-help group, in which they can learn about other women's experiences and share information with them.

10. Women who do not intend to choose abortion, irrespective of test results, should have the same access to screening and prenatal diagnosis as any other woman.

Parental Immunity Statute

1. No cause of action, whether civil, criminal, or for injunctive or other forms of equitable relief, may be brought against a parent of a child based upon the claim that:

A. The fetus was, is, or may in the future be injured by a potential parent's decision to refuse genetic testing or prenatal diagnosis
B. The child, based on information obtained from genetic testing or prenatal diagnosis, should not have been conceived, or if conceived, should not have been born or
C. The fetus was, is, or may in the future be injured by a potential parent's medical decisions, health status, employment, or personal habits prior to or during pregnancy or childbirth.

2. The provisions of this Act provide no defense to any party other than a parent.

Notes and References

1. Bowers v. Hardwick, 106 S.Ct. 2841 (1986); Roberts v. United States, 445 U.S. 552 (1980).

2. See, e.g., Thornburgh v. American College of Obstetricians and Gynecologists, 106 S.Ct. 2169 (1986); Carey v. Population Services Int'l, 431 U.S. 678 (1977); Roe v. Wade, 410 U.S. 113 (1973); Griswold v. Connecticut, 381 U.S. 479 (1965); Skinner v. Oklahoma, 316 U.S. 535, 541 (1941); Santosky v. Kramer, 455 U.S. 745 (1982); Quilloin v. Walcott, 434 U.S. 246 (1978); Wisconsin v. Yoder, 406 U.S. 205 (1971); Meyer v. Nebraska, 262 U.S. 390 (1923).

3. Eisenstadt v. Baird, 405 U.S. 438, 453 (1972) (emphasis omitted). Only four members of the Court concurred in the opinion.

4. In re Farrell, No. A-76, slip op. (N.J. Supreme Ct. June 24, 1987) (quoting Schloendorff v. Society of New York Hosp., 211 N.Y. 125, 129-30, 105 N.E. 92, 93 (1914) (Cardozo, J.)).

5. See Laurie Nsiah-Jefferson's essay "Reproductive Laws, Women of Color, and Low-Income Women" in this book.

6. Vatican Doctrinal Statement, Instruction on Respect for Human Life in Its Origin and on the Dignity of Procreation: Replies to Certain Questions of the Day, 1987.

7. One obstetrician recommends that a pregnant woman who refuses medical treatment for her fetus be charged with felony. J.R. Lieberman et al., "The Fetal Right to Live," 53 Obstetrics and Gynecology 515 (1979). Margery Shaw, a physician and jurist, suggests that a woman who withholds necessary prenatal care incurs a "prospective liability for [such] negligent acts" affecting the fetus. "The Potential Plaintiff: Preconception and Prenatal Torts," in Genetics and the Law II, edited by A. Milunsky and G. Annas (1980). John Robertson has written that "[p]renatal screening could also be directly mandated by statute, with criminal penalties for the woman who fails to obtain it." "Procreative Liberty and the Control of Conception, Pregnancy, and Childbirth," 69 Virginia Law Review 405, 407, 449 (1983). See also Stearns, "Maternal Duties During Pregnancy," 21 New England Law Review 595 (1985-86); Deborah Mathieu, "Respecting Liberty and Preventing Harm: Limits of State Intervention in Prenatal Choice," 8 Harvard Journal of Law and Public Policy 19, 54 (1985); Margery Shaw, "Conditional Prospective Rights of the Fetus," 5 Journal of Legal Medicine 63 (1984); Patricia King, "The Juridical Status of the Fetus: A Proposal for Legal Protection of the Unborn," 77 Michigan Law Review 1647, 1657 (1979); Note, "Parental Liability for Prenatal Injury," 14 Columbia Journal of Law and Social Problems 47 (1978); Note, "Recovery for Prenatal Injuries: The Right of a Child Against Its Mother," 10 Suffolk University Law Review 582, 602 (1976).

8. See Boston Women's Health Book Collective, The New Our Bodies, Ourselves (1985); Edwards and Waldorf, Reclaiming Birth (1984); Y. Brackbill, J. Rice, and D. Young, Birth Trap: The Legal Low-Down on High-Tech Obstetrics (1984); G. Cassidy-Brian, F. Hornstein, and G. Downer, Woman-Centered Pregnancy and Birth (1984); and B. K. Rothman, Giving Birth: Alternatives in Childbirth (1982).

9. M. S. Golbus and C. J. Epstein, "Prenatal Diagnosis of Genetic Diseases," 65 American Scientist 703-711 (1977).

10. See J. Adler, "Every Parent's Nightmare: A Father's Story of the Birth of a Handicapped Child," Newsweek, March 16, 1987, at 56.

11. V. McKusick, Mendelian Inheritance in Man 6 (6th ed. 1983); S. R. Stephenson and D. D. Weaver, "Prenatal Diagnosis: A Compilation of Diagnosed Conditions," 141 American Journal of Obstetrics and Gynecology 319-343 (1981).

12. Most traits that are believed to be partly inborn involve interactions among many genes and cannot be diagnosed in utero. Even when the inheritance of a trait is mediated by one gene, people who carry that gene usually show no symptoms if they have received a copy of it from only one of their parents. Such traits, and the gene that mediates them, are called recessive, and an individual who carries only one copy of a recessive gene is described as heterozygous for that gene, or as

a carrier. To exhibit a recessive trait or disease, people must inherit a copy of the corresponding gene from both parents, in which case they are said to be homozygous for the gene. Preventing the people who are homozygous and therefore manifest such diseases from having children does not appreciably decrease the frequency with which the gene occurs in the population, because the overwhelming majority of genes for recessive diseases are carried singly by people who do not know that they are carriers. For example, each of us carries between five and seven genes that would mediate a disease or disability if we had two copies of them, rather than just one. To know which genes they are, each of us would have to be screened for all possible genetic diseases. Even if this were possible, which it is not at present, it would be a gross misuse of medical resources, since genetic diseases are rare. But new forms of eugenic arguments have begun to appear, cast in terms of a so-called right to be born healthy. Based on this supposed "right," some would bar people who, because of their family history, know that they may carry the gene for a "debilitating genetic disease" from exercising their right to have children.

13. For more background on screening programs, see Ruth Hubbard and Mary Sue Henifin, "Genetic Screening of Prospective Parents and Workers: Some Scientific and Social Issues," in Biomedical Ethics Reviews - 1984, edited by James Humber and Robert Almeder (1984). See also R. H. Kenen and R. M. Schmidt, "Stigmatization of Carrier Status: Social Implications of Heterozygote Genetic Screening Programs," 68 American Journal of Public Health 1116-1120 (1978); R. F. Murray, Jr., "The Practitioner's View of the Values Involved in Genetic Screening and Counseling: Individual vs. Societal Imperatives," in Ethical, Social and Legal Dimensions of Screening for Human Genetic Disease, edited by D. Bergsma, M. Lappe, and R. O. Roblin, pp. 185-199 (1974).

14. J. C. Chang and Y. W. Kan, "A Sensitive New Prenatal Test for Sickle-Cell Anemia," 307 New England Journal of Medicine 30-36 (1982).

15. B. Culliton, "Sickle Cell," 178 Science 138-142, 282-286 (1972).

16. C. F. Whitten, "Sickle Cell Programming--An Imperilled Promise," 288 New England Journal of Medicine 318-319 (1973).

17. D. Wilkinson, "For Whose Benefit? Politics and Sickle Cell," 5(8) Black Scholar 26-31 (1974). See also Regina Kenen and Robert Schmidt, "Social Implications of Screening Programs for Carrier Status: Genetic Disease in the l970s and AIDS in the l980s," in Dominant Issues in Medical Sociology, edited by Howard Schwartz (1987).

18. H. Harris, Prenatal Diagnosis and Selective Abortion, p. 14 (1975); M. M. Adams et al., "Down's Syndrome: Recent Trends in the United States," 246 Journal of the American Medical Association 758-760 (1981); L. Holmes, "Genetic Counseling for the Elder Pregnant Woman: New Data and Questions," 298 New England Journal of Medicine 1419-1421 (1978).

19. See Barbara Katz Rothman, The Tentative Pregnancy: Prenatal Diagnosis and the Future of Motherhood (1986).

20. R. E. Magenis et al., "Paternal Origin of the Extra Chromosome in Down's Syndrome," 37 Human Genetics 7-16 (1977).

21. C. C. Nielsen, "An Encounter with Modern Medical Technology: Women's Experiences with Amniocentesis," 6 Women and Health 109-124 (1981).

22. See note 19 supra.

23. B. D. Blumberg, M. S. Golbus, and K. H. Hansen, "The Psychological Sequelae of Abortion Performed for a Genetic Indication," 122 American Journal of Obstetrics and Gynecology 799-808 (1975).

24. A "not so sci-fi fantasy" of this sort was projected by one of us several years ago. See Ruth Hubbard, "Personal Courage Is Not Enough," in Test-Tube Women: What Future for Motherhood? edited by Rita Arditti, Renate Duelli Klein, and Shelley Minden (1984).

25. The information in this section is mainly drawn from M. Sun, "FDA Draws Criticism on Prenatal Test," 221 Science 440 (1983) and G. Annas, "Is a Genetic Screening Test Ready When the Lawyers Say It Is," Hastings Center Report, Dec. 1985, p. 16. The impact of ambiguous results on attitudes toward abortion is docu-

mented in Faden et al., "Prenatal Screening and Pregnant Women's Attitudes Toward the Abortion of Defective Fetuses," 77 American Journal of Public Health 288 (1987). Women are much more likely to abort an affected fetus if the test predicts with certainty that the fetus has a condition like a neural tube defect, despite variability in the degree of disability associated with the diagnosis. Id.

26. Quoted in G. Annas, supra note 25.

27. Id. A recent study of participants in a maternal AFP screening program found substantial gaps in their knowledge, including a tendency to think that a positive test assured them of a healthy baby. Faden, "What Participants Understand About an AFP Screening Program," 75 American Journal of Public Health 1381 (1985).

28. See M. Lappe, "The Limits of Genetic Inquiry," 17 Hastings Center Report 5-17 (1987).

29. Recently published medical literature abounds with references enshrining the fetus as patient. See The Unborn Patient: Prenatal Diagnosis and Treatment (1984); The Patient Within the Patient: Problems in Perinatal Medicine, a special issue of the journal Birth Defects (Vol. 21, 1985); and articles in other journals such as Prenatal Diagnosis and Journal of Ultrasound Medicine.

30. N. S. Adzick et al., "Recent Advances in Prenatal Diagnosis and Treatment," 32 Pediatric Clinics of North America 1103 (1985).

31. Id. at 1115.

32. Id.

33. M. R. Harrison, "Fetal Surgery for Congenital Hydronephrosis," 306 New England Journal of Medicine 591 (1982).

34. New York Times, Oct. 7, 1986, §C, at 17 (report on current status of fetal surgery). See also "Report of the International Surgery Registry," 315 New England Journal of Medicine 336 (1986).

35. Margery Shaw, "The Potential Plaintiff: Preconception and Prenatal Torts," note 7 supra.

36. Margery Shaw, "Conditional Prospective Rights of the Fetus," 5 Journal of Legal Medicine 63, 100 (1984).

37. John A. Robertson, Procreative Liberty, note 7 supra at 438. George Annas counterargues that a pregnant woman has not "waived" her rights in relation to her fetus because she has chosen to carry her pregnancy to term. No knowing waiver has occurred in a legally cognizable fashion. What is more, a pregnant woman continues to have the right to terminate her pregnancy up until the time of birth, at least where her life and health are at stake. Annas, "At Law: Pregnant Women As Fetal Containers," Hastings Center Report, Dec. 1986, p. 13.

38. Robertson, note 7 supra at 449-450.

39. Id. at 450. See also J. Robertson and J. Schulman, "Pregnancy and Prenatal Harm to Offspring: The Case of Mothers with PKU," 17 Hastings Center Report 23 (1987) (postbirth sanctions preferable to prebirth seizures of pregnant women to protect fetal health).

40. See Chicago Tribune, Apr. 9, 1984, at 1, col. 4 (fetus designated ward of state due to pregnant woman's heroin addiction); Boston Globe, Apr. 27, 1983, at 8, col. 1 (news story on physician's request for court order compelling pregnant woman to be tested for drug abuse and to take "what steps are necessary to insure the fetus's proper development"). See also In re Baby X, 97 Mich. App. 111, 293 N.W.2d 736 (1980) (use of heroin by mother during pregnancy constitutes child abuse under state law).

41. Ca. Penal Code § 270 (West Supp. 1986). The statute covers both children in being and "a child conceived but not yet born." See Annas, note 37 supra; Newsweek, Dec. 8, 1986, at 87; New York Times, Nov. 16, 1986, at 24, col. 1.

42. In re Procanik, 97 N.J. 339 (1984); Harbeson v. Parke Davis, Inc., 98 Wash.2d 460, 656 P.2d 483 (1983); Turpin v. Sortini, 31 Cal.3d 220, 643 P.2d, 182 Cal. Rptr. 337 (1982).

43. Turpin v. Sortini, supra, 182 Cal. Rptr. 344, 345.

44. Curlender v. Bio-Science Laboratories, 106 Cal. App. 3d 811, 165 Cal. Rptr. 477 (1980).

45. Marjorie Shultz, "From Informed Consent to Patient Choice: A New Protected Interest," 95 Yale Law Journal 219, 267-269 (1985).

46. Curlender, supra, 165 Cal. Rptr. at 488.

47. Cal. Civil Code § 43.6.

48. For example, the Minnesota legislature passed a law forbidding "wrongful birth" and "wrongful life" suits based on the claim "that but for the negligent conduct of another, he [sic] would have been aborted." Minn. Stat. § 145.424 (1984). The Supreme Court of Minnesota upheld the constitutionality of the statute after it was challenged by the mother of a child with Down syndrome who claimed she had not been offered amniocentesis. Hickman v. Group Health Plan, Inc., 396 N.W.2d 10 (Minn. Sup. Ct. 1986). See also, e.g., Missouri Stat. Ann. § 188.130; Utah Stat. Ann. § 76.7-305.5. Five states have enacted legislation prohibiting wrongful birth actions, and 21 states have had such bills introduced. See Note, "Wrongful Birth Actions: The Case Against Legislative Curtailment," 100 Harvard Law Review 2017, 2018-2019 (1987).

49. Note 47 supra.

50. Grodin v. Grodin, 102 Mich. App. 396, 301 N.W.2d 869 (1980).

51. For example, in Stallman v. Youngquist, 152 Ill. App. 3d 638, 504 N.E. 2d 920 (1st Dist. 1987) a court permitted a child to sue her mother for fetal injuries that allegedly occurred to her because of her mother's negligent involvement in an automobile accident while pregnant. The court reasoned that the litigation, in substance, was between the child and the mother's automobile insurance carrier. See Beal, "'Can I Sue Mommy?' An Analysis of a Woman's Tort Liability for Prenatal Injuries to Her Child Born Alive," 21 San Diego Law Review 325 (1984) for a discussion of the policy reasons behind parental tort immunity and justifications for its demise.

52. Dawn E. Johnsen, "The Creation of Fetal Rights: Conflicts with Women's Constitutional Rights to Liberty, Privacy, and Equal Protection," 95 Yale Law Journal 599, 624-625 (1986). See also Dawn E. Johnsen, "A New Threat to Pregnant Women's Autonomy," 17 Hastings Center Report 33 (1987).

53. George Annas presents this compelling image in his analysis of the prosecution of Pamela Rae Stewart for child abuse because of her actions during pregnancy and childbirth. "At Law," supra note 37, p. 13. He argues that, rather than prosecuting women, "[t]he best chance that the state has to protect fetuses is through actions to enhance the status of all women." Id.

54. Roseff, Informed Consent: A Guide for Health Care Providers (1981).

55. Id.

56. Shultz, "From Informed Consent to Patient Choice," supra note 45.

57. Physicians' reluctance to accept a patient's feelings and wishes is greater than usual in obstetrics because of the tendency to perceive the pregnant woman and her fetus as a "maternal/fetal unit." See N. Rhoden, "Informed Consent in Obstetrics: Some Special Problems," 7 Western New England Law Review 67 (1987). See also R. F. Murray, "Cultural and Ethnic Influences on the Genetic Counseling Process," 20(4) Birth Defects 71-74 (1984).

58. Harris v. McRae, 448 U.S. 297 (1980).

59. For a discussion of the pressures on women to undergo amniocentesis, see Ruth Hubbard, "Personal Courage Is Not Enough," supra note 24; and Barbara Katz Rothman, supra note 19, The Tentative Pregnancy. See also A. Capron's discussion of "Defensive Medicine and the Technological Imperative" in his article "Tort Liability in Genetic Counseling," 79 Columbia Law Review 618, 666-673 (1979) and his suggestion of informed consent as the antidote.

60. The median age of onset for disabling conditions is 41 years. A significant proportion of disabling conditions is caused by accidents: 31 percent of currently disabled men and 18 percent of currently disabled women were injured in an accident. Therefore, even if prenatal diagnosis greatly reduced the number of infants born with disabilities, this would not curtail the need for social supports for disabled

individuals. D. Ferron, Disability Survey. Department of Health and Human Services, Social Security Administration Office of Policy, Research and Statistics, Report No. 56 at 24, 60-61 (1981).

61. See Adrienne Asch's essay "Reproductive Technology and Disability" in this book. See also Anne Finger's chapter "Claiming All of Our Bodies: Reproductive Rights and Disabilities" and Marsha Saxton's "Born and Unborn: The Implications of Reproductive Technologies for People with Disabilities" in Test-Tube Women, supra note 24; and Adrienne Asch and Michelle Fine, "Shared Dreams: A Left Perspective on Disability Rights and Reproductive Rights," in 18 Radical America 51-58 (1984).

62. See Ruth Hubbard, "Legal and Policy Implications of Recent Advances in Prenatal Diagnosis and Fetal Therapy," 7 Women's Rights Law Reporter 201, 210-214 (1982) and accompanying footnotes by Janet Gallagher; and Ruth Hubbard, "Personal Courage Is Not Enough," supra note 24.

63. In a study of 80 families with a disabled child, 40 percent reported that an important contributor to their stress was the lack of needed education, medical care, and other services. Holding parents responsible for the child's behavior was also cited as a major source of stress. New York Times, Aug. 11, 1986, § 4, at 17, col. 2.

64. Low-birth-weight infants are at greater than usual risk of experiencing brain hemorrhages, infectious diseases including pneumonia, and many other life-threatening conditions as well as neurodevelopmental problems, learning disorders, and behavior problems. S. Brown, "Can Low Birth Weight Be Prevented?" 17 Family Planning Perspectives 112 (1985); McCormick, "The Contribution of Low Birth Weight to Infant Mortality and Childhood Morbidity," 312 New England Journal of Medicine 82, 86 (1985).

65. D. Binsacca et al., "Factors Associated with Low Birthweight in an Inner City Population: The Role of Financial Problems," 77 American Journal of Public Health 505 (1987). Programs that provide food supplements to low-income pregnant women, such as the federally funded Special Supplemental Food Program for Women, Infants and Children (WIC), raise infant birth weight. M. Kotelchuck et al., "WIC Participation and Pregnancy Outcomes: Massachusetts Statewide Evaluation Project," 74 American Journal of Public Health 1086 (1984). Each year in the United States, 300,000 women give birth having had little or no prenatal care, and two thirds of infants who die are babies of these mothers. "Infant Mortality," New York Times, June 26, 1987, § A, at 1. Making prenatal care available to poor women significantly reduces the incidence of low birth weight and accompanying perinatal mortality. K. J. Leveno et al., "Prenatal Care and the Low Birth Weight Infant," 66 Obstetrics and Gynecology 599 (1985).

66. D. Binsacca, supra note 65.

67. "Infant Mortality," New York Times, supra note 65.

68. Id.

69. Cuts in Federal programs including Medicaid, Aid to Families with Dependent Children, and the WIC food program have left many poor families without access to prenatal and postnatal care. Dana Hughes et al., The Health of America's Children: The Maternal and Child Health Data Book (The Children's Defense Fund, 1986). See also Lori Lew, "Legislative Research Bureau Report: A Proposal to Strengthen State Measures for the Reduction of Infant Mortality," 23 Harvard Journal of Legislation 559 (1986) and references cited therein.

FETUS AS PATIENT

Janet Gallagher

Position Paper

Recent media reports and scholarly articles have forecast the emergence of the fetus as patient and a new legal doctrine of "fetal rights." It is argued, for example, that a nonconsenting pregnant woman may be forcibly subjected to a cesarean section, if doctors judge that her fetus is at serious risk;[1] that the government may place restraints on a pregnant woman's physical activities, diet, and life-style;[2] that mothers can be sued by their children for injuries caused by their "prenatal negligence";[3] and that pregnant women should be excluded from the protection of "living will" statutes.[4]

The most dramatic and widely publicized claims of "fetal rights" have been made in the context of court-ordered cesarean sections. Although there have been but a handful of reported incidents thus far,[5] the cases are of great symbolic and precedential significance. They convey a drastic message about women's moral and legal status, and they serve to legitimize a forceful--indeed physically violent--reassertion of doctors' control over pregnancy and childbirth.

In 1981, a Georgia court, acting at the request of hospital doctors, ordered that a cesarean be performed on a nonconsenting pregnant woman. The doctors testified that the woman's placenta blocked the birth canal. They said that there was a 99 percent chance that the baby could not be born alive vaginally and a 50 percent chance that the "defendant" (the pregnant woman with religious objections to surgery) would also die. The court, citing *Roe v. Wade* as authority for a state interest in the potential life of a viable but unborn child, granted temporary custody of the fetus to the local government social service agency and gave it full authority to make all decisions, including giving consent to surgical delivery.[6]

In a similar Michigan case, also arising from a diagnosis of placenta previa, the court went further: It ordered the woman to "present herself" to the hospital by a specified date and time to be delivered. The court further ordered that, if she did not appear at the specified time, the local police department would pick her up and deliver her to the doctors. Both of these women went on, despite the medical predictions, to give vaginal birth to healthy babies. The Georgia woman is said to have successfully prayed, and the Michigan woman not only prayed, but fled into hiding with her entire family until after the birth.[7]

In one shocking 1987 case, the legal department at a Washington, D.C. hospital called in a local judge who ordered that a cesarean be performed on a patient with cancer who was 26 weeks pregnant. The judge ordered the surgery, despite medical testimony that it would shorten the woman's life and over the objections of the woman, her husband, her family, her attending doctors, and the entire obstetrics/gynecology department of the hospital. The fetus, ten weeks premature and already compromised by the woman's poor health, died almost immediately. The woman died two days later. Her death certificate cited the surgery as a contributing cause of death.[8]

There are a number of other contexts in which these "fetal rights" claims have arisen. A Massachusetts husband, for example, went to court to try to force his wife to undergo a surgical procedure to prevent a miscarriage. The judge did order the surgery, but was overruled by the state's highest court.[9] A Baltimore doctor sought a court order to force one of his pregnant patients, an alleged drug abuser, to follow his order that she not take any nonprescription drugs.[10] The highest court in Michigan allowed a father, on behalf of his son, to sue the mother for "prenatal negligence" based on the claim that the drug she had taken during pregnancy had discolored the boy's teeth.[11] In 1986, a San Diego woman was jailed for six days on charges of medical neglect of her fetus.[12] A judge subsequently threw out the charges, but a

California state legislator promptly filed a bill to authorize similar prosecutions.[13]

Treatment refusals by pregnant women are rare. Women have always put themselves at risk to bring pregnancies to term. Although the advent of fetal surgery triggered a great deal of legal speculation about the power of the state to force unwilling women to undergo such procedures for the sake of their unborn children, doctors actually involved in experimental fetal surgery speak of being besieged by and having to dissuade women seeking the new therapies.[14] If anything, in fact, it may be that pregnant and birthing women are too compliant in their dealings with the medical profession, and too willing to accept cesarean sections and other invasive procedures.[15] The reports of forced cesareans have coincided with a growing concern about the number of cesarean deliveries in the United States (as high as 31.4 percent of births in some hospitals).[16]

Close analysis of actual "maternal-fetal conflict" cases, however, reveals that many of them could have been and should have been avoided. The very fact that they happen at all--whatever the outcome--usually reflects a failure: of public policy, of the organization of medical care or social services, or of simple human communication.[17] Since refusals themselves are relatively uncommon, nonconsent to truly necessary interventions will be extremely rare.

Nonetheless, fetal rights claims often arise in sympathetic contexts; doctors and social workers do sometimes find themselves confronted by agonizing and infuriating dilemmas posed by pregnant women's behavior. Granted that the Georgia and Michigan women (and a startling number of other patients for whom judges have been asked to order forced surgery) proved not to need cesareans, there undoubtedly will be some cases in which doctors' worst fears prove tragically well founded. Legal and public policy responses to our increased knowledge about, and ability to treat, the fetus are required.

By and large, judicial opinions and scholarly articles favoring fetal rights have focused most heavily on the issue of the legal status of the fetus. The fetus is presented as an independent entity, abstracted from the reality of the woman's body, much as though the commentators had encountered it upon the street. Such an isolated focus upon the fetus, however, is inappropriate from both a legal and an ethical standpoint. Given the very geography of pregnancy, questions as to the status of the fetus must follow, not precede, an examination of the rights of the woman within whose body and life the fetus exists.

Woman as Vessel

Of all the arguments for fetal rights, the most popular and seemingly secular have been those based on our increased knowledge about prenatal development and new medical advances permitting *in utero* diagnosis and treatments.[18] Close examination, however, suggests that such medical developments have simply provided contemporary scientific rhetoric for reassertion of an enduring set of deeply patriarchal beliefs. The explosion of technologies that make the fetus seem more accessible to the world at large--visually, medically, emotionally--have spurred a resurgence of powerful, largely unacknowledged social attitudes in which pregnant women are viewed and treated as vessels.

Such attitudes are far from new. Mary O'Brien, citing Aristotle's belief that "women contribute nothing to the child but . . . arrested menstrual flow," observes that "[T]his idea that men contribute spirit or soul or some other human 'essence' must have struck chords in the masculine imagination, for it lingered for centuries."[19] A 1977 study of Roman Catholic teaching on sexuality notes the shaping influence of the view of the male seed as "the active principle" and of women as "receptacles, or the seed-gardens, as it were, for human reproduction."[20] As Dr. Beverly Wildung Harrison of Union Theological Seminary has written, that "sire-centered view of embryology"[21] has a strong ideological correlation with "the assumption that children are really the fruit, even the possession of men"[22] and with male attempts to control women's reproductive power.

Contemporary preoccupation with the fetus, although couched in terms of new scientific knowledge, reflects much the same imagery. George H. Williams, for example, describes fetuses as "the unwitting and diminutive denizens of that universal and mysterious realm of maternal darkness from whence we all emerge."[23] A law review article reports that the fetus, "once a 'medical recluse,'" has emerged from its "gestational hiding place."[24]

Dr. Bernard Nathanson, an anti-abortion activist who achieved prominence as the narrator of the 1985 film *Silent Scream,* writes of the fetus as "the alien" and "the little aquanaut," in "intra-uterine exile" from the human community.[25] In such a view, women are reduced to "the mother ship," the "uterine capsule."[26] Small wonder that the woman's human rights of bodily integrity and decision making are obscured or overlooked altogether. Nathanson's fetus as patient comes to be viewed in isolation, as "a creature bricked in, as it were, behind what seemed an impenetrable wall of flesh, muscle, bone and blood."[27]

The language of the current fetal rights drive is strikingly similar to that of the nineteenth-century campaign of male "regular" doctors against abortion. Dr. Horatio Storer, a leader of that movement, also saw fetuses as *contained* within women--as "children nestling within them--children fully alive from the moment of conception that have already been fully detached from all organic connection with their parent and only re-attached to her for the purpose of nutriment and growth."[28] There are other parallels as well. Historians have documented the key role of anti-abortion campaigns in the effort of the emerging "regular" doctors to eclipse their midwife and homeopathic rivals.[29] Then, as now, the contest for power in the birthplace had both ideological and economic dimensions.[30]

Professor Carroll Smith-Rosenberg has pointed out that Victorian discussion of sex and reproduction functioned as "a metaphoric discourse in which the physical body symbolized the social body, and physical and sexual disorder stood for social discord and danger":[31]

> While Victorian women used images of marital rape to protest male social power, male physicians and legislators saw bourgeois women's rejection of motherhood as the principal source of familial and social disorder. Wild-eyed aborting matrons, hysterical young women unwisely seeking education, unmarried professional women, all bespoke male social anxieties in an uncertain world.[32]

Smith-Rosenberg notes the sense of economic insecurity triggered among doctors by women's picking and choosing among physicians, in part to ensure their access to abortion.[33] But such "consumer" freedom of choice also posed threats that were not solely financial. It engendered a "sense of low status and fear of professional impotence . . . [and] this power, especially when exerted in a concerted manner, seemed a violation of women's subservient status in society."[34] The anti-abortion campaign can be at least partly explained as an effort to achieve the redomestication of married WASP women, the physicians' primary clientele, by shaping an ideology of female chastity and maternalism.[35]

The contemporary drives against abortion and for fetal rights reflect a similar "backlash," and not just on the part of doctors.[36] Like the anti-abortion effort, the fetal rights drive is fueled by deep social unease over the changing roles of women and rapid changes in male-female relationships.[37] Both movements, although distinguishable from one another, represent attempts to reimpose a deeply held view of the natural, God-given role of woman as self-sacrificing mother.[38]

The push for legal sanctions against deviations from that role demands that government "define the public norms of morality and designate which acts violate them."[39] The fact that enforcement would be impossible or even counterproductive[40] pales before the symbolic value of the statute or court decision itself:

> The existence of law quiets and comforts those whose interests and sentiments are embodied in it . . . The fact of affirmation through acts of law and government expresses the public worth of one set of norms, of one sub-culture vis-à-vis those of others. It demonstrates which cultures have legitimacy and public domination, and which do not. Accordingly it enhances the social status of groups carrying the affirmed culture and degrades groups carrying that which is condemned as deviant.[41]

Seen in this light, proposals for judicially enforced self-sacrifice and demands for criminalization and tort liability for deviations from the idealized maternal norm are revealed as vehicles for stigmatizing the less circumscribed or less subordinate female life-styles now emerging. It is significant, for example, that so much of the drive to curb birth defects arising from reproductive hazards in the workplace has focused on excluding women of childbearing age from better paying nontraditional jobs despite clear evidence that prospective fathers' exposure may have equal or more serious impact.[42] Similarly, virtually all of the guilt-inducing publicity on the dangers posed to fetuses by smoking is aimed at pregnant women,[43] despite the fact that the risks posed to fetuses *and* pregnant women by smoking fathers are also well established.[44]

Fetal rights proposals not only attempt to impose more traditional behavioral norms by government sanctions, but they also enshrine a stereotypical and demeaning view of women. As defense attorneys in the San Diego "fetal abuse" case pointed out, "By acting as overseer to a woman's decisions during her pregnancy, the state inescapably denigrates her capacity as a decision maker."[45] And the fetal rights movement reinforces physician authority after a decade of "consumer" rebellion against the medicalization of birth and of movement toward alternatives like midwives, birthing centers, and homebirth.[46]

Nonetheless, the fetal rights drive cannot be dismissed out of hand as an antifeminist backlash or a professional "plot." Pregnant women's choices or behavior can present real dilemmas to doctors, and have devastating consequences to individual children and to society as a whole. However, we do need to expose and challenge the undercurrents of anxiety and resentment, and of professional self-interest and aggrandizement, that fuel the demand for legal intervention. We must also bring critical scrutiny to bear upon today's highly emotional, almost idolatrous, view of the fetus.

The Fetus--Symbol of Hope and Fear

For it is not just anti-abortionists and fetal rights proponents who have adopted an intense focus on the fetus. Perhaps the greatest irony of *Roe v. Wade* is that *Roe* and the earlier *Griswold* decision striking down state laws barring contraception[47] have--by creating the legal availability of birth control and abortion--made it possible for many in this generation of Americans to view pregnancy as a joyfully assumed, chosen, and celebrated human task. Happily (and sometimes determinedly), prospective parents have developed a real awareness of, an anxiety about, and even an emotional attachment to the fetus.[48]

Such fascination with the fetus extends well beyond the circle of those personally involved with pregnancy. It enjoys a much broader appeal. In fact, for some people and groups within our society, the fetus seems to have taken on the status of a veritable object of devotion and become the center of a religious cult.[49] Part of that can probably be attributed to the enormously powerful emotional associations virtually all of us have with pregnancy.

Birth and children have traditionally functioned as symbols of promise, continuity, and renewal. And the fetus, now accessible to our view through ultrasound and new photographic techniques, serves even more dramatically as a symbol of innocence and hope. For some people, the image of the fetus evokes a sense of awe. It has become a symbol of nature: a token of people's reverence for the universe, for natural processes, and for creation itself.

Such a symbolic development may be understandable, but it poses real dangers to law and public policy. It is dangerous when some people's symbol of innocence and promise is projected onto another individual's body. This view of the fetus can foster development of a bizarre and punitive attitude toward pregnant women, whatever their choices. The individual women themselves become invisible or viewed only as vessels--carriers of the infinitely more valuable being.[50]

The fetus has become a symbolic stand-in for individual anxieties about deep social change. It is perceived and exploited as a symbol of fear as much as of hope. And the fear being invoked is not simply the retrospective anxiety about one's own "wantedness" so often found in anti-abortion propaganda;[51] it is anxiety about one's safety now in a hostile and selfish world. Historian Allen Hunter observes:

> Like the family, the fetus is a condensed symbol. The fetus simultaneously stands for the desire to regain traditional society, and for hostility to feminism and freer sexuality which threaten that world . . . Further, the desire to protect the fetus--itself thematized as a miraculous meeting of nature and God--is connected with the view that the world is changing in ominous and threatening ways, ways that even deny life itself the opportunity to come into being.[52]

Kristin Luker's extensive interviews revealed that many right-to-life activists are motivated by their perception that the personhood they believed actually belonged to embryos and fetuses is being stripped away from that vulnerable group, and by their fear that other "imperfect," not "socially useful," or "inconvenient" people will be next--in other words, that they themselves are at risk.[53]

The impulse to "rescue" the fetus from the body and control of the mother seems driven by anxiety. Only powerful emotional undercurrents could account for the drastic departure from established constitutional and common law principles demanded by fetal rights advocates and illustrated by cases like that of the cancer patient in Washington, D.C. subjected to a forced cesarean.[54] Anxiety for the fetus and glorification of high-tech rescue medicine are also reflected in battles over the bodies of pregnant women who are brain-dead, or in a persistent vegetative state because of accident or illness.

Living Wills--The Pregnancy Exception

The Washington, D.C. case subjecting a gravely ill woman to a forced cesarean must be viewed in the context of the continuing legal and legislative debate over individual rights to medical decision making. In state after state, right-to-life organizations and official spokesmen for the Catholic hierarchy have made clear that they will continue to block passage of "living will" statutes that would allow dying patients to refuse intrusive and painful procedures, *unless* legislators insert provisos specifically excluding pregnant women.

Proponents of such exclusionary provisions point to several moving and highly publicized cases in which pregnant women's bodies have been maintained on life-support systems in order to allow the development and birth of their fetuses.[55] Virtually all of these cases, however, involved families or partners who felt sure that the individual woman involved would have made such a choice. The agreement reached, for example, in the 1986 "Baby Poole" case in Oakland, California[56] honored not only the wishes of the prospective father, but the very strong and clearly expressed wishes of the pregnant woman.[57] But insistence on keeping pregnant women on life support, whatever their own clear and conscientious decisions and the choices of the other people most intimately involved, takes ideology to a new level of cruelty.

The pregnancy exception proposed by right-to-life forces and the Catholic lobbying organizations would totally disregard the wishes of the woman and her loved ones. It would compel doctors and hospitals to *force* intensive and intrusive treatment upon a woman, inflicting precisely the sort of indignity upon the patient and anguish upon the family and friends that "Do Not Resuscitate" (DNR) orders and living wills are meant to protect us against. A positive medical development--an advance that might serve the needs of a family and the intentions of a pregnant woman--would be distorted into a weapon in the symbolic warfare over abortion and women's roles.

Furthermore, cases such as "Baby Poole" involved a well-advanced pregnancy with fairly optimistic prospects for the birth of a healthy baby. Life support only had to continue for seven and a half weeks. The pregnancy exceptions proposed by anti-choice groups could require continuation of life support for 6 or 7 months, even in cases were there was little chance of the fetus surviving at all, let alone ultimately being born as a healthy child. In several other cases in which the fetus did not survive, insistence on continuation of the pregnancy brought emotional trauma and invasive publicity to families.[58]

The right-to-life or Catholic Conference demands typically put forth in the living will discussions are for language such as "the provisions of this act shall not apply to a pregnant patient" or "the declaration of a qualified patient known to the attending physician to be pregnant shall be given no force and effect." The temptation to "compromise" by inserting conditions about viability or phrases like that proposed in the "Uniform Rights of the Terminally Ill Act" ("as long as it is probable that the fetus could develop to the point of live birth with continued application of life-sustaining treatment") is illusory and should be resisted. Given the current climate, such provisions are an open invitation for dispute and litigation. Hospitals and doctors may be pressured into inappropriate continuation of treatment by fear of publicity and right-to-life pickets, the specter of criminal prosecution, or even potential exposure to a whole new set of malpractice suits.

Claims that the prior, clearly expressed wishes of a woman who is brain-dead or in a persistent vegetative state have no legal or moral force must be sharply challenged. If people did not feel that they had a stake in what is done to their unconscious or dying bodies, they would not take the trouble to make living wills or discuss DNR orders. Strangers, even doctors and courts, have no legal or moral right to raid the bodies of others--against the dying individuals' clear and express wishes--for fetuses any more than they would to "salvage" organs for those in need of them. And society has traditionally granted great weight to the feelings and choices of survivors about the treatment of the bodies of the dead.[59] Mishandling or abuse of dead bodies is punishable by criminal penalties, and by tort liability for "outrage" or reckless infliction of emotional distress.[60]

For husbands or other loved ones, pregnancy exceptions virtually guarantee a prolonged, more intensely traumatic experience of loss. Not only must loved ones go through the premature death of the woman herself, but they must also be subjected to the genuinely horrifying knowledge that her body is being appropriated by strangers against her explicit wishes, used as an object in their symbolic warfare. How will fathers and family members respond toward a child born of such a situation?

Legislators have a responsibility to resist attempts by single-issue fanatics to poison and distort every public policy discussion by forcing it into the abortion debate. In the DNR and living will contexts, the demands for pregnancy exceptions represent an attempt to hold pregnant women, their loved ones, and the legislative process itself hostage to the narrow ideological preoccupations of one interest group. The dilemmas and tragedies of individual patients are exploited as propaganda fodder in the anti-abortion campaign.

There is an alternative approach, free of the objectionable features of the mandatory exception. Pregnant women and the people they would designate to make decisions about their care should have full information and an opportunity to make decisions about continuing a pregnancy under circumstances in which the pregnant woman faces health problems. Forms for living wills, health care proxy, or durable power of attorney for health care, as well as the educational and explanatory material developed for patients and doctors, should include pregnancy-related treatment among the kinds of medical care issues regarding which individuals may choose to give special instructions. Such a provision would protect both procreative decision-making rights, and the woman's continuing right to bodily integrity and self-determination. In the absence of formal instructions by a woman, doctors and loved ones should consider themselves bound by their sense of what her choice would have been.

A Distortion of Roe v. Wade

Ironically enough, one set of legal arguments for fetal rights is based on the state interest in the potential life of the fetus identified in *Roe v. Wade.*[61] Proponents of such an approach maintain that the viability "line" established in *Roe*[62] marks a point at which society may override the woman's rights.[63] Some fetal rights theoreticians use *Roe*'s viability "line" as part of a waiver argument, urging that "[once] [the woman] decides to forego abortion and the state chooses to protect the fetus, the woman loses the liberty to act in ways that would adversely affect the fetus."[64]

This attempt to use *Roe* as a legal weapon against pregnant women--to claim it as justification for detention, criminal charges of "abuse," drastic restraints on liberty, and even unconsented-to surgical invasion--stands the decision on its head, and not merely in terms of the right to abortion. *Roe v. Wade* may have its flaws, but granting open season on pregnant women after viability is not one of them. *Roe* permits states to bar abortions after viability, *if* the abortion is not necessary for the woman's life or health.[65] It does not license the state to jail women or to invade their bodies against their will. Such a reading of the 1973 decision is at odds with the Supreme Court's use of *Roe* in scrutinizing specific state and local laws regulating abortion;[66] with the now well established interpretation of the case in the area of medical decision making;[67] with the decision's historic recognition and protection of individual rights of procreative choice; and with the decision's assessment of the legal status of the fetus.[68]

Although *Roe v. Wade* does recognize that a state *may* assert a compelling state interest in the potential life of a fetus after viability,[69] it also makes clear that, even at such a relatively late stage of pregnancy, the state interest must be subordinated to the woman's life and health.[70] The Supreme Court's birth control and abortion cases have made clear that decisions about procreation are largely reserved to the individuals directly involved. In addition to that right of procreative privacy, women choosing to continue a pregnancy enjoy certain rights of familial decision making. Parental choices as to the medical care of children, although occasionally subject to being overridden in emergency cases, are nonetheless treated with great respect by doctors and by the courts.[71]

Also, there is a sharp distinction to be drawn between the limitation on a woman's legally recognized right to choose abortion in the third trimester permitted to the state in *Roe*, and the claim that the government may surgically invade a woman's body against her will or require her to subordinate herself physically to what doctors perceive to be the health needs of the fetus. A drastic surgical intrusion such as a cesarean invades well-established rights of self-determination and bodily integrity, which are distinct from the abortion right and which enjoyed substantial legal protection long before *Roe*.

Bodily Integrity and Self-Determination

The real irony of the attempt to use *Roe* as authority for the fetal rights interventions is that, in recent years, *Roe* has emerged as one of the key precedents in the development of legal protection for an individual's right to make personal decisions about medical care and to resist bodily invasion such as unwanted surgery. Far from licensing the sort of invasion and appropriation of the woman's body involved in the fetal rights proposals, *Roe* stands instead as one of the legal bulwarks of a woman's (or man's) right to resist unwanted medical procedures or restrictions.[72]

The fetal rights cases and proposals fly in the face of what has become a very consistent, powerful trend in American law: protection for individual rights of personal autonomy and bodily integrity, especially in the area of medical decision making. The most dramatic and highly publicized of these cases have involved patients' rights to forego invasive or painful procedures that might prolong, but cannot save, their lives.[73] The right to refuse treatment, however, is not limited to the terminally ill.[74] Indeed, so central is the legal tradition of respect for self-determination that the courts have developed procedures to protect the medical decision-making rights of the mentally disabled[75] and the unconscious[76] by allowing others to assert those rights on their behalf.

Court-ordered surgery on a competent,[77] nonconsenting individual is profoundly at odds with established legal principles. Surgery or other invasive treatment may be ordered only when there is a compelling state interest and when no less drastic alternative is available. The state interests, as outlined in *Superintendent of Belchertown v. Saikewicz*,[78] are: (1) the preservation of life; (2) the protection of the interests of innocent third parties; (3) the prevention of suicide; and (4) maintaining the ethical integrity of the medical profession. A state interest in the protection of third parties may sometimes overcome patient refusal rights. However, aside from the idiosyncratic caselaw concerning Jehovah's Witnesses and transfusions,[79] it is *public*--not a particular individual's--health and safety that warrant certain forced treatment.[80] A woman's refusal of treatment may, as in the placenta previa cases, place her own safety at risk. The claim, however, that a state interest in preventing suicide licenses unconsented-to treatment has been repeatedly rebuffed in recent cases.[81] When attorneys have tried to argue that state interests in the ethical positions of doctors and hospitals might outweigh the rights and choices of patients, they have met with increasingly pointed judicial disapproval.[82]

Unconsented-to medical treatment is to be avoided. Stringent requirements for administrative procedures and judicial review have been imposed on the sterilization of the mentally disabled.[83] Courts have barred the nonemergency use of psychotropic drugs on institutionalized mental patients who refuse them, unless the state proves both that the patient is incompetent to make such a decision, and that, if competent, the patient would choose to take the drug.[84] Even in the area of criminal law, judges have barred nonconsensual surgical searches[85] and placed tight restrictions on corporal punishment.[86]

Our legal tradition has consistently refused to impose physical burdens and risks on one person for the sake of another. Parents, for example, are not legally required to risk themselves to rescue their children from danger or to donate blood or organs to them.[87] The law draws a sharp distinction between ethical or emotionally imposed duties and duties imposed by legal, governmental dictate. Society may admire those prepared to undergo invasive surgery to aid relatives or others in need, but we do not *require* such self-sacrifice.[88]

So unthinkable is the notion of forced physical self-sacrifice that there are virtually no cases dealing with the issue. The only case directly on point holds squarely against coercion. In *McFall v. Shimp*, a Pittsburgh judge had to rule in a case involving a young man dying of a form of cancer that might have been arrested by bone marrow transplants from a compatible donor. His cousin, the only compatible potential donor located, would not submit to the procedure. The judge refused to order the transplants. "Morally," he wrote, "this decision rests with the defendant and, in the view of the Court, the refusal of the defendant is morally indefensible." But, the judge insisted, "to *compel* the Defendant to submit to an intrusion of his body would change every concept and principle upon which our society is founded. To do so would defeat the sanctity of the individual and would impose a rule which would know no limits and one could not imagine where the line would be drawn."[89]

Whatever one's position as to the legal or ethical status of the fetus, forced surgical intervention on its behalf cannot be viewed as any more morally or legally acceptable than would be state-compelled organ donation.[90] As one clear-sighted Washington state judge declared in refusing to order a cesarean, "I would not have the right to require the woman to donate an organ to one of her other children, if that child were dying . . . I cannot require her to undergo that major surgical procedure for this child."[91]

The Status of the Fetus

Imposing legal restraints and liabilities on women on behalf of fetuses represents a sharp departure from even the pre-*Roe v. Wade* legal and ethical views of the fetus. *Roe*'s observation that United States law has never treated the unborn as persons in the full sense[92] remains true today. Although there have been several highly publicized cases to the contrary, the overwhelming majority of United States courts follow the common law principle that the criminally caused death of a fetus is not homicide unless there has been a live birth--an existence, however momentary, independent from the mother.[93]

In addition to relying upon *Roe v. Wade*, proponents of "fetal rights" invoke court decisions allowing suits by children born with injuries caused by prenatal events such as exposure to toxic chemicals, claiming that it is inconsistent to acknowledge a right to sue, but not a "right to life" or a "right to begin life with a sound mind and body."[94] But those cases do not rely on any claim of fetal personhood. Compensation is paid to a person, born alive, who suffers from injuries traceable to prenatal causes. Some lawsuits even involve pre*conception* injuries, which occurred well before any fetus or embryo ever came into existence.[95]

The states are divided on allowing legal compensation for the wrongful death of a fetus. Many of the jurisdictions that do permit such suits require that there must have been a live birth or that the fetus must have been "viable" at the time of injury. Judicial decisions allowing recovery are really attempts to recognize the very real pain and loss suffered by a woman, and by a family, when a pregnancy is ended against her will.[96]

The prenatal tort and wrongful death cases are more accurately described as a trend toward reproductive rights, rather than as a recognition of "fetal rights." These cases protect reproductive choice by establishing a duty not to interfere with the reproductive capacity of prospective parents.[97] Suits brought by children for prenatal injuries or by parents under wrongful death statutes are often the only way to protect those rights, since reproductive injuries frequently manifest themselves in the children of those affected well after the parents would be permitted to sue. Decisions allowing such lawsuits are not at odds with *Roe v. Wade*. In fact, they recognize the complementary right to parent implicit in *Roe* itself, and they stand for the safeguarding of rights of reproductive choice.[98]

No Stopping Point

The fetal rights approach takes the limited state interest in the fetus recognized by *Roe* and distorts it beyond all recognition, requiring that the state appropriate and invade the woman's body in the service of a second, conveniently inarticulate patient. Such government-mandated imposition of physical subordination and risk upon one human being for the sake of another would be legally unprecedented, even if there were no dispute at all as to the status of the fetus.[99] It would confer upon the fetus an affirmative grant of legal rights and entitlements sweeping far beyond any enjoyed by those of us with undisputed constitutional personhood.

The Washington, D.C. case in which a pregnant woman died two days after a forced cesarean dramatically illustrates just how easily judges can careen out of control once launched on the "slippery slope" of enforcing the patienthood of the fetus. In a nightmarish scenario that would have been labeled a paranoid feminist fantasy if suggested as a possibility by opponents of fetal rights, a gravely ill pregnant woman was literally sacrificed in order to give her 26-week fetus "a better, though slim, chance."[100] The court accepted an interpretation of *Roe v. Wade* in which a woman's decision-making rights "cease at the point of viability,"[101] and pointed to an earlier Washington, D.C. case in which a cesarean was authorized to be performed on a woman at full term who had been in labor for 60 hours.[102] Court enforcement of fetal patienthood, therefore, has already led to a willingness to treat pregnant women as vessels to be used, risked, and discarded.

Legal enforcement of fetal rights would require a system of surveillance and coercion oppressive to all women of childbearing age. Given the intense emotional and symbolic associations now entangled with issues of women's rights during pregnancy, there simply is no stopping point--no "bright line"--to the fetal rights demands. Worse than the actual court cases have been the arguments and proposals showing up in legal journals and reviews. Professor John Robertson, for example, once argued that, under *Roe*, it would be legally permissible to order the forced feeding of a pregnant anorexic teenager.[103] (Fortunately, anorexic teenagers almost never become pregnant, since anorexia itself interferes with reproductive processes.)

Robertson now holds that such prebirth seizures, which he continues to regard as constitutionally valid, would be unwise for policy reasons. He insists, however, that women could, and sometimes should, be subjected to postbirth sanctions for "egregious" deviations from the maternal norm.[104] Robertson sees no reason why such sanctions should be restricted to violations of maternal duty after viability, arguing that it is the woman's intent to go to term rather than the stage of fetal development that is determinative.[105] In fact, in an earlier article, Robertson went so far as to require pregnancy testing:

> If she has reason to know she is pregnant--if, for example, she has been sexually active and has missed a period--but she has not yet had her pregnancy confirmed, it does not seem unreasonable to require her either to have a pregnancy test or to refrain from ac-

tivities that would be hazardous to the fetus if she were pregnant.[106]

Dr. Margery Shaw has also suggested that women could be held criminally liable for "negligent fetal abuse" for various prenatal harms, and subject to lawsuits by children "if they knowingly and wilfully choose to transmit deleterious genes."[107] Shaw regards drinking by pregnant women as furnishing alcohol to minors, and "wouldn't mind prohibiting bartenders from serving drinks to an obviously pregnant woman. They have to be wary of serving to minors now and could monitor pregnant women in the same way."[108] In Shaw's view, parental liability need not be restricted to actual pregnancy; it may extend back into the indefinite preconception past. Passing on unhealthy genes could be viewed as tortious behavior: "[H]aving children is a foreseeable event. Accordingly, parents should consider the quality of the gametes (eggs and sperm) that will be passed on to their children."[109]

Prospective parents' right to have some say in the conditions of birth could be drastically curtailed under the fetal rights scheme. Professor Robertson suggests, "Mothers [sic] who insist on giving birth at home after receiving a warning from the physician will have acted recklessly, and could be prosecuted for child abuse, feticide or homicide if the child were injured or died during or after the birth."[110]

Professor Angela Holder, who disapproves of the fetal rights approach, foresees even more draconian measures, asking:

> If there is a risk of some problem during delivery, could a woman be removed from a community hospital and the care of her family practitioner [and] against her will be admitted to a tertiary care hospital with a high risk obstetric unit and a neonatal intensive care unit?[111]

If pregnant women can be legally forced to undergo major surgery, then why should they not be subject to legal restrictions on prenatal diet,[112] work,[113] sex,[114] and sports?[115] After all, such restrictions are certainly less invasive or burdensome than a cesarean section. As Dawn Johnsen has argued, issues of civil liability concerning pregnancy would trigger endless deeply intrusive litigation:

> Women would be at the mercy of an undefined and ever-developing common law . . . Given common stereotypical public concepts of the "proper" role of women, par-

> ticularly pregnant women, there is very little behavior that might not be found by a jury to be "unreasonable" . . . It would not, in fact, be "unreasonable" for a pregnant woman, faced with the prospect of post-natal civil liability according to community standards of propriety, to assume that the only safe course of behavior is to lie prone for nine months.[116]

Clearly, any lawyer worth his or her salt would have a professional obligation to inform clients that continuation of any pregnancy to term, or even to viability, would subject them to open-ended criminal and civil liability. Under this theory, women who undergo a spontaneous abortion of a chosen pregnancy at any stage could become vulnerable to legal as well as emotional and physical consequences. Since it is estimated that 60 percent of all conceptuses abort (usually because of genetic flaws),[117] and since interest in embryos and fetuses has become an obsessive preoccupation of large numbers of Americans, we can anticipate a flood of child abuse reports and investigations. The fetal rights theories would, if adopted, give rise to many of the problems forecast by opponents of the Constitutional amendments proposed by anti-abortion forces during the last ten years.[118] The proposed legal doctrine is, in all but its inability to halt immediately those abortions permitted by *Roe*, a backdoor Human Life Amendment.

The proposed doctrine would also license state and private intrusions into the most intimate sphere of personal and family conduct. Parents could find themselves subjected to detailed inquiry as to their sexual relationships, since intimacy during pregnancy might be shown to have occasioned *in utero* injury or infection.[119]

Those urging that women be liable to suit by their children for prenatal negligence point out that the doctrine of parental immunity, which barred suits between family members, is eroding.[120] It is true that a good number of state courts have begun to allow such cases to proceed, but such decisions almost invariably represent judges' efforts to allow individuals with medical needs to tap household or car insurance policies.[121] Also, since the vast majority of mothers will not carry sufficient insurance to make them attractive "deep pocket" targets for lawsuits, we are likely to see plaintiffs' lawyers move toward development of legal theories allowing suits against doctors and other insured health care providers for failure to monitor or report their pregnant patients. Such cases will drive an even deeper wedge between women and their medical care providers, and lead doctors to avoid just those high-risk patients who most need prenatal care and counsel.[122]

The Washington, D.C. case is not the only incident to demonstrate the slipperiness of the "fetus as patient" slope. Other stories from around the country illustrate the dangers posed by such a doctrinal development more clearly than any set of law review hypotheticals. In late 1986, for example, a San Diego woman was arrested and jailed for six days on charges of medical neglect of her fetus. Prosecutors alleged that her disregard of doctors' instructions had caused the brain-death of her son. They charged that Pamela Rae Stewart had taken street drugs, had sexual intercourse with her husband, and failed to report promptly to the hospital when she began bleeding.[123] A municipal court judge dismissed the charges, holding that the legislature had never intended the parental support statute under which prosecutors brought the case to be used for such purposes.[124] A state legislator, incensed by the dismissal, introduced legislation applying existing child endangerment statutes to fetuses.[125] In another incident, a 16-year-old Wisconsin girl was detained in a secure detention facility rather than the foster home or treatment center mandated by the State Juvenile Code, when the local department of Social Services reported that she "lacks motivation or ability to seek prenatal care unless closely monitored by outside sources."[126]

Discriminatory Impact

Not surprisingly, the impact of fetal rights claims has fallen most heavily upon the most vulnerable groups of women. In fact, the danger of discrimination is glaringly apparent from a review of the cases thus far. A 1986 survey of doctors revealed that 81 percent of the pregnant women subjected to court-ordered interventions were black, Asian, or Hispanic; 44 percent were unmarried; 24 percent did not speak English as their primary language; and none were private patients.[127]

Other impermissible factors may play a role. As one commentator has pointed out, religious minorities have been particularly affected: "In almost all cases that have arisen to date, the woman refusing C-section is a member of some religious group or subculture which holds beliefs at variance with the majority in our country."[128] In one reported case in which doctors did not even obtain court authorization before performing a cesarean on a nonconsenting patient, the woman in question had arranged to give the child up for adoption.[129]

The gender discrimination inherent in fetal rights demands was graphically illustrated by the 1986 San Diego prosecution of Pamela Rae Stewart. One of the charges against Stewart, whose baby had been brain-damaged at birth and died some months later, was that she had disregarded doctors' orders by having sex with her husband.[130] No charges were ever leveled against the husband, despite the fact that he was also aware of the doctors' instructions.[131]

Denial of Due Process

It has also become clear that the proposed new doctrine of fetal rights cannot be applied in a considered, nonarbitrary manner. Cases reported thus far demonstrate the near impossibility of assuring even the most minimal level of due process in legal conflicts between pregnant women and their doctors. These situations usually develop very quickly. In 88 percent of the incidents reported in a major national survey, the whole process took less than six hours; "in 19 percent, the orders were actually obtained in an hour or less, at times by telephone."[132] If the pregnant woman is represented by counsel at all (and there has been no lawyer representing the woman in several cases), it is virtually always by a stranger, appointed by the court at the last minute and operating under extreme time pressure.[133] There is virtually no time for written briefs or consultation with experts familiar with the medical and legal issues. Very few cases are reviewed by appellate courts, since women who have been subjected to (or narrowly escaped) forced treatment have little inclination for further dealings with the court system.

The procedural inadequacy is particularly troublesome given the fact that what doctors or social service agencies are seeking from judges is such a drastic and unprecedented power: authorization to appropriate and invade the bodies of individual women for the supposed benefit of another patient, the fetus. The air of crisis and informal procedures militate against the careful consideration of competing evidence and evaluation of alternatives consistently mandated in other cases involving any bodily invasion. Certainly courts dealing with other medical decision-making conflicts, like surgical body searches of criminal defendants,[134] the administration of psychotropic drugs to institutionalized mental patients,[135] or the sterilization of the mentally disabled,[136] have insisted upon much more rigorous procedural standards and upon significantly more information. In those other cases, court-ordered medical treatment, even of the incompetent or of those whose legal rights are limited, is imposed only after detailed fact finding conducted through full adversarial hearings. Doctors seeking court authorization must meet an exacting standard of proof. They have to establish both the necessity of the procedure and the fact that no less drastic means are available. Their burden increases with the degree of invasiveness, risk, or indignity involved.[137]

Medical Uncertainty

Doctors would be hard pressed to meet such a legal standard for many of the procedures and restraints now proposed on behalf of the fetus. Cesarean sections are a good case in point. Although dangers to the fetus or the pregnant woman may be realistically forecast based on previous experience or studies, and although interventions might very well be *advisable* in a given case, deciding on a cesarean involves medical judgment rather than certainty. Doctors can be, and have been, wrong in their belief that surgery was required.

In at least six cases in which doctors have declared cesareans necessary and asked judges to order surgery, the women proceeded to give vaginal birth to healthy children. In the *Jefferson* case in 1981, for example, Georgia courts were told that, without a cesarean, there was a 99 percent chance that the baby would die and a 50 percent chance that the pregnant woman refusing surgery would perish.[138] The very night the Supreme Court delivered its opinion ordering surgery, a hospital test revealed that Ms. Jefferson's condition, in which the birth canal was blocked by the placenta, had corrected itself. Ms. Jefferson gave birth to a healthy baby girl three weeks later. The Georgia Medical Association headlined its story: "Georgia Supreme Court Orders Caesarean Section--Mother Nature Reverses on Appeal."[139]

Similarly ominous medical testimony was given in a 1982 placenta previa case in Michigan. The judge not only ordered the woman to report to the hospital, but instructed the police to pick her up and deliver her to the doctors if she failed to arrive. That woman fled into hiding with her entire family and, a couple of days later, gave uneventful vaginal birth to a healthy 9-pound, 2-ounce boy in a hospital that was unaware of the earlier diagnosis.[140]

In yet another Michigan case, a judge was telephoned by a hospital doctor, and asked to override the wishes of an African woman who refused a cesarean despite "failure to progress" in labor and indications of fetal distress. The judge had just indicated his willingness to order surgery when "the mother suddenly progressed rapidly to the second stage of labor and vaginally delivered an infant with Apgar scores of 8/9."[141] In the cases of the woman from Georgia and the other woman from Michigan, the women's refusals of surgery had been based on religious beliefs. Here, the woman and her graduate student husband quite sensibly pointed out that, when they returned to Africa, there would be no hospital for a repeat cesarean within more than a hundred miles. The couple was also influenced by a previous birth experience, during which doctors had also urged a cesarean and the wife had delivered vaginally.[142]

In 1984, a Washington state judge was asked to order that a cesarean section be performed on a woman who had tested positive for herpes, a condition in which active genital lesions can sometimes infect a baby during birth and cause blindness. The judge observed ruefully that, "This is a case that makes the courts wish they were doing something else for a living," but refused to order surgery.[143] The child was born without any ill effects. New York City Judge Margaret Taylor had a similar experience when she refused to override the refusal of a cesarean by a woman who said that she knew what she was doing and reminded the judge, "It's my body." That baby, too, was delivered vaginally and safely. Professor Nancy Rhoden reports a 1986 New York City case in which a woman diagnosed with hypertension and preeclampsia had religious objections to surgery. The hospital obtained a court order, but the woman never returned to the institution. Instead, she had a successful delivery at home.[144]

Welcome as all of these birth outcomes may be, they cast fatal doubt on the wisdom of lending the weight of judicial authority to medical judgment calls, especially when implementing the medical recommendations will require significant physical violence. In August of 1987, the Ethics Committee of the American College of Obstetricians and Gynecologists (ACOG), recognizing that medical "tests, judgments and decisions are fallible," concluded that resort to the courts was almost never justified.[145]

Too Many Cesareans

Skepticism about the necessity of many cesarean sections is no longer the exclusive province of the feminist health movement.[146] A 1987 study by the Public Interest Research Group (PIRG) charged that about half of the 906,000 cesareans performed in 1986 were unnecessary.[147] Although representatives of the medical profession had some criticisms of the PIRG report, there is general consensus that the rate of cesareans should be closer to 12 to 14 percent than to the current 24.1 percent.[148] Concern about the high cesarean rate has been growing for years. In 1977, a National Institutes of Health (NIH) task force issued a consensus report urging that efforts be made to reverse the trend.[149] And voices within the medical profession itself have expressed concern and called for reform.[150]

The wide variations in cesarean rates among hospitals is enough to trigger skepticism as to the necessity of many procedures. The rate can range anywhere from 31.4 percent in one suburban Boston hospital to as low as 5 percent in an Indianapolis facility.[151] And the variation in rates is not explicable in terms of risk factors. Significantly, the Indianapolis hospital with a poor, high-risk population and a low cesarean rate reported the same decline in stillbirths and neonatal deaths as occurred in the rest of the nation.[152] There are strong indications that variations in cesarean rates reflect socioeconomic factors, rather than risk factors.[153]

Unnecessary cesareans pose risks to women, to their babies, and to the relationship between them. Cesarean section carries a much higher risk of maternal death. The maternal mortality rate is "three to thirty times that associated with vaginal delivery."[154] The NIH task force review of the medical literature revealed that 10 to 65 percent of women who deliver by cesarean section experience postsurgical infections, "a rate five to 10 times the complication risk following vaginal birth."[155] Common infections affecting such women include intrauterine cystitis, peritonitis, abscess, gangrene, sepsis, urinary tract infections, and respiratory infections.[156]

Cesarean birth can also cause complications for babies. The most serious risk is that of respiratory distress--breathing difficulty caused by lung immaturity in newborns delivered too early. This problem is especially likely to occur with repeat cesareans resulting from previous surgical delivery.[157] There may also be breathing difficulties, because the infant's lungs have been deprived of the compressing effect of vaginal delivery.[158]

Surgical delivery can have a negative impact on a woman's ability to care for and establish a relationship with a newborn. As one expert has noted, "With cesarean section, appreciably more women experience postpartum illness and virtually 100 percent experience a great deal more pain, weakness, and difficulty in holding and caring for their newborns."[159] Some women report feelings of isolation, humiliation, anger, self-blame, and deep disappointment after undergoing cesarean sections.[160] The emotional consequences of a *forced* cesarean could certainly be even more devastating.

Other coercive prescriptions for fetal health should also be treated with a healthy degree of skepticism. Not only do they draw upon and reinforce punitive, controlling attitudes toward women, but they may also prove flat out wrong. Sociologist Barbara Katz Rothman has pointed out what "fetal abuse" might have looked like given the conventional medical wisdom 25 years ago:

> Janet M., a diabetic, refused her DES treatment, prescribed as especially important in the prevention of miscarriages among diabetics. Further, although she was eleven pounds overweight at the time of conception, she refused to limit her weight gain over the course of pregnancy to under thirteen pounds. She compounded the problem by not taking the diuretics prescribed, and twice refused to show up for scheduled x-rays, citing a mistrust of medications and radiation.[161]

Less Drastic and More Effective Alternatives

Government action that infringes upon an individual's fundamental rights of personal decision making and bodily integrity is constitutionally permissible only if it serves a compelling state interest and, even then, only if there is no less drastic alternative. The demands made on behalf of the "fetus as patient" cannot meet that test. *Roe* recognizes that a state interest in the fetus may be sufficiently compelling to bar "non-therapeutic" abortions after viability; it does not license seizing a woman's body and forcing her to undergo surgery or significant restraints on liberty. Also, much more often than not, there are less drastic ways to further the goals of maternal and infant health.

Although not all clashes between pregnant women and their doctors can be prevented, the number of incidents could be sharply reduced by diminishing the pressure on doctors to perform cesareans. Some observers attribute the increased reliance on cesarean sections to professional "de-skilling," which means the physicians' unfamiliarity with a range of normal labor experiences and with alternatives to cesareans.[162] Dr. Harry S. Jonas, President of ACOG, reports that "residents aren't getting the training they need in procedures that offer an alternative to cesareans. I hope we can achieve the understanding that you've got to teach these things."[163]

Anxiety created by intrusive medical procedures or an unsupportive or--worse--litigious environment may create problems leading to cesareans: "A woman may become anxious because of the confinement, the paraphernalia attached to her body, the sound of the monitor, the diversion of at least some measure of human attention from her to the apparatus."[164] A medical anthropologist commenting on the case of the African graduate student's wife who so narrowly avoided a court-ordered cesarean stressed the "possibility that the anxiety generated by the negotiations about the cesarean section contributed to arrest of labor and fetal distress."[165] Studies have shown a close relationship between psychological factors and labor difficulties.[166]

Refusals of truly necessary procedures and "noncompliance" with doctors' suggestions for healthier pregnancies could also be prevented through improvement of the communication skills of doctors and other health care workers. A 1983 study reported that treatment refusals were not only surprisingly common, but that "Refusals were generally based on factors within the physician-patient relationship, especially failures of communication and trust."[167] Previous unsatisfactory experiences with medical care and distrust based on cultural and racial differences were also factors.[168] The authors of the study concluded that "[A]ttention to those factors within the control of the medical profession, most notably communication with patients, may substantially reduce instances of refusal."[169]

Yet another strategy for reducing the likelihood of last-minute showdowns between women and their birth attendants is to require hospitals to publicize their cesarean section rate, so that women who feel strongly about not having a surgical delivery can choose a birthplace more likely to meet their needs. Since competition for maternity patients has become a major fact of economic life for many hospitals,[170] such "consumer" pressure can prove very effective. In 1982, women's health activists in Philadelphia succeeded in obtaining changes in hospital practices simply by making it known to the hospitals that they were collecting information for a guide to "Childbirth Choices."[171]

Legislation requiring publication of cesarean section rates (see Appendix A, *infra*) passed both houses of the New York State legislature in 1987, but was vetoed by Governor Mario Cuomo at the behest of medical and hospital associations. Governor Cuomo argued that statistics can be misleading and promised action by the State Department of Health. The Health Department has since undertaken a comprehensive effort to reduce the state's cesarean rate, including development of a consumer's guide to birthing practices that will cite hospital cesarean section rates and a study of hospitals with particularly high rates.[172] An ACOG doctor heading the newly formed state task force declared that he had already established a process at his own hospital under which every cesarean birth would be subject to clinical review.[173]

Reducing Doctors' Fear of Lawsuits

One of the key factors leading to doctors' overreliance upon cesareans is their enormous and realistic fear of malpractice suits if a child is born with a condition that might arguably have been prevented by surgical delivery. Not only are obstetricians subject to more frequent malpractice lawsuits than most other specialists,[174] but damages in obstetrical cases, especially the "brain-damaged baby" cases, can be spectacularly high.[175] A study of damage awards and settlements paid out in 1986 by the New York City Health and Hospitals Corporation, for example, revealed that, although malpractice claims involving childbirth were only 11 percent of the suits, they accounted for 41 percent of all costs.[176]

The causes of the "malpractice crisis" may be sharply disputed,[177] but the impact--financial, professional, and emotional[178]--on doctors has been enormous. A 1983 survey by the American College of Obstetricians and Gynecologists disclosed that 9 percent of the respondent members had quit obstetrics because of the risk of lawsuits, 10 percent were delivering fewer children, and another 18 percent were refusing more high-risk obstetrical cases.[179] And malpractice pressures inevitably affect the way that doctors practice medicine. Dr. Helen Marieskind found in her 1979 study that the "[t]hreat of a malpractice suit if a Caesarean were not performed and the outcome was 'less than a perfect infant' was the most frequent reason for the increase in the Caesarean section rate given by physicians during interviews."[180] A professional newsletter circulating among obstetricians confirms this picture, reporting that: "Ninety percent of malpractice suits brought against obstetricians involve failure to perform a cesarean section or the improper use of forceps. So, among the suing population, if the physician does a cesarean section he does his best job."[181] An observer of the spiraling rate of cesarean sections has noted that given the pervasive fear of lawsuits, "Doctors are willing to accept a lesser and lesser degree of risk for the fetus."[182]

Doctors and hospitals confronted with a woman's refusal of a cesarean, therefore, find themselves in a dilemma at once ethically, emotionally, and financially threatening. The Colorado hospital attorneys who called in a judge in one of the first reported court-ordered cesarean cases related that they sought court action in part because they were concerned about institutional liability.[183] ACOG stated about the recent cesarean imposed upon a dying woman in Washington, D.C., "This demonstrates what happens when decisions are being driven by fear of legal action rather than by medical judgment."[184]

Reliance upon the tort system to ensure care for children born with disabilities can all too often result in injustice and in a defensive medical stance that is distorting women's health care during pregnancy and childbirth.[185] Universal health insurance and a society more hospitable to the disabled would help lessen the problem. In the meantime, judges and juries must make clear that doctors and hospitals are under no duty to disregard a competent woman's rights by overriding her informed refusal of a cesarean section or other invasive medical treatment (*see* Model Jury Instructions and Commentary, Appendix B), and that invoking judicial authority is not only not required, but dubious medical practice.[186]

Coercion: Unjustifiable and Counterproductive

There are less drastic and less discriminatory alternatives to the "fetal rights" approach. They are mandated not only by constitutional restraints on government power, but by realistic concern for maternal and infant health. The punitive, controlling approach of forced treatment and civil or criminal liability is not just legally impermissible; it will not work.

Calls for court-ordered surgery and for legal punishment for negligent behavior in pregnancy or "fetal abuse" are unacceptable given the unavailability of even the most basic prenatal and obstetrical health care for many pregnant women. Between 1980 and 1985, the infant mortality rates in the United States tied for the worst ranking among 20 industrialized nations.[187] According to the *Boston Globe*, a 1985 Massachusetts study found glaring inadequacies in prenatal care in that state, which no doubt are typical of those that affect women nationwide:

> In New Bedford, Fall River, Chicopee, Lynn, Fitchburg and Lowell, fewer than seven women out of 10 who gave birth in 1983 received adequate prenatal care. The proportion of women getting adequate prenatal care was lowest in Holyoke, where only 45 percent of women who gave birth in 1983 received adequate prenatal care . . . [T]he state should require all health insurers to cover maternity benefits . . . None of the Blue Cross-Blue Shield nongroup insurance policies for individual subscribers cover maternity benefits . . . and most group policies do not cover maternity benefits for teenage dependents.[188]

Such drastic shortcomings in the health care system raise serious questions as to the motivation behind, and likely effectiveness of, fetal rights demands. A society that has made so little attempt to meet the needs of pregnant women and their children cannot level heroic demands on individual women, and should not move to exact the gift of life by force of law and violence. Professor Angela Holder has stated:

> Only when every impoverished woman in this country who wants it has access to adequate prenatal care, diet, and the option of delivering her baby in a hospital can the government claim that its concern about the infants born in our country is sufficiently genuine so that it has some moral right to prosecute and control women it believes to be engaging in fetus abuse.[189]

The uproar over the Pamela Rae Stewart prosecution had at least one benefit: it focused public attention on the insufficiency of medical care for poor pregnant women in the San Diego area. Three days after "medical neglect" charges against Stewart were dismissed, the local newspaper reported new countywide statistics showing that "many pregnant women--mostly those who are poor, many who work in farms and fields in rural areas or those who hold minimum wage jobs--never see a doctor or midwife until they are in labor."[190]

Proposals that poor nutritional habits trigger punitive state action against pregnant women[191] or that vitamins be forcibly administered[192] seem a bitter joke in the context of federal policies in which serious consideration is given to declaring catsup a vegetable, and public funding limitations have blocked assistance to more than half the pregnant women and children eligible for supplemental nutrition programs.[193] Indeed, in the context of the current realities of maternal and child health in the United States, the fetal rights drive is a destructive distraction, diverting medical and legal attention from the affirmative programs of environmental protection, public education, and health care reform genuinely required to ensure healthy pregnancies and well babies.

Until, for example, more stringent restrictions are placed upon smoking by the general public, harsh policies against pregnant women are both unfair and ineffective. It simply does not make sense to single out fetuses for protection and then only vis-à-vis the prospective mother. Restrictions on smoking by the woman's spouse or companion, other family members, and coworkers are also necessary to protect fetal health.[194] Public education about the dangers of smoking could be made less gender-specific.[195] An end to tobacco subsidies, higher taxes on tobacco, and legal curbs on smoking in public places would be useful and less constitutionally dubious steps.

Suggestions for the criminalization of drinking by pregnant women[196] must be subjected to similarly critical evaluation. Again, general education and deterrence campaigns seem more likely to vindicate the state's interest without transgressing constitutional norms. The government should move to correct the drastic shortage of treatment programs for alcoholic women. A June, 1987 article in the *New York Times* quoted a Massachusetts judge describing this deplorable situation: "Nationally there are 5557 treatment programs for men, 375 for women. In Massachusetts, like most states, 30 to 50 percent of the alcoholics are women, yet they are less than 20 percent of those treated. For them, society sees alcoholism as a moral weakness, not an illness."[197] And residential treatment programs must offer child care if they are to be of use to women.[198]

Deterring Prenatal Care

Targeting pregnant women for punitive and controlling policies is not only constitutionally dubious, it is all too likely to backfire. Women fearful of forced cesareans or of legally "policed" pregnancies may avoid prenatal care altogether. The Michigan woman who fled a court-ordered cesarean gave birth in another hospital,[199] but in one New York case in which a hospital obtained a court order authorizing surgery, the woman decided not to return to *any* hospital and gave vaginal birth, safely, at home.[200] Not every case of women driven away from medical assistance will necessarily turn out so successfully. Coercive measures may lead to concealed pregnancies, nonassisted births, an upsurge in abandonment, and even infanticide.

It was concern about the deterrent effect of legal intervention on women's use of the health care system that led to the outspoken medical opposition to the prosecution of Pamela Rae Stewart for "medical neglect" of a fetus. The President of the California Medical Association declared, "[P]rosecution is counterproductive to the public interest as it may discourage a woman from seeking prenatal care or dissuade her from providing accurate information to health care providers out of fear of self-incrimination."[201] Also, the statement of August 1987 from the American College of Obstetricians and Gynecologists that physicians should rarely, if ever, seek court-ordered treatment of pregnant women was grounded in the recognition, not only that medical judgments are fallible, but that resort to government authority and physical force has a devastating impact on women's relationships with hospitals and doctors.[202]

Defining fetuses as children for the purposes of neglect and abuse laws must be seen as self-indulgent posturing in light of the crisis already engulfing child abuse agencies.[203] In fact, in 1984 the Illinois Department of Children and Family Services went into court to fight a judge's order that it intercede on behalf of the fetus of a pregnant drug addict.[204] Additionally, as lawyers for Pamela Rae Stewart pointed out, putting women in prison for fetal abuse or neglect is unlikely to further maternal or infant health; they cited a pending federal lawsuit alleging that pregnant women prisoners at the California Institution for Women were being deprived of adequate prenatal and postpartum care.[205]

Demands for forced institutionalization of pregnant substance abusers[206] are similarly counterproductive. Care-avoidance patterns have already been widely noted among pregnant addicts: "The reasons for their reluctance to seek out prenatal services are attributed not only to their overall lifestyle of self-neglect, but also to their fears that Child Welfare agencies will be contacted and their newborns and/or other children will be taken away from them."[207]

With pregnant addicts, as with other women confronting high-risk pregnancies, supportive programs that emphasize and build upon the woman's desire to meet the needs of her children are more likely to have a positive effect. We have not made those programs available to many pregnant women. The Pamela Rae

Stewart case, for example, highlighted shortcomings in San Diego County's services for female substance abusers: "Women addicts must wait up to six months for one of just 26 slots in residential rehabilitation programs where they can live with their children--the programs experts say may offer the best chance for success."[208] Models do exist. Voluntary, innovative Pregnant Substance Abuser programs have been established at five New York City municipal hospitals to encourage prenatal care. Women are provided with continuous and supportive prenatal medical care (including home visits and escort services to clinic appointments), as well as counseling, advocacy assistance, and referral to other agencies and services. Staff are specially trained to deal with the special needs of pregnant substance abusers.[209]

Conclusion

Treatment refusals by pregnant women are quite rare, and the cases in which truly necessary interventions are refused will be fewer still. But even in those very rare cases in which the danger to the fetus is real, the benefits are indisputable, and the risks to the women are minimal, even then we should honor those refusals. Mistakes will be made in pregnancy and birth as in every other human context. They may be made by doctors; they may be made by patients and their families. While the fetus is within the body of a woman, the power and responsibility to choose must rest with her. To ground decisions elsewhere--with doctors or lawyers or judges--is to treat pregnant women as seed gardens or vessels, to deny them the bodily integrity and self-determination specific to human dignity.

The debate over the "fetus as patient" has less to do with abortion or with the status of the fetus than with the rights of women to carry and bear children with dignity. We need not agree on the question of whether the fetus is a person once we recognize that a woman is. Pregnancy, even after viability, does not divest a woman of her rights of bodily integrity and self-determination, nor of her right to be free from physical invasion or appropriation on behalf of another. The fetal rights demands *are* a symbolic affront to all women: a dismissal of our moral agency and full citizenship, reducing us to the status of potential "motherships," or carriers of "precious cargo." Supposed gains in neonatal survival and health must be weighed against the individual and societal costs. Women have been driven into hiding, tied down to operating tables, and cut open against their will. They have been frightened away from hospital birth and prenatal care. Legal enforcement of the primacy of the "fetus as patient" threatens to create a climate in which pregnancy and birth are laced with intimidation, coercion, and raw physical violence.

In the spring of 1987, New York City newspapers reported on a neighborhood's resistance to the opening of a home for "boarder babies"--infants left in hospitals, some of them ill with AIDS and some born to mothers addicted to drugs or otherwise unable to cope with parenting. A local woman, an opponent of the proposed home, declared, "We love babies. We just hate drug addicts." Our society simply cannot get away with such a stance. The exclusive focus on the fetus, or even on babies, is a cheap emotional distraction from the more difficult and complicated tasks at hand. We cannot effectively love babies and treat their mothers badly.

Appendix A

Bill A. 3821, New York State

The People of the State of New York, represented in Senate and Assembly, do enact as follows:

Section 1. The public health law is amended by adding a new section twenty-eight hundred three-i to read as follows:

§2803-i. Maternity patients informational pamphlet. 1. The commissioner shall require that every hospital shall distribute at the time of pre-admission directly to each prospective maternity patient and, upon request, to the general public an informational pamphlet. Such pamphlet shall be prepared by the commissioner and shall contain material briefly describing maternity related procedures performed at such hospital and such other material as deemed appropriate by the commissioner.

2. Such pamphlet shall also include statistics relating to the annual percentage of maternity related procedures performed at each hospital, including but not limited to the following:

(a) the annual rate of caesarean sections, primary and repeat, performed at such facility;

(b) the annual percentage of women with previous caesarean sections who have had a subsequent successful vaginal birth;

(c) the annual percentage of deliveries in birthing rooms in such facility;

(d) the annual percentage of deliveries by certified nurse-midwives;

(e) the annual percentage of births utilizing fetal monitoring listed on the basis of electronic internal, external and auscultation;

(f) the annual percentage of births utilizing forceps, listed on the basis of low forceps delivery and mid forceps delivery;

(g) the annual percentage of births utilizing breech vaginal delivery;

(h) the annual percentage of vaginal births utilizing analgesia;

(i) the annual percentage of births utilizing anesthesia including general, spinals, epidurals, and peracervical;

(j) the annual percentage of births utilizing induction of labor;

(k) the annual percentage of births utilizing augmentation of labor;

(l) the annual percentage of births utilizing episiotomies; and

(m) the percentage of mothers breast feeding upon discharge from such facility.

3. Compilation of the statistics set out in subdivision two of this section shall be the responsibility of the commissioner provided, however, that the statistics set out in paragraphs (b), (c) and (l) of such subdivision shall be compiled by each facility and submitted to the commission for inclusion in the informational pamphlet until such time as the commissioner deems otherwise.

4. Statistical information contained within the informational pamphlet shall be presented in a three year aggregate with each of the years included in the aggregate listed separately.

5. The commissioner shall review and update the informational pamphlets on an annual basis.

6. The cost of preparing such pamphlets shall be paid by the department from monies appropriated therefor.

§2. This act shall take effect on the one hundred twentieth day after it shall have become a law.

Appendix B

Model Jury Instructions

A doctor who respects the informed refusal of surgery or other diagnostic or therapeutic procedures, regimens, medications, or treatments by a competent patient who is pregnant is not liable for injuries to the woman or to her later born child caused by failure to administer the treatment refused. The doctor has neither a legal right nor a legal duty to override the refusal of a competent patient. In fact, doctors who disregard the treatment refusal of a patient and perform an unconsented-to operation or procedure upon her commit what is known to the law as assault and battery and may not only be sued for damages but could also find themselves answerable to criminal charges. It is no defense that the operation [procedure or treatment] which was performed was a medically sound procedure; it is the patient's right to decide whether she wishes to consent to the surgery [procedure or treatment].

Doctors *are* under a duty to make an understandable disclosure to the patient of what it is that they propose to do. The doctor must explain in terms understandable by a reasonable person the reason for the proposed surgery [procedure or treatment], what the risks and benefits to the patient's health or life and to the potential life and health of the fetus may be in performing the surgery [procedure or treatment], what the risks may be if no operation [procedure or treatment] is carried out, whether the surgery [procedure or treatment] is one that is ordinarily done under the same conditions, whether other or different operations [procedures or treatments], if any, are used, and the manner in which the alternative [procedures or treatments] are performed, and the nature and extent of the risks involved in the alternative [procedures or treatments]. The choice or decision is the woman's, but she should be provided with information as to the risks and benefits to her and to the fetus.

If you find that the doctor told the pregnant woman all those factors reasonably required to be given to a patient in her condition at the time the operation [procedure or treatment] was refused, or that any information you find was omitted would not have materially affected her decision to submit or not to the proposed surgery [procedure or treatment], and that the injuries to the woman or to the after born child were caused by the patient's refusal, then your verdict must be for the defendant. I charge you, as a matter of law, that you are not to consider the failure of the defendant[s] to carry out the operation [procedure or treatment] refused by the patient as an act of negligence.

Commentary

The patient's informed refusal of treatment negates physician liability for malpractice in failing to carry out the proposed procedure. *Steele v. Wood*, 327 S.W. 2d 187, 196 (Mo. 1959); *Parker v. Goldstein*, 78 N.J. 472, 189 A.2d 4441 (1963); *Hunter v. United States*, 236 F. Supp. 411 (M.D. Tenn. 1964); Rozofsky, *Consent to Treatment*, § 1.10-1.10.2 (1984) at 35-43; 61 *American Jurisprudence* 2d Physicians and Surgeons § 175; 50 *American Law Reports* 2d 1043 § 4 at 1054 (1956).

The right to refuse treatment is grounded in both common law and constitutional sources. Bodily integrity and self-determination enjoy longstanding common law protection. *Union Pacific Railway Co. v. Botsford*, 141 U.S. 250 (1891); *Mohr v. Williams*, 95 Minn. 261, 104 N.W. 12 (1905); *Schloendorff v. Society of New York Hospital*, 211 N.Y. 125, 105 N.E. 92 (1914). In re *Conroy*, 98 N.J. 321, 346-48, 486 A.2d 1209, 1210-12 (1985); *Rogers v. Com'r of Dept. of Mental Health*, 390 Mass. 489, 498, 458 N.E. 2d 308, 314 (1983); In the Matter of *Eichner*, 52 N.Y. 363, 376, 438 N.Y.S. 2d 266, 272, 420 N.E. 2d 64, 70 (1981). In recent years, courts have sometimes also cited constitutional privacy grounds in upholding patient refusal rights. In re *Quinlan*, 70 N.J. 10, 38-42, 355 A.2d 647, 660-663 (1976). While the right to refuse treatment has been most widely publicized in recent cases involving the "right to die with dignity," it is by no means confined to the terminally ill. *Bartling v. Superior Court*, 163 Cal. App. 3d 186, 209 Cal. Rptr. 220 (App. Ct. 1984) and In re *Conroy*, 98 N.J. 321, 355, 486 A.2d 1209, 1226 (1985). See generally Cantor, "A Patient's Decision to Decline Life-Saving Medical Treatment: Bodily Integrity Versus the Preservation of Life," 26 *Rutgers Law Review* 229 (1973); Clarke, "The Choice to Refuse or Withhold Medical Treatment: The Emerging Technology and Medical-Ethical Consensus," 13 *Creighton Law Review* 795 (1980); Annas, Glantz, and Katz, *The Rights of Doctors, Nurses and Allied Health Professionals* 71-97 (1981); 93 *American Law Reports* 3d 67 (1979). Pregnant women are not excluded from the protection of the doctrine. In fact, their individual rights as patients are enhanced by the protection accorded to procreative and familial decision making. Gallagher, "Prenatal Invasions and Interventions: What's Wrong with Fetal Rights," 10 *Harvard Women's Law Journal* 9 (1987).

In *Jefferson v. Griffin Spaulding County Hosp. Auth.*, 247 Ga. 86, 274 S.E. 2d 457 (1981), the Georgia Supreme Court authorized performance of a cesarean on a nonconsenting woman. In its per curiam opinion the Court relied upon *Roe v. Wade*, 410 U.S. 113 (1973), *Raleigh-Fitkin Memorial Hosp. v. Anderson*, 42 N.J. 421, 201 A.2d 537, cert. denied, 377 U.S. 985 (1964); and *Strunk v. Strunk*, 445 S.W. 2d 145 (Ky. Ct. App. 1969).

Invocation of *Roe v. Wade* as authority for disregarding the woman's objection to undergoing surgery demands serious distortion of that landmark case. Such a reading of the 1973 decision is at odds with the Supreme Court's use of *Roe* in scrutinizing specific state and local laws regulating abortion [see, e.g., *Collauti v. Franklin*, 439 U.S. 379 (1979)]; with the now well established interpretation of the case in the area of medical decision making [Gallagher, "Fetal Personhood and Women's Policy," in Sapiro (ed.), *Women, Biology, and Public Policy* 91, 97-100 (1985)]; with the decision's historic recognition and protection of individual rights of procreative choice; and with the decision's assessment of the legal status of the fetus [*Roe v. Wade*, 410 U.S. 113, 157-162 (1973)].

Although *Roe v. Wade* does recognize that a state may assert a compelling state interest in the potential life of a fetus after viability, 410 U.S. 113, 163-64, it also makes clear that even at such a relatively late stage of pregnancy that state interest must be subordinated to the woman's life and health. See also *Thornburgh v. American College of Obstetricians and Gynecologists*, 476 U.S. 747 (1986).

The right to refuse treatment is not absolute. It may be overridden by a compelling state interest in, for example, public health and safety. *Prince v. Massachusetts*, 321 U.S. 158 (1944). But the state may not subject a competent, unconsenting adult to surgery or other invasive procedures for the benefit of another patient, fetal or born. See, e.g., *McFall v. Shimp*, No. 78-17711 in Equity (C.P. Allegheny County, Pa., July 26, 1978) (refusing to order marrow transplant by unconsenting individual to relative regarded by doctors as having no other chance to survive); Regan, "Rewriting *Roe v. Wade*," 77 *Michigan Law Review* 1569 (1979). *Strunk v. Strunk*, 445 S.W. 2d 145 (Ky. Ct. App. 1969) represents an effort to implement the wishes of, and serve the best interests of, an incompetent sibling by preserving the life of a close relative. It does not provide a precedent for unconsented-to surgery performed on a competent woman.

There have been a number of court-ordered blood transfusions performed on Jehovah's Witnesses. Application of the *President and*

Directors of Georgetown College, 331 F.2d 1000 (D.C. Cir. 1964). Such orders have sometimes been described as having been issued for the sake of minor dependents or, in the case of pregnant women, for the sake of the fetus. *Raleigh-Fitkin Memorial Hospital v. Anderson*, 42 N.J. 421, 201 A.2d 537, cert. denied, 377 U.S. 985 (1964). The blood transfusion cases, however, are most realistically analyzed as having turned on issues of patient competence or on the judge's conviction that--by assuming the spiritual responsibility for the decision--he honored the real wishes of the patient. Gallagher, "Prenatal Invasions and Interventions," 10 *Harvard Women's Law Journal* 9 (1987). Cases involving fetuses are subject to the same narrow limitations as the other transfusion cases and must also now be scrutinized in light of the significant expansion of procreative and medical decision-making rights since *Raleigh-Fitkin* was decided. *Roe v. Wade*, 410 U.S. 113 (1973); In re *Quinlan*, 70 N.J. 10, 355 A.2d 647, cert. denied, 429 U.S. 922 (1976); In re *Grady*, 85 N.J. 235, 426 A.2d 467 (1981).

The fetus's legal status is not the issue posed by a woman's treatment refusal during pregnancy or birth. Whatever duty doctors may have to the fetus under other circumstances, see, e.g., *Hughson v. St. Francis Hosp. of Port Jervis*, 92 A.D.2d 131, 459 N.Y.S. 2d 814 (N.Y. App. Div. 1983), they have no duty to disregard a competent patient's refusal of consent to surgery or other invasive procedures proposed on behalf of that fetus. No patient--whatever his or her legal status--can be so favored at the expense of another. *McFall v. Shimp, supra.* Nor may the state delegate to doctors a power it may not exercise itself. *Planned Parenthood v. Danforth*, 428 U.S. 52 (1976). The duties and liabilities imposed by state tort law are subject to constitutional constraints. *New York Times v. Sullivan*, 376 U.S. 254, 265 (1964).

Deliberate disregard of an explicit patient refusal is battery. "An unauthorized operation is a wrongful and unlawful act for which the surgeon will be liable in damages." *Pugsley v. Privette*, 263 S.E. 2d 69, 74 (Va. 1980). Absent consent, even nonnegligent, harmless, or concededly beneficial treatment may constitute a tort. *Id.* at 74-75; *Estate of Leach v. Shapiro*, 469 N.E. 2d 1047, 1051 (Ohio App. 1984); *Marino v. Ballestas*, 749 F.2d 162, 167 (3d Cir. 1984) (applying Pa. law). Even a blood test done in contravention of a patient's stated wishes can result in liability for assault and battery. 61 *American Jurisprudence* 2d Physicians and Surgeons § 197 (1981), citing *Bednarik v. Bednarik*, 18 N.J. Misc. 633, 16 A.2d 80 (N.J. Ch. 1940).

Exceptions to the informed consent rule, for emergencies or in cases of implied consent, are specifically not applicable when there is an explicit refusal. "Even in emergencies, however, it is held that consent will not be implied if the patient has previously stated that he would not consent (*Restatement of Torts* 2d, § 62, Illustration 5; Powell, Consent to Operative Procedures, 21 *Md. L. Rev.* 189, 199; Bryn, Compulsory Life-saving Treatment for the Competent Adult, 44 *Fordham L. Rev.* 1, 15, n. 64)." In re *Eichner*, 420 N.E. 2d 64, 70 (N.Y. 1981). See also Raines, "Editorial Comment on Cesarean Delivery for Fetal Distress without Maternal Consent," 63 *Obstetrics and Gynecology* 596, 598 (1984); and Rozofsky, *Consent to Treatment*, § 1.14.3 at 70 (1984). Subsequent patient unconsciousness does not relieve the health provider's responsibility to honor the refusal. Rozofsky, § 2.1.2 at 91.

The law presumes competency. Annas and Densberger, "Competence to Refuse Medical Treatment: Autonomy vs. Paternalism," 15 *Toledo Law Review* 561, 575 (1984); 41 *American Jurisprudence* 2d Incompetent Persons § 129; Annotation, 25 *American Law Reports* 3d 1439 (1969) at 1440. Even in cases involving institutionalized mental patients, doctors must prove that a patient is incompetent for the purpose of consenting to treatment. See, e.g., *People v. Medina*, 705 P.2d 961, 970 (Ariz. 1985); In re *Boyd*, 403 A.2d 744, 747, n. 5 (D.C. Ct. App., 1979). A plaintiff alleging that the woman had been incompetent at the time of treatment refusal, therefore, would have to overcome the presumption. A medically "irrational" decision by a patient does not necessarily give rise to an inference of incompetence. *Lane v. Candura*, 376 N.E. 2d 1232 (Mass. App. 1978); 93 *American Law Reports* 3d 67, § 3[b] at 73 (1979).

Annas and Densberger, *supra*, at 572, define competency in this context as "the capacity to understand and appreciate the nature and consequences of one's acts." The doctor has a duty of full disclosure to the pregnant woman. A pregnant patient should be fully apprised of the possible consequences--to her and to the child to be born--of her refusal. She should be informed of possible alternatives to the proposed treatment. *Marino v. Ballestas*, 749 F.2d 162 (3d Cir. 1984) (applying Pa. law). Reasonable alternatives are to be defined in the circumstances of each case, in accordance with the individual patient's situation and concerns. Rozofsky, *Consent to Treatment*, § 1.11.2 at 48 (1984). And if it is claimed that there has been a negligent failure of disclosure, the adequacy of the information provided to the patient is a question for the jury. *Steele v. Wood*, 327 S.W. 2d 187 (Mo. 1959).

Notes and References

1. Jefferson v. Griffin Spaulding County Hosp. Auth., 247 Ga. 86, 274 S.E.2d 457 (1981).

2. See, e.g., Parness, "The Duty to Prevent Handicaps: Laws Promoting the Prevention of Handicaps to Newborns," 5 Western New England Law Review 431 (1983); Robertson, "The Right to Procreate and In Utero Fetal Therapy," 3 Journal of Legal Medicine 333 (1982).

3. Shaw, "Conditional Prospective Rights of the Fetus," 5 Journal of Legal Medicine 63 (1984); Doudera, "Fetal Rights? It Depends," Trial, Apr. 1982, at 38.

4. E.g., Ill. Ann. Stat. ch. 110 1/2, Sect. 3(c) (Smith-Hurd 1984).

5. Kolder, Gallagher, and Parsons, "Court-ordered Obstetrical Interventions," 316 New England Journal of Medicine 1192-1196 (May 7, 1987).

6. Jefferson v. Griffin Spaulding County Hosp. Auth., 247 Ga. 86, 274 S.E.2d 457 (1981).

7. Gallagher, J., "Prenatal Invasions and Interventions: What's Wrong with Fetal Rights," 10 Harvard Women's Law Journal 9, 47 (1987).

8. Greenhouse, "Wide Appeal Filed on Forced Caesarean Delivery," New York Times, Nov. 25, 1987, p. A-15, col. 1; In re A.C. (No. 87-609) D.C. Court of Appeals, Nov. 10, 1987; Greenhouse, "Appeals Court Vacates Forced Caesarean Ruling," New York Times, March 22, 1988, p. A-17, col. 1.

9. Taft v. Taft, 388 Mass. 331, 446 N.E.2d 395 (1983).

10. No ruling on the issue was ultimately required, since the woman agreed to enter a drug program as a condition of regaining custody of an already born child.

11. Grodin v. Grodin, 102 Mich. App. 396, 301 N.W.2d 869 (1980).

12. Chambers, "Dead Baby's Mother Faces Criminal Charges on Acts in Pregnancy," New York Times, Oct. 9, 1986, p. A-22, col. 1.

13. Carson, "Bill Offered Based on Pamela Rae Stewart Baby Case," San Diego Union, Mar. 7, 1987, p. A-3, col. 1.

14. Address by Dr. William H. Clewell, American Society of Law and Medicine Conference, Boston (Oct. 29, 1984). See also Jameson, "I Risked My Life To Have A Baby," Chatelaine, Mar. 1985, at 52 and Distelheim, "Jeanette Percival's Brave Decision," Good Housekeeping, May 1985, at 36.

15. See, e.g., Cohen and Estner, Silent Knife (1983).

16. Knox and Karagianis, "Caesarean Births: High Rates, Impassioned Debate," Boston Globe Magazine, October 21, 1984, at 10, 11.

17. See discussion of physician education, infra, in section on "Less Drastic and More Effective Alternatives."

18. These arguments have been employed in both the legal and popular debates. See Justice O'Connor's widely noted dissent in Akron, 462 U.S. 416, 442 (1983) and Dr. Bernard Nathanson on "fetology" in the 1985 anti-abortion film, Silent Scream.

19. O'Brien, M., The Politics of Reproduction 48 (1981).

20. Kosnik, A., Carroll, W., Cunningham, A., Modras, R., and Schulte, J., Human Sexuality: New Directions in American Catholic Thought 59 (1977).

21. Harrison, B. W., Our Right to Choose: Toward a New Ethic of Abortion 137 (1983).

22. Id. at 134.

23. Quoted in Harrison at 145.

24. Haliday, "The Fetal Patient and the Unwilling Mother: A Standard for Judicial Intervention," 14 Pacific Law Journal 1065, 1065 (1983).

25. Nathanson, B., The Abortion Papers 124, 119, 117 (1983).

26. Id. at 124, 123.

27. Id. at 119.

28. Quoted in Milbauer, The Law Giveth: Legal Aspects of the Abortion Controversy 118 (1983).

29. Ehrenreich and English, Witches, Midwives, and Nurses: A History of Women Healers (1973); Gordon, Woman's Body, Woman's Right (1976); Mohr, Abortion in America (1978).

30. Katz Rothman, In Labor: Women and Power in the Birthplace (1982).

31. Smith-Rosenberg, Disorderly Conduct 40 (1985). Smith-Rosenberg explains this symbolic transformation:

> "Public language" exists to convey socially shared experiences in an affective but deliberately distorted manner. Driven to discuss what is too painful or too political to be discussed overtly, societies as a whole, or specific groups within them, develop metaphoric or mythic systems that cloak real meanings behind symbolic masks. The most "public language" thus can be decoded to reveal the interplay of social experiences and emotional realities. Sociological realities and political motivations will then emerge, not stripped bare by analysis, but enveloped in the feelings that constitute one of their most central components. Id. at 45.

32. Id. at 46.

33. Id. at 232.

34. Id.

35. Petchesky, Abortion and Woman's Choice: The State, Sexuality, and Reproductive Freedom 78-82 (1984).

36. Smith-Rosenberg reports that mid-nineteenth-century men "molded the twin themes of birth control and abortion (always defining them as women's decisions) into condensed symbols of national danger and decay." Smith-Rosenberg, supra n. 31, at 180. For recent parallels, see Jacoby, "Be Fruitful or Be Sorry" (review of Ben J. Wattenberg's The Birth Dearth), New York Times Book Review, July 12, 1987, p. 9 and King, "Robertson Urges New Policy to Increase U.S. Birth Rate," New York Times, Oct. 24, 1987, p. A-9, col. 1.

37. Sociologist Kristin Luker's landmark study of pro-choice and anti-abortion activists, Abortion and the Politics of Motherhood (1984), demonstrates that, whereas the ostensible issue of debate is the status of the fetus, the abortion battle also represents a clash between starkly contrasting views of women. See especially 158-192.

38. Anxiety over changing sex roles is not solely a male or even an antifeminist experience. Some feminist critiques of alternative forms of reproduction, particularly of surrogate motherhood, reflect a deep fear that the power of motherhood is being snatched away. Elsewhere, I have examined this response:

> Recent U.S. legal trends awarding joint or even sole custody to fathers after divorce have created a widespread uneasiness among women, reflected not only in the legal literature (Polikoff, 1982), but in the popular culture as a recurring motif in television soap operas and as the theme of best-selling novels like Susan Miller's The Good Mother. Nonetheless, this anxiety over the possible snatching away of the role of motherhood sometimes manifests itself as a self-consciously feminist variant of the same insistence on a specifically female sphere of nurturance, a strict emotional and social division of labor along gender lines, that Kristin Luker (1984, pp. 158-163) found in her study of Right to Life activists and that Beverly Harrison (1983, pp. 79-84) discerns among otherwise liberal opponents of abortion.
>
> Ironically enough, this fear of the loss of motherhood emerges contemporaneously with an obsessive male fear about the loss of mother, a belief that women's demand to use technology (abortion) to choose whether and when we will bear children presages the very end of nurturance (Willis, 1983). These oddly complementary fears rest upon and reinforce a static, polarized vision of gender roles and may spring from anxiety gen-

erated by changes in gender relations and expectations. Gallagher, "Eggs, Embryos and Foetuses: Anxiety and the Law," in Stanworth, M. (ed.), Reproductive Technologies: Gender, Motherhood and Medicine (1978).

39. Gusfield, "Moral Passage: The Symbolic Process in Public Designations of Deviance," 15 Social Problems 175, 175 (1965). Gusfield's study of the Prohibition movement, which he characterizes as a middle-class Protestant effort to reassert symbolic social power over Catholic immigrants with very different cultural systems, presents illuminative points of comparison with the contemporary drive against abortion and for fetal rights. See Luker, supra n. 37, at 158-215. Luker points out that anti-abortion activists "want a human life amendment to the Constitution (or a federal law) primarily in order to make a moral statement about abortion and only secondarily in order to prevent all abortions in practice . . . Their desire to see a moral affirmation of the wrongness of abortion outweighs their concern about the problems of implementing a human life law." Id at 234. Professor John Robertson's insistence on postbirth sanctions for "egregious" cases of deviations from a reasonable maternal standard of behavior would seem to fall into this category. See Robertson and Schulman, "Pregnancy and Prenatal Harm to Offspring: The Case of Mothers with PKU," Hastings Center Report, August/September 1987, p. 23.

40. See infra, section on "Coercion: Unjustifiable and Counterproductive" and ff.

41. Gusfield, supra n. 39, at 310.

42. See the position paper by Joan Bertin, "Reproductive Hazards in the Workplace," in this book. See also Stellman, J. M., and Henifin, M. S., "No Fertile Women Need Apply," in Hubbard, Henifin, and Fried (eds.), Biological Woman: The Convenient Myth (1982) and Williams, W., "Firing the Woman to Protect the Fetus," 69 Georgetown Law Journal 641 (1981).

43. See, e.g., "Babies Don't Thrive in Smoke-filled Wombs," advertisement in The Tablet, newspaper of the Roman Catholic diocese of Brooklyn, New York, June 20, 1987, p. 15.

44. "'Passive Smoking' Study Identifies Risks to Infants, Pregnant Women," Chronicle of Higher Education, Nov. 26, 1986, p. 7. Although as American Civil Liberties Union attorney Janet Benshoof said of the San Diego prosecution, "The next thing you know women will be prosecuted because they failed to leave their homes because their men were smoking, and smoking isn't good for fetuses." See also, Chambers, "Dead Baby's Mother Faces Criminal Charge on Acts in Pregnancy," New York Times, Oct. 9, 1986, p. A-22, col. 1.

45. Memorandum for Defendant, at 30, California v. Stewart (February 23, 1987) (No. M508197). San Diego B/P brief, p. 30. Professor Lawrence Tribe has pointed out that the singling out of "less powerful sub-groups of the population, imposing on them a higher standard of mandatory self-protection than the government imposes on the population as a whole" stigmatizes them as less responsible, and more in need of restraint by the state. Tribe, L., American Constitutional Law § 15-12 (1978).

46. Starr, The Social Transformation of American Medicine 391 (1982); Green, "Birthing Alternatives: A Matter of Choice and Turf," Medical World News, May 28, 1984, p. 42.

47. Griswold v. Connecticut, 381 U.S. 479 (1965).

48. Note, for example, the wave of books geared toward prospective parents: The Safe Pregnancy Book, Caring for Your Unborn Child, Nourishing Your Unborn Child. See, e.g., Goldstein and Crichton, "Books on Pregnancy, Childbirth and Childrearing: A Checklist," Publishers Weekly, April 22, 1983, at 34.

49. The most disturbing manifestations of such cultlike treatment of the fetus occur within the right-to-life movement: a Connecticut clergyman erects a 30-foot poster of a fetus atop his church and marches it through the streets as part of a local parade; another carries an aborted fetus about the country in a small coffin, displaying it as "Baby Choice"; an artist creates a "reliquary" of fetal remains in formaldehyde. Gallagher, "The Cult of the Fetus," unpublished manuscript (1986), p. 1.

50. Such attitudes can be as funny as they are horrifying. See, e.g., liberal columnist Nat Hentoff's lament: "If only the pro-choice left could think of the fetus as a baby seal, in utero." Hentoff, "How Can the Left Be Against Life," The Village Voice, July 16, 1985, p. 1.

51. Right-to-life propaganda deliberately invites and plays upon individual fearful identification with the fetus. See, e.g., Bergel, "When You Were Formed in Secret/Abortion in America," a double "Intercessors for America" booklet (1986), which gives readers a dreamy, personalized version of embryonic and fetal development: "This is an account about you and your life before birth . . . [In week one,] your mother had no idea you had 'nested' into her womb . . . [By week two,] . . . you were able to move with a delightfully easy grace in your buoyant world. By the end of the month you could swim." The booklet's flip side juxtaposes pictures of fetuses and of old people, exclaiming, "Who knows what the next domino might be, as we systematically cheapen human life and undermine the family structure . . . ?"

52. Hunter, "In the Wings: New Right Ideology and Organization," 15 Radical America 113, 132 (Spring 1981). See also Petchesky, "Fetal Images: The Power of Visual Culture in the Politics of Reproduction," 13 Feminist Studies 263 (Summer 1987).

53. Luker, supra n. 37, at 156-157.

54. See infra, section entitled "No Stopping Point."

55. See, e.g., "Out of Death, A New Life Comes," Newsweek, April 11, 1983, p. 65; Shrader, "On Dying More than One Death," Hastings Center Report, Feb. 1986, p. 12.

56. Davis, "Brain-Dead Mother Gives Birth Today," Oakland Tribune, July 30, 1986, p. A-1.

57. Davis, "Friends Say Baby Was Important to Teacher," Oakland Tribune, June 25, 1986, p. A-5. ("This is what Odette would have wanted.")

58. See Palmer, "Baby Born to Brain-Dead Woman Dies," Atlanta Constitution, August 17, 1986. ("The baby's mother . . . was found unconscious in a public restroom with a needle hanging from her arm, apparently the victim of a drug overdose . . . [T]he woman's husband, had asked earlier that her life support equipment be disconnected. But another man . . . claimed to be the baby's father and asked that she be kept alive.") See also Burke, "Baby at Center of Legal Controversy Dies After Birth," Hartford Courant, April 9, 1987. (A 1-pound, 14-ounce baby, delivered by a cesarean performed on a woman comatose since her jail-cell suicide four months earlier, died within 15 minutes of birth. A judge had refused a request from the mother of the woman that an abortion be performed, triggering intense publicity. The grandmother suffered a stroke attributed to stress.)

59. See Larson v. Chase, 47 Minn. 307, 50 N.W. 238 (1891).

60. "Family Is Awarded $380,000 in Cemetery's Burial Mix-up," New York Times, July 27, 1987, p. A-15, col. 6.

61. Robertson, "Procreative Liberty and the Control of Conception, Pregnancy, and Childbirth," 69 Virginia Law Review 405, 437 (1983). In his most recent article, Professor Robertson disclaims the label "fetal rights," insisting that "the offspring's right is contingent upon live birth" and that his position does not threaten the woman's procreative rights. Professor Robertson, however, defines a woman's procreative rights quite narrowly, limiting them to the right to abort before viability. Robertson and Schulman, "Pregnancy and Prenatal Harm to Offspring: The Case of Mothers with PKU," Hastings Center Report, August/ September 1987, p. 23.

62. In its 1973 opinion recognizing a woman's right to choose abortion, the Supreme Court laid out a three-part scheme of pregnancy. In the first trimester, the woman could, in consultation with her doctor, freely choose to terminate her pregnancy. In the constitutional balance established by the Court, during the early stage of pregnancy, a woman's right to privacy outweighs any possible state interest in regulating abortion. In the second stage of pregnancy, the government could impose certain regulations on abortion, but only those "reasonably related" to the protection of the woman's health. In the third stage of pregnancy, described by the court as following "viability" (see Nan Hunter, "Time Limits on Abortion," in

this book), the government is permitted to make regulations protective of its "compelling state interest" in the "potential human life" and may even forbid those abortions not necessary to preserve the life or health of the woman. Roe v. Wade, 410 U.S. 113, at 164-65.

63. See, e.g., Jefferson v. Griffin Spaulding County Hosp. Auth., 247 Ga. at 87, 274 S.E.2d at 458. For detailed legal examinations and rebuttals of these and other fetal rights arguments, see Gallagher, "Prenatal Invasions," supra n. 7; Rhoden, "The Judge in the Delivery Room: The Emergence of Court-ordered Cesareans," 74 California Law Review 701 (1986); Nelson, Buggy, and Weil, "Forced Medical Treatment of Pregnant Women: Compelling Each to Live as Seems Good to the Rest," 37 Hastings Law Journal 703 (1986); and Johnsen, "The Creation of Fetal Rights: Conflicts with Women's Constitutional Rights to Liberty, Privacy, and Equal Protection," 95 Yale Law Journal 599 (1986).

64. Robertson, supra n. 61, at 437; Shaw, supra n. 3, at 88.

65. Roe v. Wade, 410 U.S. 113, 164 (1973).

66. See, e.g., Collauti v. Franklin, 439 U.S. 379 (1979).

67. See infra, section on "Bodily Integrity and Self-Determination"; see also Gallagher, "Fetal Personhood and Women's Policy," in Sapiro (ed.), Women, Biology, and Public Policy 91, 97-100 (1985).

68. Roe v. Wade, 410 U.S. 113, 157-162 (1973).

69. 410 U.S. 113, 163-64.

70. 410 U.S. 113, 157-162. See also Thornburgh v. American College of Obstetricians and Gynecologists, 476 U.S. 747 (1986).

71. See Gallagher, "Prenatal Invasions and Interventions," supra n. 7, at 29-31.

72. In re Quinlan, for example, the New Jersey Supreme Court relied upon Roe v. Wade when it recognized a comatose and irreversibly ill woman's right to privacy and allowed removal of her artificial life support. 70 N.J. 10, 355 A.2d 647 (1976). For additional discussion of Roe v. Wade as a bodily integrity and self-determination precedent, see Gallagher, "Prenatal Invasions and Interventions," supra n. 7, at 16-18.

73. See, e.g., Bartling v. Superior Court, 163 Cal. App.3d 186, 209 Cal. Rptr. 220 (Cal. Ct. App. 1984).

74. See, e.g., Bouvia v. Superior Court, 225 Cal. Rptr. 297, 179 Cal. App.3d 1127 (1986); In re Conroy, 98 N.J. 329, 486 A.2d 1209 (1985); John F. Kennedy Hosp. v. Bludworth, 452 So.2d 921 (Fla. 1984).

75. Superintendent of Belchertown State School v. Saikewicz, 373 Mass. 728, 370 N.E.2d 417 (1977).

76. In re Quinlan, 70 N.J. 10, 54-55, 355 A.2d 647, 664, cert. denied 429 U.S. 922 (1976).

77. There are cases of forced treatment on incompetent pregnant women. In 1983, a Massachusetts Probate and Family Court judge appointed a temporary guardian for the purpose of consenting to a cesarean section for a woman who was two to three weeks postterm and had revoked her earlier consent to induction of labor, pulled out her intravenous needle, and attempted to leave the hospital. Immediate delivery was regarded as necessary for her safety and that of the unborn child. The judge found that "she is not competent to make decisions about her labor and delivery or about anything else. She has had psychotic episodes in the past and appears psychotic at this moment." The patient's husband was present and assented to the proposed surgery. Franklin Medical Center v. Linda T., Franklin County Probate and Family Court, July 28, 1983. [Unpublished order and findings in the possession of the author.]

Another case, reported in the psychiatric literature, involved court-ordered detention of a woman far advanced in pregnancy in order to ensure prenatal care and a medically safe delivery. Soloff, Jewell, and Roth, "Civil Commitment and the Rights of the Unborn," 136 American Journal of Psychiatry 114 (1979). Although the authors, nonattorneys, contend that the court's order rested on the fetus's "right to be well born," the decision seems to have turned instead on an implicit finding of incompetence. The patient had completely denied the pregnancy, had been admitted through an involuntary emergency commitment following a physical attack on her mother, and had been diagnosed as schizophrenic during a similar admission two years earlier.

78. 373 Mass. 728, 744, 370 N.E.2d 417, 427 (1977).

79. Fetal rights advocates point to cases authorizing blood transfusions to Jehovah's Witnesses as authority for forced medical treatment. But the Jehovah's Witness cases most frequently cited, Raleigh-Fitkin Memorial Hospital v. Anderson, 42 N.J. 421, 201 A.2d 537, cert. denied 377 U.S. 985 (1964) and In re President of Georgetown College, 331 F.2d 1000 (D.C. Cir. 1964), cannot carry the weight of the fetal rights proposals. Decided in 1964, neither case reflects the current emphasis on respect for individual self-determination and bodily integrity in the area of medical decision making. Also, Jehovah's Witness cases seem most frequently decided on the basis of the judge's conviction either that the patient is not really competent or that court-ordered treatment is an acceptable vehicle for honoring the patient's religious scruples without real risk to life. For a more detailed discussion of the blood transfusion cases, see Gallagher, "Prenatal Invasions," supra n. 7, at 34-37.

It may be that court-ordered transfusions to Jehovah's Witnesses--pregnant and not--have become such a well-established, seemingly acceptable pattern that many lower court judges will continue to authorize them on a regular basis. Whatever one may think of such a practice, it is vital that hospitals and judges recognize that the Jehovah's Witness blood transfusion precedents do not support other forced treatment on unconsenting patients.

80. See Jacobson v. Massachusetts, 197 U.S. 11 (1905) (upholding compulsory vaccination); Myers v. Commissioner of Corrections, 379 Mass. 728, 399 N.E.2d 452 (1979) (institutional security may override prisoner/patient refusal).

81. See, e.g., In re Conroy, 98 N.J. 321, 350-51, 486 A.2d 1209, 1224 (1985).

82. See, e.g., Bartling v. Superior Court, 163 Cal. App.3d 186, 195, 209 Cal. Rptr. 220, 225 (1984); In re Conroy, 98 N.J. 321, 352-53, 486 A.2d 1209, 1225 (1985).

83. In re Grady, 85 N.J. 235, 261, 426 A.2d 467 (1981).

84. See, e.g., Rogers v. Comm'r of Dept of Mental Health, 390 Mass. 489, 458 N.E.2d 308 (1983).

85. Winston v. Lee, 470 U.S. 753 (1985).

86. In addition to applying the explicit constitutional ban on "cruel and unusual punishment" of the Eighth Amendment, courts have proved reluctant to allow so-called organic therapies such as psychosurgery. See Kaimowitz v. Dep't of Mental Health, 2 Prison L. Rep. 433 (1973), which relied in part upon Roe v. Wade in prohibiting experimental psychosurgery on a mental patient. In 1985, the Supreme Court of South Carolina barred the castration of three convicted rapists who had chosen that sentence option over 30-year prison terms. State v. Brown, 284 S.C. 407, 326 S.E.2d 410 (1985).

87. Regan, "Rewriting Roe v. Wade," 77 Michigan Law Review 1569 (1979). Close analysis of treatment refusal cases belies the notion that third-party interests are determinative. A review of the treatment refusal cases collected at 9 American Law Reports 3d 1391 and in a New York case, Randolph v. City of New York, Supreme Court-Manhattan Trial Term, New York Law Journal, Oct. 12, 1984, p. 4 (doctor cannot be sued for honoring patient's refusal, but may be liable for deviation from standard of care if he does begin unconsented-to treatment), reveals that judicial deference to the allegedly compelling state interest in innocent third parties (most commonly dependent children whose parents refuse care) has been largely pro forma. While a number of opinions recite such an interest, almost all of those do so in order to stress the nonexistence of such a concern in the case at hand and cite that distinction as the basis for failing to follow the cases in which judges order the refused procedure. See, e.g., In re Melideo, 88 Misc.2d 974, 975, 390 N.Y.S.2d 523, 524 (Sup. Ct. 1976) and In re Brooks Estate, 205 N.E.2d 435, 442 (Ill. 1965). And in a number of treatment refusal cases in which there are minor children, judges honor the refusal anyway. See In re Osborne, 294 A.2d 372 (D.C. 1972) and Mercy Hospital v. Jackson, 489 A.2d 1130 (Md. 1985). Those relatively few cases

in which treatment is ordered over the objection of the patient seem to turn--not on the existence of minor children, although that factor may be cited as a justification--but on the judge's belief that the patient would accept treatment if the court shoulders the spiritual responsibility for the decision. See Gallagher, "Prenatal Invasions," supra n. 7, at 34-37.

88. Compare the lionization afforded Senator Jake Garn when he donated a kidney to his daughter with the cavalier appropriation of an unconsenting woman's body found in the forced cesarean cases. Greenhouse, "Garn to Give Kidney to Diabetic Daughter, 27," New York Times, Sept. 10, 1986, p. A-13, col. 1. It is hard to escape the implication that a woman's self-sacrifice in childbirth is not only taken for granted, but will be exacted by physical force if unforthcoming. Such coercion distorts our moral discourse. See Harrison, supra n. 21, at 198 ("Women, like men, may indeed elect self-sacrifice, but self-giving is no virtue apart from free choice.")

89. McFall v. Shimp, 127 Pitts. Leg. J. 14 (Allegheny Cnty., July 26, 1978).

90. Advocates of fetal rights attempt, inappropriately, to invoke two cases involving organ transplants from incompetent donors (one a child of seven and the other a resident of a state school for the "feeble-minded") as authority for performing forced cesareans on competent adult women. Reliance on such precedents may indicate an unstated assumption that pregnancy, or perhaps gender itself, renders a woman incompetent per se. See Gallagher, "Prenatal Invasions," supra n. 7, at 26-28.

91. Kolder, Gallagher, and Parsons, "Court-ordered Obstetrical Interventions," 316 New England Journal of Medicine 1192, 1195 (May 7, 1987).

92. Roe v. Wade, 410 U.S. 113, at 157-58, 162.

93. See, e.g., State of Minnesota v. Soto, 378 N.W.2d 625 (Minn. 1985).

94. Shaw, supra n. 3, at 95, quoting Smith v. Brennan, 31 N.J. 353, 157 A.2d 497 (1960). In fact, Shaw would go so far as to subject parents to liability for failure to abort a "defective" fetus. Id. at 110.

95. See, e.g., Renslow v. Mennonite Hospital, 40 Ill. App.3d 234, 351 N.W.2d 870 (1976).

96. See, e.g., Dunn v. Roseway, 333 N.W.2d 830, 833 (Iowa 1983): "The parents' loss does not depend on the legal status of the child; indeed the absence of the child is the crux of the suit." This interpretation of the wrongful death cases is strengthened by the fact that several jurisdictions that do not apply their wrongful death statutes to fetuses do allow recovery for tortious fetal death under parental claims of emotional distress--either as a distinct cause of action or as an element of damages in a prospective parent's own cause of action for negligent injury. See, e.g., Graf v. Taggert, 43 N.J. 303, 204 A.2d 140, 144 (1964); Endresz v. Friedberg, 24 N.Y.2d 478, 487 (1969); Sesma v. Cueto, 181 Cal. Rptr. 12 (Ct. App., 4th Dist. Div. 1, 1982); and Johnson v. Superior Court of Los Angeles County, 177 Cal. Rptr. 63, 65 (Ct. App., 2d Dist. Div. 5, 1981): "Whether or not a fetus is a person for wrongful death purposes is not determinative of whether a prospective parent can have a relationship with the fetus which would sustain a Dillon [emotional distress] action . . . It is . . . patently clear that a mother forms a sufficiently close relationship with her fetus during pregnancy so that its stillbirth will foreseeably cause her severe emotional distress."

97. A more explicit recognition that the injury to be compensated and deterred is one to the prospective parents could allow judges to jettison the dishonest and confusing biological entity theory of Dietrich v. Northampton, 138 Mass. 14 (1884), under which recovery for the tortious death or injury of a fetus is conditioned on judicial ability to posit somehow a distinct physical being. Bonbrest v. Kotz, 65 F. Supp. 138 (D.D.C. 1946) and subsequent opinions that allowed recovery by reasoning that the fetus biologically assumes a separate juridical personality at viability may have improved upon the earlier case's harsh result, but they remain trapped within the same ultimately unsatisfactory terms. Abandonment of the outmoded and strained reliance on concepts like viability and juridical personhood would also free judges to develop a more coherent, direct approach to preconception and prenatal injuries.

98. See "Right to Choose v. Byrne: Brief Amicus Curiae," 7 Women's Rights Law Reporter 285, 293-96 (1982).

99. For a discussion of this nonsubordination argument, see Gallagher, "Prenatal Invasions and Interventions," supra n. 7, at 23-28.

100. In re A.C. (No. 87-609) D.C. Court of Appeals, Nov. 10, 1987, slip op. at 6.

101. Statement by guardian ad litem for fetus, In re A.C., transcript of trial court proceedings at 70.

102. In re Madyun Fetus, No. 189-86 Daily Washington L. Rep. 2233 Co. B (Sup. Ct. D.C. Civ. Div. July 26, 1986).

103. Robertson, "The Right to Procreate and In Utero Fetal Therapy," 3 Journal of Legal Medicine 333, 356 (1984).

104. Robertson and Schulman, "Pregnancy and Prenatal Harm to Offspring: The Case of Mothers with PKU," Hastings Center Report, August/September 1987, pp. 23, 28.

105. Id. at 25.

106. Robertson, "Procreative Liberty," supra n. 61, at 447, n. 129. Woe to the woman who does not keep accurate track of her menstrual cycle and tidy records of sexual activity to disprove liability 18 or 21 years later.

107. Shaw, supra n. 3, at 111.

108. Shaw, quoted in a forum on "Maternal v. Fetal Rights," in At Issue, a newsletter for obstetricians and gynecologists, May 1986, p. 4.

109. Shaw, supra n. 3, at 81.

110. "Procreative Liberty," supra n. 61, at 454, n. 157.

111. Holder, "Maternal-Fetal Conflicts and the Law," 10 Female Patient 80, 86 (1985).

112. See Robertson and Schulman, "Pregnancy and Prenatal Harm to Offspring: The Case of Mothers with PKU," Hastings Center Report, August/September 1987, p. 23.

113. "Should Your Pregnant Patient Work in Her Third Trimester," Modern Medicine, Oct. 1982.

114. "Sex Infection Called Biggest Risk to Fetus," Chicago Tribune, June 13, 1981, p. 10, col. 1.

115. Gauthier, "Guidelines for Exercise during Pregnancy: Too Little or Too Much?" The Physician and Sportsmedicine, April 1986, p. 162.

116. Johnsen, "The Creation of Fetal Rights: Conflicts with Women's Constitutional Rights to Liberty, Privacy, and Equal Protection," 95 Yale Law Journal 599, n. 68 (1986).

117. Shaw, supra n. 3, at 89.

118. See, e.g., Copelon, "Danger--A Human Life Amendment Is On The Way," Ms., February 1981; Pilpel, "The Collateral Legal Consequences of Adopting a Constitutional Amendment on Abortion," 5 Family Planning/Population Reporter 44 (1976).

119. See, e.g., "Sex Infection Called Biggest Risk to Fetus," Chicago Tribune, June 13, 1981, p. 10, col. 1.

120. See, e.g., Robertson, "Procreative Liberty," supra n. 61, at 439-442.

121. See, e.g., Stallman v. Youngquist, No. 86-0315, slip op. (Ill. App. Ct., Feb. 11, 1987) in which the Court declared, "[I]n reality, the sought after litigation is not between child and parent but between child and parent's insurance carrier . . . "

122. Mary Sue Henifin, drafter of this book's position paper "Prenatal Screening," has suggested that legislation barring parental liability for "wrongful life" could also extend immunity for parental choices and behavior before conception and birth. That broader statute would provide:

1. No cause of action, whether civil, criminal, or for injunctive or other forms of equitable relief, may be brought against a parent of a child based upon the claim that:

A. The fetus was, is, or may in the future be injured by a potential parent's decision to refuse genetic testing or prenatal diagnosis

B. The child, based on information obtained from genetic testing or prenatal diagnosis, should not have been conceived, or if conceived, should not have been born or

C. The fetus was, is, or may in the future be injured by a potential parent's medical decisions, health status, employment, or personal habits prior to or during pregnancy or childbirth.

2. The provisions of this Act provide no defense to any party other than a parent.

Whether or not states choose to approach this issue through legislation, judges must be sensitized to the constitutional and policy issues, or the evolution of intrafamilial tort law may result in serious infringements on women's rights and health.

123. Warren, "Woman Is Acquitted in Test of Obligation to an Unborn Child," Los Angeles Times, Feb. 27, 1987, p. 1.

124. "Doctors Aren't Policemen . . . ," San Diego Tribune, Feb. 28, 1987, p. C-3, col. 1.

125. Carson, "Bill Offered Based on Pamela Rae Stewart Baby Case," San Diego Union, Mar. 7, 1987, p. A-3, col. 1.

126. "Girl Detained to Protect Fetus," Wisconsin State Journal, Aug. 16, 1985, p. 2, col. 3.

127. Kolder, Gallagher, and Parsons, "Court-ordered Obstetrical Interventions," 316 New England Journal of Medicine 1192-1196 (May 7, 1987).

128. Introductory Comments by Howard Brody, M.D., Ph.D., in "Medical Ethics Case Conference: Ethical and Legal Issues in a Court-ordered Caesarean Section," Medical Humanities Report issued by the Medical Humanities Program, Michigan State University (Winter 1984).

129. Jurow and Paul, "Cesarean Delivery for Fetal Distress Without Maternal Consent," Obstetrics and Gynecology 596 (1984).

130. Chambers, "Dead Baby's Mother Faces Criminal Charge on Acts in Pregnancy," New York Times, Oct. 9, 1986, p. A-22, col. 1.

131. Bonavoglia, "The Ordeal of Pamela Rae Stewart," Ms., July/August 1987, pp. 92, 201.

132. Kolder, Gallagher, and Parsons, "Court-ordered Obstetrical Interventions," 316 New England Journal of Medicine 1192-1196 (May 7, 1987) at 1195.

133. See Gallagher, "Prenatal Invasions and Interventions," supra n. 7, at n. 204.

134. See, e.g., Winston v. Lee, 470 U.S. 753 (1985).

135. See, e.g., People v. Medina, 705 P.2d 961 (Ariz. 1985).

136. See, e.g., In re Grady, 85 N.J. 235 (1981).

137. See Gallagher, "Prenatal Invasions," supra n. 7, pp. 20-23.

138. Jefferson v. Griffin Spaulding County Hosp. Auth., 247 Ga. 86, 274 S.E.2d 457 (1981).

139. 70 Medical Association of Georgia 451 (1981).

140. Flanigan, "Mom Follows Belief, Gives Birth in Hiding," Detroit Free Press, June 28, 1982, at 3-A.

141. "Medical Ethics Case Conference: Ethical and Legal Issues of a Court-ordered Caesarean Section," Medical Humanities Report issued by the Medical Humanities Program, Michigan State University (Winter 1984).

142. Id.

143. Unpublished opinion, No. 84-7 500060 (Superior Court of Benton County, April 20, 1984).

144. Rhoden, "The Judge in the Delivery Room: The Emergence of Court-ordered Cesareans," 74 California Law Review 1951, 1960 (1986).

145. "Patient Choice: Maternal-Fetal Conflict," Committee on Ethics, American College of Obstetricians and Gynecologists, Aug. 11, 1987.

146. See, e.g., Corea, The Hidden Malpractice (1984); Arms, Immaculate Deception (1975); Cohen and Estner, Silent Knife (1983).

147. Shabecoff, "Panel Says Caesareans Are Used Too Often," New York Times, Nov. 3, 1987, p. C-5, col. 1.

148. Id.

149. The Cesarean Birth Task Force, "National Institutes of Health Consensus Development Statement on Cesarean Childbirth," 57 Obstetrics and Gynecology 537 (1981).

150. See, e.g., "Dystocia Most Cited Indication for Cesarean; Held Overdiagnosed," Ob/Gyn News, Aug. 15-31, 1984; "Calls for Stronger Peer Review to Cut Rate of Cesareans," Ob/Gyn News, June 1-14, 1984.

151. Knox and Karagianis, "Caesarean Births: High Rates, Impassioned Debate," Boston Globe Magazine, Oct. 21, 1984, at 10, 11. Article is continued in issue of Oct. 28, 1984 at 13.

152. Pearson, "Cesarean Section and Perinatal Mortality: A Nine-Year Experience in a City/County Hospital," American Journal of Obstetrics and Gynecology 156 (Jan. 15, 1984).

153. Sociologists have found that "middle and upper class women are at higher risk for cesareans than lower class women." Hurst and Summey, 18 Social Science and Medicine 621 (1984) at 621. A study of California vital records data for the years 1978-80 revealed much lower average cesarean rates for hospitals staffed with prepaid or salaried physicians than for private hospitals. "Publishing Cesarean Rates May Aid M.D. Accountability," Ob/Gyn News, May 15-31, 1984. The PIRG report and other critical reviews suggest that some cesareans can be attributed to the fact that they are "more profitable and more convenient" for doctors. Shabecoff, supra n. 147. Cesarean delivery has been estimated to be 47 to 85 percent more expensive than vaginal birth. Hurst and Summey, supra, at 625. "The length of stay for a Caesarean section is approximately double the length of stay for a vaginal delivery, with a corresponding length of stay for the infant." Marieskind, H., An Evaluation of Caesarean Section in the United States 64 (1979). Each 1 percent increase in the cesarean section rate adds more than $54 million to the cost of hospital care in the United States. "Doctor Says Increasing Rate of Cesareans is Alarming," Boston Globe, Dec. 21, 1984, p. 42, col. 1.

154. Gilfix, "Electronic Fetal Monitoring: Physician Liability and Informed Consent," 10 American Journal of Law and Medicine 31, 42 (1984).

155. Knox and Karagianis, supra n. 151, at 58.

156. Brackbill, Y., Rice, J., and Young, D., Birth Trap: The Legal Low-Down on High-Tech Obstetrics 25 (1984).

157. "NIH Consensus Statement," supra n. 149, at 542.

158. Marieskind, supra n. 153, at 59.

159. Gilfix, supra n. 154, at 42-43.

160. Knox and Karagianis, supra n. 151, at 56.

161. Rothman, B.K., "Case Studies Commentary," Hastings Center Report, February 1986, p. 25.

162. Marieskind, supra n. 153, at 96-97. One Boston doctor interviewed in Knox and Karagianis, supra n. 151, lamented that "[O]bstetrical trainees can go through a three-year residency without spending an entire labor at a woman's side."

163. "Dr. Jonas Points Out Why Lingering Fears Over VBAC Exist Among Physicians, Patients," American College of Obstetrics and Gynecology Newsletter, April 1987, 1, at 2.

164. Gilfix, "Electronic Fetal Monitoring: Physician Liability and Informed Consent," 10 American Journal of Law and Medicine 31, 45 (1984).

165. Medical Humanities Report, supra n. 141, at 4.

166. Sosa et al., "The Effect of a Supportive Companion on Perinatal Problems, Length of Labor, and Mother-Infant Interaction," 303 New England Journal of Medicine 597 (Sept. 11, 1980); "Psychological Factors, External Stress May Necessitate Cesarean," Ob/Gyn News, Nov. 15-30, 1982.

167. Applebaum and Roth, "Patients Who Refuse Treatment in Medical Hospitals," 250 Journal of the American Medical Association 1296, 1301 (Sept. 19, 1983). See also Goleman, "Physicians May Bungle Key Part of Treatment: The Medical Interview," New York Times, Jan. 21, 1988, p. B-10, col. 1.

168. Id. at 1298. See also Andrews, "Taking Care of the Doctor-Patient Relationship," 1981 American Bar Foundation Research Journal 251, 260-265 (Winter 1981).

169. Applebaum and Roth, supra n. 167, at 1301.

170. See, e.g., Lewin, "Hospitals Pitch Harder for Patients," New York Times, May 10, 1987, Sect. 3, p. 1.

171. Interview with Liz Werthan of Choice, July 21, 1987. The pamphlet included information on hospital practices such as fetal monitoring and episiotomies as well as cesarean sections.

172. Gilgoff, "State to Investigate High C-Section Rate," Newsday, Nov. 20, 1987, p. 17.

173. Sullivan, "New York Starts Effort to Slow Cesarean Rate," New York Times, Nov. 24, 1987, p. B-4.

174. "The Rising Rate of Caesarean Sections," Newsday, May 1, 1987, p. 3. ("More than a quarter of all ob-gyn specialists are sued for malpractice at least three times during their careers . . . ")

175. See, e.g., "$9,000,000 Verdict--Medical Malpractice--Alleged Negligent Failure of Defendant Ob/Gyn to Timely Respond to Signs of Fetal Distress During Labor . . . ," New England Jury Verdict Review and Analysis, Feb. 1987, p. 1; Fox, "Court's $6 Million Award Record for Brain-Damage Case," New York Law Journal, March 3, 1987, p. 1.

176. "City Pays $147 Million in Malpractice Claims," New York Amsterdam News, Dec. 5, 1987, p. 52.

177. See, e.g., Bellotti, Van de Kamp, Thornburg, Mattox, Brown, and LaFollette, "An Analysis of the Causes of the Current Crisis of Unavailability and Unaffordability of Liability Insurance," prepared for the National Association of Attorneys General (May 1986); Boyd, "Reagan to Seek Changes in Laws to Limit Awards in Liability Suits," New York Times, April 1, 1986, p. A-23, col. 1; "Malpractice: A Crisis of Doctors' Making," National Law Journal, Sept. 16, 1985; Kristof, "Insurance Woes Spur Many States to Amend Law on Liability Suits," New York Times, March 31, 1986, p. A-1, col. 2; "Statement of James M. Shannon, Attorney General, Commonwealth of Massachusetts, on behalf of the National Association of Attorneys General, before the Senate Antitrust, Monopoly, and Business Rights Subcommittee on S. 80, a Bill to Repeal the McCarran-Ferguson Act, and for Other Purposes," Feb. 18, 1987.

178. See Eisenberg, "A Doctor on Trial," New York Times Magazine, July 20, 1986, p. 26.

179. Lehman, "The Cost of Doing Business," Boston Globe, Aug. 26, 1985, p. 43, col. 2.

180. Marieskind, H., supra n. 153, at 82.

181. "Are Primary Cesarean Section Rates Too High?" Ob/Gyn News, Nov. 15, 1981.

182. Slatalla, "The Rising Rate of Caesareans," Newsday, May 11, 1987.

183. Bowes and Selgestad, "Fetal versus Maternal Rights: Medical and Legal Perspectives," 58 Obstetrics and Gynecology 209, 211 (1981).

184. American College of Obstetricians and Gynecologists, "Statement on Court-ordered Cesarean for Dying Woman," Nov. 24, 1987. See also Rhoden, supra n. 144, at n. 284 on fears of malpractice playing a role in a New York City case.

185. For an excellent, detailed discussion of this issue, see Young (ed.), Proceedings of the Forum on Malpractice Issues in Childbirth: 1985, International Childbirth Education Association (1985).

186. "Patient Choice: Maternal-Fetal Conflict," Committee on Ethics, American College of Obstetricians and Gynecologists, Aug. 11, 1987.

187. "Maternal and Child Health Data Book," Children's Defense Fund (1987). Conditions for the poor and minorities are worse. In 1986, 34 of the nation's 54 largest cities posted infant death rates above the national average. "Infant Death Rates," Newsday, Sept. 10, 1987, p. 10.

188. Knox, "Disparity Widens in Infant Deaths," Boston Globe, May 28, 1985, pp. 1, 10.

189. Holder, "Maternal-Fetal Conflicts and the Law," 10 Female Patient 80, 90 (1985).

190. Curran-Downey, "'No Care' Babies a Major County Problem, Say Officials," San Diego Union, Feb. 28, 1987, p. B-3, col. 1.

191. See, e.g., "Baby Placed in Foster Home; Doctor Claims Prenatal Abuse," Des Moines Register, April 3, 1980, p. 11-a.

192. Robertson, "The Right to Procreate," supra n. 2, at 356.

193. "Wisdom in WIC," editorial, New York Times, Oct. 14, 1987.

194. See "Fathers Smoking May Harm Fetuses," New York Times, Jan. 18, 1983, p. C-2, col. 1.

195. "'Passive Smoking' Study Identifies Risks to Infants, Pregnant Women," Chronicle of Higher Education, Dec. 26, 1986, p. 7.

196. See Shaw, supra n. 3, at 74.

197. "For Alcoholic Women, a Place to Go," New York Times, June 14, 1987, p. 51, col. 1.

198. Id.

199. Flanigan, "Mom Follows Belief, Gives Birth in Hiding," Detroit Free Press, June 28, 1982, p. 3-A.

200. Rhoden, supra n. 144, at 1960.

201. Declaration of Gladden V. Elliott, M.D., President, California Medical Association, appendix to Memorandum for Defendant, California v. Stewart (Feb. 23, 1987) (No. M508197).

202. "Patient Choice: Maternal-Fetal Conflict," Committee on Ethics, American College of Obstetricians and Gynecologists, August 11, 1987.

203. "Abused-Child Deaths Up 29% in Reports of 24 States for '86," New York Times, Jan. 27, 1987, p. A-15, col. 3. ("'Caseloads are getting too high,' she said. 'Some families who might have been helped by treatment didn't get it. Death is one of the outcomes.'")

204. "'Addicted' Fetus Sparks Court Battle," Chicago Tribune, April 9, 1984.

205. See Barry, "Quality of Prenatal Care for Incarcerated Women Challenged," VI Youth Law News 6:1-4 (November/December 1985).

206. Rindaldo, "Court Case Weighs Rights of Pregnant Woman, Fetus," Staten Island Advance, March 30, 1987, p. A-10, col. 1.

207. Declaration of Lydia Roper, appendix to Memorandum for Defendant, California v. Stewart (Feb. 23, 1987) (No. M508197).

208. Schacter, "Help is Hard to Find for Addict Mothers," Los Angeles Times--San Diego County, Dec. 12, 1986, p. 1.

209. "Prenatal Program for Pregnant Addicts," correspondence with Suzanne Halpin of New York City Health and Hospitals Corporation, dated March 14, 1986. The New York City programs are modeled on the PAAM Program developed at New York Medical College. See Brotman, Hutson, and Suffet, Pregnant Addicts and Their Children: A Comprehensive Care Approach (1985).

Commentary

"HARD CASES" AND REPRODUCTIVE RIGHTS

Jeannie I. Rosoff

The issues raised in the position papers are deeply troubling, for a variety of reasons. They deal with "hard cases"--like late abortion and fetal abuse--that are visible and numerous enough to cause considerable public discomfort. These hard cases evoke sharp tensions between various rights and values--tensions that may never be resolved to anyone's satisfaction. Indeed, their preferred resolution will be viewed, in most cases and by most people, as the lesser of two evils. The hard cases call into question the appropriateness of society's responses and the priorities given to different interventions. Finally, and I believe this to be crucial, they affect subgroups of the population that, in some manner and to some degree, are not considered like "us," the majority. These subgroups may be composed of individuals whose mental or physical capacity and functioning may differ from the norm, and who will be viewed sympathetically, but perhaps condescendingly, or they may consist of persons whose behavior will be viewed by the great majority of us as incomprehensible at best or at least irresponsible--if not quasi criminal.

We identify with the needs and the rights of a woman having a second-trimester abortion, because the fetus she carries is found to be affected with Tay-Sachs disease or even Down syndrome. She is likely to be the object of our sympathy or, at least, our empathy. Her choice, if we are to believe public opinion polls, is one most of us would make. However, it must be kept in mind that the number of abortions performed after the 13th week following the last menstrual period (LMP) has ranged for the last ten years or so between 130,000 and 160,000. Of this number, probably less than 5000 can be attributed to genetic reasons. What then of the others? The nationwide legalization of abortion and the increased availability of abortion services have contributed greatly to a reduction in the proportion of all abortions that are performed after 13 weeks LMP. That proportion fell from 18 percent in 1972 when abortion was legal in only a few states to 10 percent in 1977, but it has remained at roughly that level ever since. Some women still have difficulty obtaining an abortion in a timely fashion, and some other factors such as age, race, and marital status play a role in delaying a pregnancy termination. However, careful research shows that the most important factors are individual, personal characteristics such as psychological conflict, maturity, or moral quandary, and biological factors such as a history of irregular menses. The needs of the women who fall into these categories are no less intense than those of the majority, and their rights are just as valid, but they fall outside the normal experience. Therefore, we tend to believe that, if we try hard enough, if we educate enough, and if we counsel enough, we can make the "problem" of late abortion--and society's discomfort--eventually disappear.

The inability or unwillingness of a woman to change her behavior and subjugate her own needs to those of a fetus she has presumably decided to carry to term evokes conflicts between several sets of rights: those of the woman in terms of her autonomy and bodily integrity, those of the fetus who is rapidly approaching birth and biological independence, and those of the onlookers--professionals, family members, society at large--who have a justifiable concern for the health and welfare of the infant about to be born. The cases of women who refused, but in some instances have been forced, to undergo cesarean sections, again, readily engage our understanding. Many people would be willing (perhaps should be willing) to undergo a major operation on behalf of a loved one, but many would not. Certainly, all of us would object to being coerced to do so, but what of a woman who consistently and despite urgent warnings persists in behaviors almost certain to harm her baby *in utero*? Those dealing with such a woman--physicians, social workers, even her family--will alternate between compassion and anger. They will try to counsel and cajole, and eventually will seek to threaten and coerce her into acceptable behavior; namely, that of a "good" mother, selfless and willing to subordinate her needs to those of her infant. The frustration and anger are understandable, for many of these women will not only persist in their behavior, but they will behave in such manner repeatedly. They will give birth to baby after baby who comes into the world addicted to drugs, mentally or physically impaired, or infected with AIDS. Most of them lead lives so chaotic and out of control that they are unable to practice contraception effectively but, when they do become pregnant, will not consider having an abortion. Many express deep religious feelings, although to the onlooker, the lives they lead would hardly reflect their beliefs. These are women caught in turmoil and entrapped in behavior they are unable to control. For some, the sense of renewal and hope that the coming of a baby may inspire will provide sufficient motivation to curb previously uncontrolled or uncontrollable habits. For many more, the pregnancy will just aggravate destructive, perhaps deadly, behavior that, after all, is first directed against the self.

We may find it easy to believe that society should intervene in the lives of these women, and forcibly protect their yet-unborn babies at the cost of doing violence to their right of self-determination and their bodily integrity. Many professionals who work with pregnant women infected with the AIDS virus will urge and, indeed, pressure them to have an abortion. Yet AIDS-infected women stand only a 30 percent to 50 percent chance of passing on the virus to their infants, who then *may* or *may not* develop the disease. On the other hand, some women who discover during pregnancy that their babies *will* be born with a deadly condition or one certain to cause severe pain or impairment, will decide to carry the pregnancy to term anyway. Few would suggest that they be forced to have an abortion or be pressured to do so even though that baby too is likely to suffer and to have to depend on society's resources.

We recognize and respect the fact that some people's religions or moral beliefs will require them to have their baby, even if they know in advance that the baby will die soon after birth, or that their own lives will be radically changed by the birth of a severely impaired infant who may need years and years, even decades, of constant and arduous care. We also recognize that some people are more nurturing than others, or have more internal resources in dealing with illness and adversity than others. Also, we know that virtually all will care for and love a less than "perfect" infant once it is born. Nevertheless, few parents would freely choose, if they had the means to make that determination in advance, to have a severely disabled baby. Prospective parents do not seek to have flawless infants; they just wish that these infants will be born under the best conditions they, as parents, can secure. They will usually delay having children until they are married, until they have enough financial resources to provide for them, or until they feel emotionally ready. Also, some will have an abortion if these conditions are not met. The availability of the new genetic screening technologies simply enlarges their control over the circumstances under which they will choose to become parents.

It is understandable that some may fear that the availability and the wide use of genetic screening technologies might serve to reinforce existing prejudices against disabled persons. Yet, whether an individual is discriminated against or stigmatized because his/her mental or physical capacity is impaired does not depend on whether the impairment is the result of faulty genes or an automobile accident, or whether it was acquired *in utero* or in adolescence. Prejudice is blind and must be fought on its own merits (or demerits). Fear of prejudice, however, should not become the basis for the denial of medical techniques that enlarge human freedoms.

Finally, much has been said about what we--society--should do: Counsel and provide support to individuals so that they make more appropriate decisions; redirect priorities away from investment in high-tech services towards greater investment in maternal and child health programs. There is no question that those things should be done and that they are not done adequately now. However, I feel obliged to point out that there is no way that a societal trade-off between genetic screening services and better maternity care for the poor could be made. Private health expenditures are not offset by public expenditures or vice versa. There are no "line items" in medical budgets, not even in societies with universal insurance coverage and much greater government control over health care expenditures and allocations than in the United States. Better maternity care for the poor must be fought for and obtained because it is good and right, not because it could or should be offset by some other cost. In the same way, the rights of individuals must be defended and secured, not because these individuals are in the mainstream and "worthy" but because the "unworthy," by whatever definition we apply, deserve those rights just as we do. In defending them against society's intrusion, we defend ourselves.

Commentary

DISABILITY RIGHTS PERSPECTIVES ON REPRODUCTIVE TECHNOLOGIES AND PUBLIC POLICY

Deborah Kaplan

My comments are founded in a social movement that promotes radical changes in the way we as a culture think about disability and disabled people. It is often referred to as the disability rights movement. Its major leaders are disabled themselves, and we are engaged in very broad social reform.

One of our goals concerns definitions, cultural and personal. We seek to change our cultural identity, and along with that effort, we are constantly struggling with self-definition and self-perception. The way we are perceived by others directly affects the extent to which society is willing to make room for us to live fully and freely. The manner in which we perceive ourselves has a great deal to do with whether we are willing to speak up for ourselves, and assert our rights to be a part of society and to contribute to the broader good.

Disabled Women and Reproductive Rights

In this context, let us look at some of the implicit definitions or perceptions of disability and disabled people in the public discourse about reproductive rights issues. It is significant and revealing that disabled women are very rarely acknowledged as a distinct group with their own reproductive rights issues. Historically, we have been regarded as "perpetual children" or "pure" because of our disabilities. This unrealistic attitude denies us our sexuality.

It also ignores the fact that many disabled women are likely to experience difficulties in finding an effective and safe method of contraception, and that many disabled women will have "problem" pregnancies with medical complications, possibly resulting in late abortions. The attitudes of medical professionals towards disabled women and reproduction have often also been based on myth rather than fact.

An illustration of these realities for disabled women can be found in the person of the adopting mother in the "Baby M" litigation concerning "surrogate" mothering. She has multiple sclerosis. News reports have indicated that she decided to enter into the agreement with the birth mother of "Baby M" because of her fears of pregnancy's medical and physical consequences for her.

Although available statistics indicate that as high a proportion as 15 percent of our population may have some type of disability, little if any research has been conducted on disability and reproductive issues. We do not know, for instance, whether pregnancy has any effect on multiple sclerosis. Physicians have counseled women with multiple sclerosis not to have children without much scientific evidence to support their advice; they have so counseled women with many other types of disabilities, as well, because it has seemed "obvious" that disabled people would not make good parents and that they need to be protected by others. Although there is anecdotal evidence from disabled women that would challenge these assumptions, research is still not conducted. Even though disabled women make up a significant part of the larger group of women in their childbearing years, and there is reason to expect that, as a group, they experience more difficulties than average in all aspects of reproduction, their problems are regarded as not significant or interesting enough to study. Old attitudes die hard, and the disability rights movement has a long way to go in challenging society's perceptions in this area.

Disability and Prenatal Screening

In the context of prenatal screening, the implicit and explicit perceptions about disability and disabled people are worth examining. Literature and promotional materials focus on the benefits to prospective parents of knowing that they will have a "normal, healthy" baby. Implicit is that a child with a disability is neither normal nor healthy. Words such as "defective" and "deformed" abound. At a recent conference in San Francisco of a national organization of abortion providers, a physician referred to disabled newborns as "gorks."

Needless to say, the disability rights movement does not seek to perpetuate such a perception or definition of disability. It points to the environment, rather than the individual, as the source of the problem. A person who uses a wheelchair is not "hopelessly confined" in a wheelchair-accessible building. A deaf person is not isolated from the outside world by his or her deafness so much as by the unavailability of sign language interpreters, telecommunication devices for the deaf (TDDs), and closed (or open) captioning on television. Disabled people are unemployed in larger numbers proportionately than any other group not for lack of talent or skills, but because of employer fears and lack of appropriate educational opportunities. The disability alone does not determine a person's ability to function or succeed. It is possible for an environment and society to support and accommodate people with disabilities, and to recognize that it is "normal" for a certain percentage of the population to be disabled.

Social Policy Towards Disability

Our society's social policies towards disability and disabled people reflect a national schizophrenia; we cannot seem to make up our minds what we want to do. Historically, the old line has been that disability is inherently tragic, and that disabled people are a burden on their families and society. Large-scale institutions were created as a response to this perception. Disabled people did not hold any unique legal rights, and indeed, had fewer rights than others. "Ugly" laws were enacted, prohibiting disabled people from being seen in public. Many disabled people were forcibly incarcerated and/or sterilized. Disabled children were not entitled to a public education. The Jerry Lewis Muscular Dystrophy Telethon is a holdover from that history and serves as evidence that those attitudes are still very much alive.

The new policies are represented by state and federal statutes prohibiting discrimination based on disability, and stipulating that disabled people are entitled to equality of opportunity and should be accommodated by social service systems, schools, and employers in order to participate fully in society. Examples are the Rehabilitation Act of 1973 (29 U.S.C. Sections 701 *et seq.*), The Education for All Handicapped Act of 1974 (20 U.S.C. Sections 1401 *et seq.*), and the Developmentally Disabled Assistance and Bill of Rights Act of 1975 (42 U.S.C. Sections 6001 *et seq.*).

In enacting these new laws and related others, Congress was specific about its perceptions and intent:

> The Congress finds that . . . it is essential . . . to assure that all individuals with handicaps are able to live their lives independently and with dignity, and that the complete integration of all individuals with handicaps into normal community living, working and service patterns be held as the final objective. White House Conference on Handicapped Individuals Act, 29 U.S.C. Section 701n.

Also typical of the new social policy is the decision of the California Supreme Court in *In re Marriage of Carney* 598 P.2d 36 (1979), a custody dispute in which the mother contested the father's custody of their two sons based on his disability, quadriplegia:

> . . . it is erroneous to presume that a parent in a wheelchair cannot share to a meaningful degree in the physical activities of his child, should both desire it. On the one hand, modern technology has made the handicapped increasingly mobile, as demonstrated by William's purchase of a van and his plans to drive it by means of hand controls . . .
>
> At the same time the physically handicapped have made the public more aware of the many unnecessary obstacles to their participation in community life. Among the evidence of the public's change in attitude is a growing body of legislation intended to reduce or eliminate the physical impediments to that participation . . .
>
> Both the state and federal governments now pursue the commendable goal of total integration of handicapped persons into the mainstream of society . . . No less important to this policy is the integration of the handicapped into the responsibilities and satisfactions of family life, cornerstone of our social system. *Id.* at 43-45.

Much of our society's thinking about prenatal screening is based on the old line and the stereotypes about disability that accompany it. It should be recognized that these issues invite a quick, emotional, gut-level response. In this case, such a response evokes our cultural aversion to disability, and our fears of becoming disabled or of having a disabled child. The often irrational thinking that accompanies these fears leads people to ignore the facts: that there are many severe disabilities that cannot be predicted through prenatal screening and that prenatal screening cannot predict the severity of the disability. In fact, many of the genetic conditions that are now included in prenatal screening are never severe. The inability to distinguish between anencephaly and mild spina bifida, for instance, results in a tendency to paint the picture with a very broad brush of hopelessness and tragedy. In spite of much evidence in the world around us that living with a moderate or mild disability is not necessarily tragic or burdensome, prenatal screening may require people to behave as though all disabilities are by their nature terrible.

Prenatal screening encourages people to overlook the social policy that is based on the fact that disability can be managed through technology, early intervention programs for very young disabled children, social support systems, and social change. The availability of prenatal screening influences social policy because, up until the present, the literature and materials of its proponents have unabashedly reinforced and strengthened negative attitudes towards disability and disabled people. It has also created the public misperception that soon we will be able to prevent almost all disabilities, no matter how unrealistic that myth may be. As a result, women and couples who go through prenatal screening may become less accepting of disabilities and disabled people. As prenatal screening becomes more widespread and available for younger women, the potential social impact should not be ignored.

Reconciling Disability Rights and Reproductive Rights

How can we talk about or take advantage of prenatal screening without further stigmatizing disabled people? Can we? The point of this discussion is not to make a statement about prenatal screening itself, but rather to focus on the way that we talk about and promote it from the perspective of disability rights and the reproductive rights of disabled people. The relative lack of involvement of disabled people in this discussion to date is particularly startling when one considers that prenatal screening involves our perceptions of disability and disabled people in a central way. More broadly, disabled women are a significant group with a stake in specific and unique reproductive rights issues. One way to start making positive changes is to bring more disabled women and disabled people into the public policy debate.

The key to change, however, lies in attempting to reconcile the way we define and perceive reproductive rights issues with the new public policies that recognize and promote the civil rights of disabled people. It should be possible to talk about prenatal screening without assuming that disability is tragic, painful, and burdensome at all times for all people. Those who are involved in this field as theorists, practitioners, or consumers could, with few exceptions, stand to become better acquainted with existing services for disabled people and with the reality of having a disability. How this might be accomplished could be the subject of a conference in itself. It will not be accomplished without involving and becoming involved with disabled people at a much more intimate level than the current relationship of occasional contact, if any. Part of the process of change will of necessity involve enhancing the value given to disabled people and their lives.

"Wrongful Life" and Disability Rights

As a personal injury plaintiffs' lawyer, I have a few comments about the "wrongful life" and "wrongful birth" lawsuits that are mentioned in the position papers. These lawsuits are based on the legal theory that the physician or other medical professional acted negligently in not informing the mother about the availability of prenatal screening, that if she had known in advance of her child's disability she would have had an abortion, and that she or the child has been injured as a result.

These cases are different from other personal injury lawsuits in that other tort cases involve a preexisting human being before the injury. Here, the injury is caused by an omission well before birth, and the injury is the birth itself. In the complaint, the document that initiates the lawsuit, it is essential that the mother alleges that, if she had been informed of the fetal disability, she would have had an abortion.

These lawsuits have a potential impact on the individual level as well as the social level. For the individual disabled child, often struggling to overcome a negative self-image and stigmatized by others, eventual knowledge of the lawsuit could be devastating, especially the fact that a public document declares that she/he should never have been born. On a broader social level, these lawsuits further the stereotype that being disabled is a completely hopeless, pathetic situation. The only redeeming feature of these lawsuits is that they potentially bring resources to families with disabled children that are not routinely provided, but nevertheless are needed.

Conclusion

No one can seriously dispute that the world is an inequitable place for disabled people. For those disabled people and their supporters who are actively engaged in challenging and reforming those inequities, the issues involved in the debate over reproductive rights policy are directly relevant. Disabled women are engaged in personal and social struggles over their own sexual freedom and reproductive choices. Although they are more likely as a group to have their reproductive choices limited by their disabilities or by other people, their problems have yet to be recognized by reproductive rights advocates. Prenatal screening and the way that we talk about it involve our fundamental attitudes towards disability and disabled people; the challenge is to involve disabled people in developing a vocabulary and ethic that complement the activities of the disability rights movement.

As a first step, the reproductive rights community needs to recognize that consideration of disability issues is central, not peripheral, to the discussion. Disabled women should be present to speak of their own experiences. Disabled theorists and activists should be engaged in the debate. The infusion of these new voices and perspectives will deepen the analysis, and make it more real and vital.

Commentary

THE FETUS IS A PATIENT[1]

Alan R. Fleischman

As a pediatrician and neonatologist, I am confronted on a daily basis with issues concerning fetal viability, fetal therapy, and genetic screening. I am often asked by families to assist them in decision making prior to, during, and after a pregnancy. Most of my commentary will focus on the issue of the fetus as a patient and fetal therapy. At the end, I will add some brief comments on genetic screening and the problem of the time limits on abortion.

The field of fetal therapy includes fetal assessments, fetal treatments, and interventions to enhance the well-being of the fetus. It can be divided into three areas: first, treatment of the mother in order to assist the fetus; second, direct medical and surgical treatments of the fetus; and third, the use of cesarean section in order to improve fetal outcome. In each of these areas, a number of different parties have interests in the outcome: the pregnant woman, the father of the fetus, the fetus itself, physicians and other health care personnel, and the society. We have heard a great deal about the legal analysis of these problems utilizing the language of "rights" to identify who should be able to make decisions concerning fetal therapy, and whether or not there ought to be limitations on those choices. An ethical analysis, different from a legal analysis, might attempt to determine what is right rather than to determine whose rights will prevail.

As a physician, I utilize two leading principles of bioethics in these situations. The first principle is known as "respect for persons." This principle incorporates two ethical convictions: that individuals should be treated as autonomous agents and that persons with diminished autonomy are entitled to protection.[2] This principle supports the right of the woman to determine what happens to her body, including the fetus, which is undeniably a part of her body. The "respect for persons" principle might also be viewed as granting the fetus protection because of its diminished autonomy. This application of the principle is far more controversial. Some might argue that the fetus is the type of entity that does not possess autonomy, and therefore, the concept of diminished autonomy ought not to apply. Alternatively, one could suggest that the fetus has the potential to develop autonomy and that this future potential ought to be respected. According to the "respect for persons" principle, the physician has a clear moral obligation to respect the autonomy of the mother and a somewhat less clear obligation to protect the fetus with its diminished or potential autonomy.

The second applicable precept of bioethics is the principle of "beneficence." This principle states that persons are treated in an ethical manner not only by respecting their decisions and protecting them from harm, but also by making efforts to secure their best interests or well-being. Beneficent actions attempt to maximize possible benefits and minimize possible harms. Specifically in the case of fetal therapy, the physician who acts according to this principle seeks to protect and promote the best interests of both the mother and the fetus. This requires an objective assessment of the various therapeutic options, and the implementation of those that protect and promote the best interests of both parties. The outcome should secure the greatest balance of benefits over harm.[3]

Physicians are comfortable with this approach to weighing benefits and risks, and offering therapeutic choices that have a favorable balance. It is far more complex to weigh benefits and risks for two patients simultaneously, but that has been the nature of obstetric and midwifery practice for centuries. Health care practitioners who treat pregnant women have always realized that there were ethical concerns and that their obligations were multiple.

The principle of beneficence should apply as well to the decisions and actions of the pregnant woman. She has a moral obligation to act in the best interest of her fetus, at least to the extent that she has decided to allow the fetus to come to term as a wanted offspring and that her actions will not place her at an undue risk. In this situation, the interests of the fetus do not stem from its moral status as an independent entity, but derive from its future standing as an infant and child, as well as a future member of the moral community.

Rather than discussing the entire area of fetal therapy, I will focus my comments on the issue highlighted by Ms. Gallagher--forced cesarean section--merely one of many areas of fetal assessment and therapy that could be analyzed. When a pregnant woman in labor refuses to consent to a recommended cesarean section for fetal indications, perhaps the most troubling aspect is the multiple conflicts that this creates. This problem has arisen in recent years because of the development of techniques in fetal monitoring during labor that enable the physician to assess fetal well-being during the process of labor and delivery. The techniques of fetal monitoring are by now well established, and although their predictive power is not perfect, they cannot be considered experimental. Furthermore, the advanced stage of fetal life that is consistent with a high likelihood of survival at the time these techniques are utilized suggests that the fetus has significant moral standing. The fetus about to become an infant is still unquestionably a fetus, given its direct physical dependence on the mother; yet, it clearly has interests, whether or not one wants to equate those interests with the interests of an already-born neonate.

Fetal monitoring includes the continuous measurement of fetal heart rate and rhythm, as well as uterine contraction. The measurement is taken externally by means of a belt placed around the woman's abdomen or internally by means of an electrode placed on the scalp of the fetus to monitor the fetal heart, and a catheter placed in the uterus to monitor contractions. These continuous measurements have been in use for over 25 years. They have resulted in careful analysis and the ability to determine patterns of heart rate and uterine contraction that serve to predict risks for the development of acidosis, asphyxia, and ultimate hypoxic-ischemic encephalopathy or irreversible brain damage. Techniques have also been developed to sample fetal blood *in utero* by pricking the scalp through the dilated cervix to ascertain fetal acid-base status and, thus, increase the data upon which prediction is based about ultimate fetal outcome.

It is important to point out that these assessment tools have been developed with the laudable aim of seeking to prevent fetal and ultimate neonatal compromise, resulting in the birth of healthy newborns. Recommendations to clinicians on how to use the fetal monitoring data are biased in support of cesarean delivery to prevent fetal compromise before it is irreversible. A significant percentage of fetuses delivered by cesarean section based on fetal monitoring criteria will not require it and would suffer no irreversible damage if delivered vaginally. However, the quality of the assessment tools does not allow for differentiation of this group of "false-positive" candidates for cesarean section. This more careful fetal assessment during labor has resulted in many infants being delivered prior to suffering irreversible brain damage, but has at the same time resulted in an increase in the number of cesarean sections in virtually every hospital in the United States.

A development of another sort has increased the pressure to recommend cesarean delivery to prevent fetal damage. Over the last ten years, there has been a dramatic increase in medical malpractice suits against obstetricians for the birth of brain-damaged babies. Many of these lawsuits result in awards to the plaintiff and parenthetically to the attorneys of millions of dollars. Thus, physicians aware that a malpractice lawsuit is a probable outcome if the fetus is irreversibly damaged are recommending many more cesarean sections, in order to protect the fetus from damage and to protect themselves from criticism and liability for not doing everything possible for the fetus.

This is the background against which the pregnant woman must make a decision when her physician recommends cesarean section. She may realize the fallibility of the data the physician has used to make the recommendation, and she probably knows that cesarean section rates have risen to almost 30 percent of all deliveries in large university teaching hospitals. Appropriately, she is reluctant to place herself at a fourfold increased risk of dying during childbirth, in addition to the significant pain and potential morbidity of the operation and the anesthesia. She also may be told that the cesarean can adversely affect her future reproductive life, making rupture of her uterus in future pregnancies more likely, and repeat cesarean section for each future delivery far more probable.

This complex weighing of this difficult decision is typically performed during active labor. The woman is lying in bed in at least intermittent pain and probably constant discomfort; she is attached to several devices, including an intravenous infusion and a fetal monitor; and she may have received medication for pain relief that can affect her ability to analyze alternatives. This situation may result in the woman not being able to make a clear, rational, and informed choice. Most frequently, she respects the advice of her physician and concurs with the proposed plan presented as the best course for her baby. The atypical woman who questions the certainty of the recommendation, voices concern about her own well-being, or raises a question about the motivation of the physician concerning future malpractice protection is viewed by the physician and other caregivers as a difficult, noncompliant patient who wishes to hurt her baby. The physician views the recommendation of a cesarean section as a minor surgical procedure with few risks for the woman and obvious benefits for the fetus. We should quickly point out the significant distinction between minor and major surgery. Minor surgery is surgery on someone else; major surgery is surgery on you.

Another group of mothers refuses cesarean section based on strongly held religious or moral grounds. These patients believe that surgery under any circumstance is unacceptable. For them, surgery to protect the fetus is no more acceptable than any other medical technological intervention. These patients often do not come to the hospital for their deliveries and choose to deliver at home instead. However, some desire to be hospitalized, but will not consent to surgery or other invasive treatments. These patients have the same concerns as the former group of patients, and in addition, they harbor a general mistrust of the values, goals, and actions of the medical profession. They are far less subject to persuasion and explanation, or to being convinced of the appropriateness of the recommended surgery. This is a very troubling group for health care professionals. Some professionals believe that, because there is a religious basis for the refusal of surgery, it should hold more weight, but others feel that religious beliefs should be given no weight at all.

Once the woman in labor has refused to consent to cesarean section, the caregivers are faced with a terribly difficult ethical dilemma. It is important to point out, however, that these cases will never reach a court if not brought there by one of the individuals in the hospital setting and if not supported by expert physician testimony. Thus, health care deliverers may wish to keep these types of conflicts out of the legal system by setting out some basic guidelines and principles by which the physicians might choose to act in these difficult cases.

Physicians might choose to analyze this situation not by elevating the status of the fetus to that of a full bearer of rights, but rather by analyzing the interests of the two patients. We have heard the anger generated when a pregnant woman is referred to as a "container" or "incubator" for the fetus, and the fear that presuming the fetus has rights will potentially allow the pregnant woman to be controlled by others acting for the fetus. This argument points out a common problem in using a rights-based ethical analysis. One's antecedent position on a moral issue often determines whether an entire class of individuals should bear certain rights. There is often no clear antecedent basis for claiming that a class of individuals has those rights other than the claimant's desire to bring about a better state of affairs.[4] In the legal domain, there may well be no other alternative than to balance the rights of the fetus with maternal rights when they come into conflict. But in the moral sphere where questions of rights are much less settled, a debate resting on rights claims may either reach a stalemate or a rhetorical pitch.

An alternative approach is to bypass the problematic talk of fetal rights and of personhood, and refer, instead, to the interests of the parties involved. This forms the basis of a consequentialist approach, in which the right course of action is the one that maximizes good consequences. In medical contexts, this perspective takes the form of an analysis in terms of risks and benefits. Construed somewhat more broadly, consequences can be viewed along utilitarian lines, embodying much more than simply the risks and benefits of treatment for the parties concerned, but also the consequences for others and the society at large. However, even if approached in the narrower framework, the risks and benefits that need to be taken into account include those that accrue to the pregnant woman as well as to the fetus. The risks to each party should not be minimized in the eagerness to describe the potential benefits to one.

As familiar as the consequentialist approach is to clinicians and researchers in medicine, this approach is not without its problems. The difficulty of predicting outcomes can be formidable. This uncertainty as well as the level of risk can be overlooked or minimized by physicians eager to offer benefits to their patients. Despite physicians' enthusiasm and arrogance in using fetal assessments during labor as absolute predictors of fetal distress and bad outcome, it must be emphasized that the diagnostic and therapeutic procedures applied to the fetus involve significant uncertainty and clear risks to both the fetus and the woman. The uncertainty about outcome is only one problematic feature of a consequentialist approach to the ethics of fetal therapy. Not only do the risks to the fetus need to be balanced against the potential benefits, but also the risks to the mother must be taken into account at the very least by balancing them against the risks to the fetus of nonintervention.

In my opinion, one great advantage of using a consequentialist approach for ethical analysis of fetal interventions is that it is less likely to make mother and fetus adversaries in battles over their respective rights. To pit the rights of the mother against those of the fetus is a divisive tactic, which has the prospect of heightening the difficulties, rather than enhancing and promoting harmony in an effort to arrive at the best outcome for all concerned. My personal analysis of this complex area results in the belief that a pregnant woman has a moral obligation to act in a manner that promotes the best interests of her fetus once she has determined that the fetus will be carried to term. The physician, I believe, has a similar duty to act in the best interests of that fetus and should, therefore, try to convince the woman to comply with recommended treatments deemed to enhance fetal well-being. The physician's duty, however, does not extend to initiating legal actions to ensure compliance. The physician may find it difficult to be sympathetic to a woman's refusal to act in the recommended manner and should use all available forms of moral persuasion to change the woman's mind, but should not be party to overriding the woman's autonomy by forcibly restraining her. To be specific concerning the case of cesarean sections recommended for fetal indications, I believe that a woman's informed refusal should be honored. A significantly increased risk of her dying during childbirth is sufficiently high to respect as binding her weighing of risks to herself from the surgery as greater than the potential risks to her fetus from foregoing the cesarean section. Although medical risks and benefits can be objectively determined, a patient's assessment of the degree of risk she is willing to assume for the sake of predicted benefits is a wholly subjective matter. Even when the degree of risk can be ascertained with some accuracy, reasonable people disagree on the question of which of life's risks are worth taking to attain specific benefits.

Let us turn now to a few comments concerning other important reproductive issues. Preventive interventions of any kind, including genetic screening, stigmatize people by calling them "at risk" for a certain problem. This labeling phenomenon is a negative consequence of a program that has the potential of many positive outcomes. In my opinion, prenatal screening is a valuable service that has been desired by families for many years. I believe it should not be considered something that is done "to women," but rather "for the woman and her family." Genetic screening has made great strides in increasing reproductive choice for women and their families by giving them more information about their fetuses without obligating the woman to obtain an abortion or continue her pregnancy to term.

Without question, the professionals involved in genetic screening and counseling should be knowledgeable and sensitive, so that the data obtained can be conveyed in a nonjudgmental and clear manner. Screening programs should never be forced on women, nor should there be determinations made concerning the desirability of abortion at the time of the screening procedure and before the data are available. It is undoubtedly correct that economic and social issues in our society and lack of access to prenatal care pose the greatest threat to future fetuses and children, but the assertion that monies spent for prenatal screening programs could be reallocated to enhance prenatal care in general seems naive. We must simultaneously seek to ensure access to prenatal care and adequate social and medical help for those women in need, while at the same time supporting genetic and other prenatal screening programs that can enhance the care of all women and their fetuses. It is important to note that, after all of the rhetoric and inconclusive arguments in the paper concerning genetic screening, the public policy recommendations are quite sound. So, here again, I concur with those conclusions, as I did with the conclusions of the "Fetus as Patient" paper, but I have reached this point by a very different route than the authors.

Finally, a brief comment on the question of the limits on abortion. I believe that advocating no legal limits on a woman's right to request abortion throughout pregnancy will result in polarizing our society to an even greater extent than exists today and runs the risk of converting those who have moderate views concerning abortion to oppose all abortions in the future.

In my opinion, women, fetuses, and the reproductive process are an extremely important part of our societal makeup, and deserve careful and sincere efforts at developing a reasonable consensus on public policy in our pluralistic society. The fetus should not be used as a political pawn or viewed as an adversary of its mother, but rather should be respected as an important entity with clear interests and the potential for future autonomy.

Notes and References

1. Based on an article: Fleischman A, Macklin R: Fetal Therapy: Ethical Considerations, Potential Conflicts, in, Ethical Issues at the Outset of Life, eds: William Weil and Martin Benjamin, Blackwell Scientific Publications, Boston, 1987.

2. National Commission for the Protection of Human Subjects. The Belmont Report, U.S. Government Printing Office, Washington, D.C., 1979.

3. Chervenak FA, McCullough LB: Perinatal Ethics: A Practical Method of Analysis of Obligation to Mother and Fetus. Obstetrics and Gynecology 66:442-446, 1985.

4. Macklin R: Moral Concerns and Appeals to Rights and Duties. Hastings Center Report 6:31-38, 1976.

Commentary

RECONCILING OFFSPRING AND MATERNAL INTERESTS DURING PREGNANCY

John A. Robertson

The papers and the policy proposals by Janet Gallagher, by Mary Sue Henifin, Ruth Hubbard, and Judy Norsigian, and by Nan Hunter raise a variety of issues concerning autonomy during pregnancy, and state action to limit or promote autonomy. In their view, neither a late stage of pregnancy nor a decision to go to term justifies limits on the autonomy of pregnant women.

Since a brief comment does not permit the in-depth scholarly analysis that these issues deserve, I will concentrate on more general points about prenatal obligations during pregnancy to expected offspring, addressing issues raised by Gallagher and Henifin et al. I will then address two issues that Hunter raises in her discussion of late-term abortions.

Maternal Obligations to Offspring during Pregnancy

The growing ability to prevent the birth of disabled infants has raised new issues about the scope of reproductive freedom--issues of special concern to women. Most women at risk welcome knowledge about treatments or behaviors that will prevent the birth of a child with disabilities. They avoid risky behavior, and accept medical treatments or surgery that will ensure a healthy birth. If a healthy birth is not possible, they may avoid conception or terminate the pregnancy.

Yet not all women know of the dangers that certain behaviors pose or of the treatments available to minimize congenital impairments. Even if they do know, they may lack access to the prenatal screening and treatment that would prevent the disabilities from occurring.

Sometimes, however, women with access ignore the knowledge or refuse treatment, and by their action or inaction cause children who could have been born healthy to be born disabled. What is the ethical status of such conduct? What should public policy toward it be?[1]

Moral Obligations to Unborn Children during Pregnancy

Questions of prenatal obligations to offspring are ethically complex because of the prenatal timing of the harmful conduct and the unborn child's location in the mother's uterus when the harmful conduct occurs. Meeting obligations to the unborn child may require placing limits on the mother's conduct that would not arise if she were not pregnant. Thus, the mother's interest in autonomy and bodily integrity must be balanced against her baby's welfare.

Yet it is not unreasonable to regard her as having a moral duty to the baby she is choosing to deliver. All persons have obligations to refrain from harming children after birth. Similarly, they have obligations to refrain from harming children by prenatal actions.[2] There is no reason why the mother who has chosen to go to term should not also have a duty to prevent harm when she may reasonably do so. The timing of the conduct does not affect the duty to avoid causing harm.

It is important to recognize that the interests of actual offspring to be free of prenatally caused harm rather than the right of the fetus to complete gestation are at issue. Duties to fetuses in themselves and duties to fetuses because they will come to term must be distinguished (and seldom are in this debate). Protecting offspring against prenatally caused harm in no way diminishes the woman's right to terminate pregnancy, for the issue arises only if the woman decides to continue the pregnancy. No right of the fetus to be brought to term is at issue.

Thus, the tendency of physicians to speak of the fetus as a "patient" should be clarified.[3] The fetus going to term is a "patient" by virtue of the expectation that it will be born alive, and not because physicians have an independent duty to bring all fetuses to term regardless of the mother's wishes.

Prenatal duties owed the planned offspring may even arise before viability. The mother's plans, and not the state of fetal development, are determinative, for even first-trimester conduct could culpably injure children who are subsequently brought to term. A woman who is undecided or ambivalent about a first-trimester pregnancy may still be morally obligated to act as if she will carry the fetus to term if first-trimester conduct poses serious risk to a baby that is born. Although she is free to terminate the pregnancy later, she is not free to injure offspring prenatally if she is uncertain about whether to continue the pregnancy, and eventually decides to do so.

Ethical analysis must balance the mother's interest in freedom and bodily integrity against the offspring's interest in being born healthy. This balance will vary with the burdens of altering the mother's conduct and the risk of prenatally caused harm to offspring. Depending on the balance of risk, benefits, and burdens, prenatal conduct may be morally discretionary, advisable, prudent, or even obligatory. The evaluation depends upon the reasonableness (as viewed by reasonable persons) of avoiding the harm, in light of its severity, certainty, and the difficulty of avoiding it.

It should also be noted that moral analysis produces a different outcome when no healthy birth is possible, because of genetic or other factors, such as prenatal transmission of AIDS, or because *in utero* injury that was once avoidable has already occurred. Delivery in those situations does not harm offspring, for they have no alternative but an impaired birth.[4] If the birth is truly wrongful, harm can be minimized by withholding all treatment. Birth, however, may harm those persons who bear the burdens of rearing the now unavoidably impaired offspring. The ethical issue is whether persons have duties to avoid conception or delivery of an unavoidably disabled offspring in order to avoid imposing rearing costs on others. Preventing the imposition of rearing costs has different ethical, legal, and policy implications than does preventing avoidable harm to offspring. As I have shown elsewhere, the case for coercive public policies in this situation is considerably weaker than when avoidable harm to offspring is at issue.[5]

Policy Options: Voluntary Compliance or Compulsion?

Several policy options are available to influence the behavior of women and others during pregnancy for the sake of planned offspring, ranging from voluntary compliance through education and access to services to coercive sanctions and seizures. Relying on voluntary compliance is the most desirable policy, since it raises fewer civil liberties and privacy issues, and is more likely to be effective. Most women will welcome such knowledge and act accordingly. If they have not been able to avoid the damaging conduct, many will choose abortion rather than bring the impaired fetus to term. The main need here is to assure that women are adequately informed and have access to treatments that can avoid the harm to offspring. A society truly concerned about prenatal harm to offspring can do much in the way of education and services to prevent such harm from occurring.

However, women who will not or cannot comply with proper conduct will end up injuring a child who could be born healthy. Should the state go beyond education and penalize irresponsible maternal behavior during pregnancy by imposing civil or criminal sanctions when actual harm to offspring has occurred? Should it prevent the harm before it occurs by incarcerating or forcibly treating pregnant women?

Gallagher and Henifin et al. take the position that coercive measures are never appropriate to prevent avoidable harm to expected offspring. Although I agree with them with regard to prebirth seizures, I see a plausible case for use of postbirth sanctions in egregious cases. I thus call into question their policy recommendations that women never be held legally accountable for conduct during pregnancy, even when it culpably causes serious injury to offspring.

Postbirth Civil and Criminal Sanctions

The law has long recognized that actions or omissions during pregnancy can be as harmful to children as actions or omissions after the child is born. Since the sixteenth century, prenatal actions that cause a child to die after live birth have been prosecuted as homicide.[6] Similarly, under the civil law, damages have been awarded for injuries that occur during pregnancy or before conception, when a child who could have been born healthy is born impaired.[7] Recent developments allowing family members to sue each other now permit such suits by children against parents if the latter have culpably caused the children avoidable injury.[8] Since these duties arise only if the woman chooses to continue a pregnancy that she is legally free to end, penalizing culpable maternal behavior that unreasonably harms offspring does not conflict with *Roe v. Wade*.[9]

In theory, a child who is severely retarded as a result of culpable prenatal conduct could sue the mother. However, suits by impaired offspring against mothers will rarely be brought. The theoretic possibility of civil suit is not likely to deter harmful prenatal conduct.

The state might pursue criminal prosecution for culpable prenatal conduct that causes severe impairment to offspring. Although only a few prosecutions for prenatal child abuse have been reported and the applicability of current child abuse and neglect laws to prenatal conduct is uncertain,[10] this avenue is constitutionally within state authority. It may turn out to be an effective tool for demonstrating society's protection of children and deterring egregiously harmful prenatal conduct in certain cases.

Gallagher and Henifin et al. assert that state sanctions would never be an acceptable tool of public policy for preventing avoidable prenatal harm to offspring. However, they do not distinguish between the rights of fetuses to be brought to term and the rights of those fetuses that will be brought to term to be treated with reasonable care; nor do they recognize the crucially important difference between postbirth sanctions and prebirth seizures as coercive policies.

If the debate were narrowed to take account of these distinctions, the desirability of public use of sanctions should depend on the gains to children relative to the harms that might arise from such a policy. The mere fact that a policy focuses on prenatal conduct and that it limits how a pregnant woman behaves or treats her body is not itself determinative of the question. The argument against use of postbirth civil or criminal sanctions would have to rest on a claim that the possibility of error in assigning responsibility and causation is so great, and the chances that women would be unfairly treated so substantial, that egregiously culpable behavior by pregnant women should go unpunished, even though equally culpable behavior done by a third party to the offspring during pregnancy or after birth would clearly be punished.

As Justice Stone reminds us in *Butler v. United States*, the possibility of abuse of a power is not an argument against its exercise in nonabusive circumstances.[11] Gallagher and Henifin et al. have presented no empirical data or persuasive reasons for thinking that prenatal child abuse laws would unjustly limit the conduct of pregnant women or others to whom they apply. I am not convinced that a slide down a slippery slope to arbitrary, loosely justified limits on prenatal conduct is likely to follow from extending child abuse laws to culpable maternal conduct during pregnancy.

Egregious cases of culpable prenatal conduct causing substantial harm and suitable for prosecution can be distinguished from less egregious cases, just as is done with allegations of postnatal child abuse. Fears that obstetricians will become pregnancy policemen are no more valid than fears that pediatricians will become "childrearing policemen" under statutes requiring the reporting of postnatal child abuse. For example, the fear that child abuse reporting requirements would deter abusive parents from bringing children to doctors for medical treatment has not proven true. It is not clear why women would be any more deterred from seeking prenatal care.

To refuse to draw any lines about prenatal conduct by claiming that pregnant women have total autonomy regardless of the avoidable effects on offspring ignores the interests and rights of children who are born injured by maternal conduct. Such injuries would not be tolerated if caused after birth, or if caused prenatally by third parties. The speculative danger of abuse is no reason to exempt children from legal protection against pregnant women who unfortunately are not able to meet reasonable community standards about safe conduct during pregnancy.

Postbirth sanctions occupy a small but important niche in preventing conduct during pregnancy that culpably injures children who would otherwise have escaped injury. Despite the legal availability of criminal sanctions for prenatal maternal child abuse, however, I want to emphasize that the main policy approach should be education and making needed services available. Voluntary compliance, rather than coercive sanctions, remains the most desirable policy.

Prebirth Seizures

The most extreme and controversial policy option is incarceration or forced treatment of pregnant women who are unlikely or unwilling to avoid the behavior that is harmful to offspring. From the perspective of the child at risk, this approach is preferable to punishing after the harm occurs, since it prevents the harm altogether.

Direct intervention on the mother, however, is the most troubling option, because it involves bodily seizures of varying duration and risk without the woman's consent. The right to be free of seizure and forced bodily intrusion is a basic right. Few cases would meet the high standards necessary to justify a direct seizure for the benefit of unborn offspring.

In general, I agree with Gallagher and Henifin et al. that prebirth seizures are not likely to be effective or fair and, thus, are not a desirable tool of public policy. Although there may be some role for postbirth sanctions, the role for prebirth seizures in preventing avoidable impairments is extremely narrow. Policy efforts are better spent informing women of risks and making treatment available than seizing recalcitrant women during pregnancy for the sake of their expected offspring.

I differ with them, however, in finding that seizures are objectionable only on grounds of policy and not on grounds of principle. In principle, seizures and forced treatment are within state power if a compelling need that outweighs the burdens of the seizure can be shown.[12] Although direct bodily seizures are rare in the law, they are not unknown. They occur in civil commitment, prison sentences, capital punishment, the draft, forced treatment of adults for the sake of minor children, and blood tests and surgery to recover evidence of crime. Their validity depends on a sufficient state interest to justify the intrusion on protected personal interests in bodily integrity, liberty, and privacy.

Although this standard is purposely high and difficult to meet, there may be rare situations in which prenatal protection of offspring satisfies it, because the benefits to offspring clearly outweigh the burdens of the intrusion. However, prenatal seizures for the benefit of offspring would have to be specifically authorized by statute and accord the woman procedural due process, including judicial review of the need for the seizure.[13] Authority for bodily seizures should also extend to comparable situations of risk, benefit, and assumed duty outside of pregnancy.

What situations would justify seizure or forced treatment of the pregnant woman (or father) to prevent harm to expected offspring? The strongest case is when the intrusion or seizure is minimal in length and the harm to be prevented is certain and substantial. Even then, considerable doubt and controversy remain about whether the power should be used. The case for seizure weakens rapidly as the length, risk, and burdens of the seizure increase, and the benefit to the offspring diminishes.

Forcing medical treatment on pregnant women would be justified only in very exceptional cases. The strongest case for forced treatment is a one-time intervention of minimal risk to the mother: administering the drug, blood, or surgery to avert severe disability in her offspring. A one-time surgical procedure without high risk and with great benefit, such as Rh transfusion or even an established fetal surgery, could meet these standards in particular cases.[14]

In practice, such cases will be very rare. Other than Rh transfusion, few procedures have such clearly established benefits. The initial hopes for *in utero* fetal surgery to repair congenital hydronephrosis or hydrocephalus have generally not been satisfied, and presently would not justify forced treatment.[15] When safety and efficacy are clearly established, few mothers are likely to refuse the procedure.

Ordering a cesarean section over the mother's refusal is the most difficult case, since it forces an unwilling person to undergo general anesthesia, abdominal surgery, and the risk of infection and other complications. Yet the benefits to the offspring also appear substantial--avoidance of anoxia and severe brain damage.

As a matter of principle, mandating cesarean section in certain cases might be within the power of the state, but it appears to be a power that should be used rarely, if at all. Risk of error, lack of due process, and magnitude of intrusion counsel against resort to such a remedy in all but very exceptional circumstances.[16]

No physician should be required to seek a mandatory cesarean section. Informing the mother of the risks and reporting harmful refusals to child welfare authorities satisfy the physician's duty. The benefit to the few children who would avoid injury seems to be outweighed by the errors likely to occur under forced treatment policies. Sanctions after birth, however, may be imposed for culpable refusals that caused serious impairment to offspring. Refusals resulting in stillbirth could not be punished in the roughly 25 states that have no laws requiring that postviable fetuses be brought to term.

In sum, prebirth seizures may fall within state power in a narrow class of compelling cases, yet rarely if ever should be sought. The likelihood of error in predicting benefit, the difficulty in assuring due process, and the burden of forced treatment and incarceration make such an extreme remedy, except in a few exceptional cases, a dubious avenue to reduction of impaired births. Given the risks of such an intrusive policy, postbirth sanctions should be the sole coercive remedy used by the state, even though exceptional cases meeting the very high standards for seizures can be met. Voluntary compliance and access to services remain the preferable policy.

Sanctions, Seizures, and Prenatal Screening Programs

The ethical, legal, and policy issues analyzed above would apply to prenatal screening obligations. The main policy should be voluntarism, encouraged by information, public education, and access to essential services. A law informing women of screening options, such as the California law requiring that women be informed of alpha-fetoprotein screening, is desirable.

Mandatory screening, backed by criminal or civil penalties, may also be justified when the costs and burdens of screening are low and a great benefit to the offspring or others can be shown. When such situations exist is, of course, subject to debate.

Forcing screening to occur, like forcing any treatment or making any bodily seizure, is the most difficult of all to justify. Yet if an extreme enough situation presented itself, such a policy would be within the constitutional authority of the state. As a general rule, such situations will be rare, since it will require a strong showing to overcome the presumption against seizures as a way to prevent harm to offspring. Also, screening only identifies a problem. Although it may enhance the possibility, it does not assure that the desired preventive action will then occur.

Time Limits on Abortion

I want to comment briefly on Nan Hunter's paper "Time Limits on Abortion." A very different set of issues is presented here--the question of obligation or duty to bring a fetus to term once a certain point in the pregnancy has been reached. To a large extent, the issue is symbolic, since 98 percent of abortions occur in the first 14 weeks. Yet enough late abortions occur and their consequences are significant enough to warrant comment.

As a policy matter, it is clearly preferable to have early rather than late abortions. Therefore, I agree with Nan Hunter that noncoercive policies of education and access to services should be adopted to reduce as much as possible the number of late-term abortions that do occur. The dangers and suffering that women experience in late abortions will also be reduced by prenatal diagnostic techniques that permit decisions about continuing the pregnancy in the case of genetic impairment to be made in the first trimester or earlier.

My comments on Hunter's paper concern the ethical and legal positions she takes on decision-making authority over late-trimester fetuses, and her failure to unpack some of the ethical and legal issues that arise.

Should Late Abortions Be Permitted?

One such area concerns the acceptability of banning or allowing late-term abortions--whether time limits on abortion are justified at all. The Supreme Court in *Roe v. Wade* permits the state to ban abortions after "viability" if it chooses. Interestingly, about half of the states have not enacted legal prohibitions on late abortions, though permitted to do so. Without intending to argue that such laws should be enacted, I want to comment on a possible justification for such a ban and, thus, push Nan Hunter to confront some distinctions that do not appear in her paper.

Although the Supreme Court's viability line for proscribing late abortions has never been adequately explained, such a line can be justified if one acknowledges that the advanced development of the fetus by 22-24 weeks of gestation makes moral demands on us that do not exist at earlier, less well developed stages.[17] Since the brain and nervous system by that point are well developed, it is very possible that abortion could inflict suffering on fetuses that would not be imposed on humans or animals without good reason. A moral obligation to respect the fetus for its own sake, as we respect newborns, rather than for its potential or likelihood of developing further, may arise at that point to limit the availability of abortion.[18]

Let me emphasize, however, that protection of the fetus in its own right at 22-24 weeks does not lead to protecting it in its own right at 12-14 weeks or earlier, when the evidence for a sentient nervous system is not persuasive. Although it does not define viability, sentience may capture some of the concerns raised by that concept.[19]

Yet even this ground for respecting advanced fetuses would not mean that late abortions could never be justified. Rather, a stronger justification may have to be shown than is necessary to abort at an earlier stage. Threats to health, genetic impairment, and impossibility of aborting earlier may all argue for permitting such an abortion to occur. However, there may be a duty to abort earlier rather than later where possible, to protect both advanced, sentient fetuses and society's symbolic commitment to the worth of human life.

In sum, questions about the permissibility or impermissibility of time limits on abortion require much greater analysis and justification than do decisions at earlier stages. The issue is not resolved by mouthing shibboleths about "control of one's body" or "potential human life" any more than it is at earlier stages. Perhaps careful, nuanced analysis of the competing interests at late stages of pregnancy will help bring reasoned analysis to bear on the conflicts that arise at earlier stages.

Late Abortion Techniques, Fetal Survival, and the Importance of Genetic Parentage

Nan Hunter also argues that, if late abortions do occur, abortion techniques that maximize the chance of fetal survival should not be required (even when they do not increase the risks to the mother).[20] Hunter is, in effect, arguing that the right to terminate the pregnancy should also include the right to cause the death of the fetus, and not merely to relieve the woman of gestational burdens by removing the fetus from the body.

Such a concept of abortion accords with the general understanding of what abortion does. Yet the general concept breaks down in late abortions. Depending on the technique (prostaglandin or hysterotomy versus saline infusion), an advanced fetus could survive the termination procedure and be treated like any other premature infant. Indeed, the famous case of *Commonwealth v. Edelin* made clear that physicians undertaking a legal abortion have a legal obligation to save fetuses that emerge alive, from the abortion procedure.[21] Once the fetus emerges alive, it is a legal person or subject and must be accorded all the rights of other premature newborns.

Hunter argues that late-trimester abortion policies should permit abortion techniques that will assure the death of the fetus. Prohibiting such techniques in order to assure fetal survival will interfere with procreative liberty, for it will lead to unwanted genetic offspring. Even if no rearing duties attach, the existence of genetic offspring is a significant reproductive event and, thus, should be reserved to the individual. Since the woman will also have gestated for several months, the reproductive impact of unwanted fetal survival is even greater, and should be left to individual discretion.

Although feminists have usually decried the importance that men have placed on creating biologic descendants, it is difficult to deny that avoiding genetic parentage *tout court* might have great psychosocial importance for men and women. Hunter's clear articulation of this issue in the context of late abortions helps us consider the importance of genetic parentage *tout court* in other reproductive situations.

After a live birth, the avoidance of genetic and gestational parentage that involves no rearing is easily overridden by the needs of the infant. The mother is free to relinquish social parentage, but she may not kill her child to avoid the psychosocial burdens of having relinquished her child for adoption.

Similarly, late-term abortions can be prohibited altogether because of the injury that would occur to the late-term fetus, even though the mother will suffer the burdens of further gestation and relinquishment. The advanced stage of fetal development gives the state a significant interest in protecting the fetus for its own sake.

Although the fetus can be protected altogether at that stage, it is no contradiction if the state chooses to limit fetal protection for the sake of the pregnant woman. Finding that women should be relieved of gestation but not the psychosocial pains of being a relinquishing mother is a rational and legally acceptable position to take (though one might oppose its wisdom as policy). The state may choose to protect the welfare of fetuses over the interest of women undergoing abortions in being relieved both of gestational and relinquishment burdens.[22]

A different answer arises when we examine the question of mandating the donation of unwanted embryos in an in vitro fertilization (IVF) treatment program, because of the much less developed status of the extracorporeal embryo.[23] Many American and foreign IVF programs require that all embryos be transferred to a uterus, so that they might have a chance to implant and come to term. Only Louisiana has a law that requires that all embryos be transferred to an available uterus.[24] Since this issue has not been extensively analyzed or debated, it is worth exploring here.

Do persons providing the gametes from which the embryo is formed have the right to discard their embryos and, thus, prevent transfer of the embryo to a uterus and eventual birth? May the state override objection to transfer by a law requiring that embryos be transferred to the woman providing the egg or to a willing recipient who relieves the couple of all rearing rights and duties in offspring? The question of the couple's right to discard excess embryos formed in the course of IVF treatment squarely focuses the issue of the genetic tie *tout court*, for no gestational tie yet exists, and the embryo is so rudimentary in development that no obligation to it for its own sake can be maintained.

A mandatory embryo donation law serves symbolic purposes only. It denotes or symbolizes a special commitment to human life. Many persons might find meaning in such a policy. However, such a policy is not required by the obligations of justice. Since no person will exist if discard occurs (if it is accepted that the embryo is not a person), no person's rights are violated by a refusal to bring him/her into being. The potential for live birth does not itself require that all possible births occur.

Whether society could adopt such a purely symbolic policy depends on the burdens it poses for reproductive choice. In this case, the reproductive burden posed is the imposition of a genetic tie *tout court*. The psychosocial burdens of genetic parentage will no doubt exist, for genetic linkages have meaning for male and female alike. A couple wishing to discard embryos has a reproductive interest in avoiding biologic offspring, even if no rearing occurs, just as women and men may not wish their oocytes or sperm removed for other purposes to be used to conceive a child. This interest exists even if they have no rearing rights and duties, and remain anonymous.

Does avoidance of this burden qualify for protection as part of the fundamental right not to procreate? The answer depends on an evaluation of the personal importance of such a tie to individuals. The Supreme Court has not yet faced this question, and is generally leery of creating fundamental rights based on psychosocial burdens alone. It may not extend protection to this interest when no gestational or rearing burdens attach.[25] If this prediction is correct, the state would be free to symbolize its respect for human life by enacting mandatory embryo donation laws and preventing couples from discarding embryos formed from their gametes.

Yet the case for making this symbolic statement is not a strong one, even if it can be constitutionally made. At a time of widespread legal abortion, the symbolic gains from transferring all embryos when only a small percentage will survive seem slight. The burden on persons who do not want a genetic connection is significant, even if not constitutionally protected. The symbolic gains are easily outweighed by the loss to the individuals involved.

Conclusion

Important questions arise in disputes about public policy in reproductive matters. Unfortunately, the debate has too often been so polarized that careful distinctions among issues and interests have not been made. Indeed, right-to-life groups have made a fetish of refusing to yield an inch in compromising the worth of any fertilized egg, a position that greatly undermines their credibility.

Unfortunately pro-choice activists and many feminists have engaged in a similar fetishism from the opposite direction, finding any assertion that procreative liberty may be limited during pregnancy to be politically incorrect. Although this stance was perhaps necessary at earlier stages of the struggle, it may now be counterproductive.

The papers commented on make useful contributions to public policy concerning reproductive choice, but they would have been more convincing if they had made more distinctions and acknowledged a greater complexity to these issues. The need to stand up for women's rights does not justify overlooking important distinctions that must be taken into account in arriving at defensible public policies.

Recognizing that women may have moral and legal obligations to offspring during pregnancy does not turn women into reproductive vessels, as described by Margaret Atwood in *The Handmaid's Tale*. In my view, the best way to protect reproductive freedom is to move beyond worst case scenarios to the nuanced, reasoned analysis that the complexity of human procreation deserves.

Notes and References

1. The comments that follow have been treated at greater length in Robertson and Schulman, "Pregnancy and Prenatal Harm to Offspring: The Case of Mothers with PKU," Hastings Center Report 23-33 (August, 1987).

2. Joel Feinberg, Offense to Others (New York: Oxford University Press, 1985), pp. 149-151.

3. William Ruddick and William Wilcox, "Operating on the Fetus," Hastings Center Report 12:5 (October, 1982), 10-14.

4. Robertson, "Embryos, Families and Procreative Liberty: The Legal Structure of the New Reproduction," 59 Southern California Law Review 939, 987-990 (1986).

5. See note 1 supra at 25, 31.

6. 3 E Coke, Institutes 50 (1648).

7. Horace Robertson, "Toward Rational Boundaries of Tort Liability for Injury to the Unborn," 1978 Duke Law Journal 1401.

8. Grodin v. Grodin, 102 Mich. App. 396, 301 N.W. 2d 869 (1980).

9. 410 U.S. 113 (1973).

10. See note 1 supra at 27-28.

11. 297 U.S. 1 (1936).

12. Schmerber v. California, 384 U.S. 757 (1966); Winston v. Lee, 470 U.S. 753 (1985).

13. Few states have statutes that explicitly extend such protection. However, several state courts have interpreted statutes protecting children to include post-viable fetuses. See note 1 supra at 27.

14. Robertson, "The Right to Procreate and In Utero Fetal Therapy," 3 Journal of Legal Medicine 438-439 (1982).

15. "Report of the International Surgery Registry, Catheter Shunts for Fetal Hydronephrosis and Hydrocephalus," 315 New England Journal of Medicine 336-340 (1986).

16. In this respect, I agree with the conclusions reached against forced cesarean sections by Kolder, Gallagher, and Parsons, "Court-ordered Obstetrical Interventions," 316 New England Journal of Medicine 1192 (1987) and Rhoden, "The Judge in the Delivery Room: The Emergence of Court-ordered Cesareans," 74 California Law Review 1951 (1986).

17. Rhoden, "Technology and Trimesters: Revamping Roe v. Wade," 95 Yale Law Journal 639, 669-673 (1986).

18. Glover, "Matters of Life and Death," New York Review of Books, May 30, 1985, at 19, 22-23 takes a similar position.

19. An advanced fetus is also a more powerful symbol of human life than is an unimplanted embryo or fetus at an earlier stage of pregnancy. Thus, even if there is no basis for respecting the fetus in its own right, the basis for respecting it as a symbol is stronger at advanced stages as well.

20. If there are health risks, the conflict between bodily integrity and offspring welfare that underlies the prenatal conduct debate would arise.

21. 371 Mass. 497, 359 N.E. 2d 4 (1976).

22. The Supreme Court in Roe v. Wade upheld a right to avoid procreation up to the point of viability in pregnancy, but it never adequately explained the basis for such a right. Was it genetic, gestational, or unwanted rearing and relinquishment burdens that were being protected? The Court's discussion dwelled on the rearing burdens entailed by an unwanted child, and paid little attention to unwanted gestational burdens (which in retrospect seem to be the most cogent basis for the decision). In any event, it never addressed the question of the genetic tie alone, and the extent to which one would have a right to prevent such a tie when the state chose to impose it. See 410 U.S. 113 (1973).

23. Another context in which the importance of women's and men's genetic bond tout court arises is their relationship to offspring born of egg or sperm donation. Do male or female gamete donors have a sense of having procreated through

such donation? Is this sense important and substantial enough that donation agreements that try to protect all the parties involved recognize it? Is it strong enough to warrant overriding an agreement giving up rearing rights and duties, as some have argued is justified in the case of surrogate mothers who have changed their minds and refused to relinquish custody? In the case of egg donation, is it so important for offspring that they will be motivated to seek out their genetic mother, even though they know their gestational mother? How will it affect relations with gestational but not genetic mothers? Analysis of these questions is reserved to another publication.

24. La. Rev. Stat. Ann. #9: 121 et seq. (1987 Supp.).

25. Robertson, note 4 supra at 977-981.

2. Reproductive Choice: Hazards and Interference

REPRODUCTIVE HAZARDS IN THE WORKPLACE

Joan E. Bertin

Position Paper

The American Cyanamid Company plant is in Willow Island, West Virginia, on a stretch of the Ohio River crowded with chemical plants. The surrounding area is economically depressed, and the population is heavily dependent on the chemical industry for employment. Between 500 and 600 workers have been employed at American Cyanamid. Their union scale wages lift them substantially above the standard of living of many of their neighbors. Until 1974, this plant employed no women in production work.

By 1978 approximately 25 women worked at the American Cyanamid plant. Early that year, they were called into meetings in small groups and were informed about a new company medical policy. The purpose of the policy was to protect the fetus from exposure to hazardous chemicals if a woman worker became pregnant. As a result of the policy, women would be excluded from most jobs in the plant. Those with the lowest seniority faced a likely possibility of termination. A company official reportedly said that the policy was supported by the federal government and that the other chemical plants in the region would adopt similar policies.

The policy applied to all women between the ages of 15 and 50 regardless of their marital status, sexual orientation, or child-bearing intentions. The only ones not affected would be those who could present medical verification of sterility. Birth control, even a spouse's vasectomy, would not suffice. In response to questions, a company doctor told women that they should not get sterilized in order to secure their employment. Nonetheless, tubal ligations were described, and the women were assured that their insurance would cover the costs. No one offered any other suggestions as to how any of these women could secure their employment and their economic well-being.

Subsequently, the scope of the policy was substantially narrowed, but by then five women had already submitted to surgical sterilization in order to protect their right to jobs. A year later, the department where five women had undergone sterilization to secure their employment was closed, allegedly for "business reasons."[1] These events occurred at a facility that has been cited for violations of federal occupational safety and health law,[2] and whose own records revealed exposures to levels of toxic substances that federal regulators and independent health professionals generally agree pose a risk to adults of both sexes.[3]

* * *

In 1979 these facts captured public attention. Since then, the question of occupational reproductive hazards and "fetal protection" policies in particular have been the subject of Congressional hearings, scholarly commentary, several court decisions, and studies by the United States Congress Office of Technology Assessment, the Council on Environmental Quality, and others. Regulatory guidelines were attempted, but abandoned.[4] Notwithstanding all this activity, the questions posed by these policies have yet to be resolved satisfactorily, while the need for solutions persists: in January 1987, some manufacturers of semiconductor chips announced an intention to limit employment of women workers in response to a study purportedly showing an increased risk of miscarriage in such employment. Unlike the earlier situations, which mostly involved male-intensive industries in which women were viewed as a marginal element of the workforce, this most recent example involves an industry substantially populated by women. The earlier cases invited speculation that "fetal protection" policies forced those women who wished to function in a "male" world literally to sacrifice their "femaleness." The newer situation, in contrast, suggests that women are always marginal workers, and that their rights and interest in employment must always yield to their paramount responsibility as childbearers.

Initially, public attention focused on exclusionary policies of the American Cyanamid version because of the cruel choice they posed for women workers--your fertility or your job. But these policies also raise crucial scientific and legal questions: what evidence supports the proposition that women (or the fetus) are uniquely susceptible to the harmful effects of toxic chemicals or are susceptible at lower levels of exposure? Did Congress and state legislatures, when they enacted broad prohibitions against sex discrimination and an expansive right to a safe and healthy workplace, intend that pregnant workers were not to be covered to the same extent as all other workers?

To date, these questions have been addressed only in part, and in a not wholly satisfactory way. This position paper will describe the nature and extent of the problem posed by exclusionary policies and the judicial and other responses to them, and will then propose legislative actions and public policy initiatives to enhance workers' rights to reproductive freedom, occupational safety, procreative health, and nondiscrimination in employment.

I. Scope of the Problem

Hundreds of thousands of lucrative jobs are closed to women workers as a result of exclusionary policies. Many of the country's largest corporations have adopted such policies,[5] and in some companies the policies may affect large numbers of jobs. The Lead Industries Association has publicly opposed the employment of fertile women since 1974,[6] and almost a million jobs are at stake in that industry alone.

Many of the policies bar all women of childbearing age or capacity from certain jobs, a rule that affects women for most of their working lives. However, the average woman bears only two children, and most working women plan the timing of those births. Many blue-collar women bear children early in life, before they begin work.[7] Thus, concern for specific workplace effects on pregnancy cannot rationally be directed at the entire female workforce, and women cannot be viewed as permanently potentially pregnant or as powerless to control conception and childbirth.

II. The Premise of Fetal Hypersusceptibility--Fact or Stereotype?

"Fetal protection" policies assume that the fetus is uniquely susceptible to injury from hazardous workplace exposures, or that it is susceptible at lower exposure levels than those which would threaten adult workers. Some companies have adopted policies solely on the basis of this unexamined assumption, rather than on the basis of an actual examination of the biological effects of specific chemicals or conditions. A review of the scientific literature on the reproductive and other health effects of occupational exposures reveals that the assumption of heightened or unique fetal risk, while sometimes ultimately validated, frequently relies more on stereotypes than on facts,[8] and that the adverse health effects of toxic chemicals are rarely confined to the fetus *in utero*.[9]

In 1977, male workers at an Occidental Chemical Company plant in California reported an unusually low birth rate in their families. Subsequent testing showed that 14 of 25 tested men were either sterile or had extremely low sperm counts. The affected men had all worked with the pesticide dibromochloropropane (DBCP).[10] Subsequently, DBCP was banned for almost all uses, based on the human data showing "low and zero sperm counts, on animal test data indicating DBCP was a carcinogen in laboratory animals, and on laboratory studies demonstrating DBCP caused heritable genetic damage."[11]

The following year, the Occupational Safety and Health Administration examined the health effects of workplace exposure to lead, from which women are commonly excluded, and concluded with regard to reproductive health:

> Germ cells can be affected by lead which may cause genetic damage in the *egg or sperm* cells before conception and which can be passed on to the developing fetus. The record indicates that genetic damage from lead occurs prior to conception in *either father or mother*. The result of genetic damage could be failure to implant, miscarriage, stillbirth, or birth defects.[12]

Similarly, the United Nations Scientific Committee on the Effects of Atomic Radiation (UNSCEAR) concluded that, in addition to risks posed by *in utero* radiation exposure, preconception *paternal* irradiation could result in "from 2 to 10 congenitally malformed liveborn children, per rad of paternal radiation, with about five times this number of recognizable abortions and about 10 times the number of losses at the early embryonic stage."[13] The Environmental Protection Agency's Teratology Policy Workgroup recently conducted an internal study of 19 pesticides regulated because of fetal effects, and five industrial chemicals. EPA concluded that "if a pesticide is extensively tested for health effects, it is likely to show additional positive effects other than teratogenicity . . . [and] teratogenicity was never the most sensitive effect, except in the case of one pesticide."[14]

Recent concern over miscarriage rates in the semiconductor industry demonstrates the problem regarding selective assessment of hazards. Among the chemicals widely used in the industry are glycol ethers, which are known to cause reproductive injury in male animals.[15] A recent compendium identified the wide variety of toxic substances used in the industry and documented numerous hazards that have been recognized for a number of years.[16] Whether employment in this industry poses a risk of miscarriage to pregnant women is still unknown--the much-touted study commissioned by the Digital Equipment Corporation has been criticized as preliminary, incomplete, and inconclusive, if not flawed.[17] However, the existence of other occupational health hazards is well established, and these risks are not addressed by policies excluding or limiting the employment of prospective mothers.

Selective concern for female aspects of reproduction is particularly peculiar in light of the record of indifference to workplace hazards evidenced by some industries with exclusionary policies. For example, in 1973, the Manufacturing Chemists Association kept the link between vinyl chloride exposure and angiosarcoma, a fatal form of liver cancer, "confidential in order 'to minimize unwarranted speculation.'"[18] At the same time, some industries barred women from jobs involving vinyl chloride exposure.[19] Similarly the Lead Industries Association (LIA) and many of its members maintained at OSHA hearings that the permissible blood lead level for all workers should be 80 micrograms per 100 grams of blood,[20] a level which OSHA found would result in "severe lead intoxication."[21] At the time of the OSHA hearings members of LIA were in general noncompliance with OSHA standards,[22] and many enforced exclusionary policies against fertile women.

These and other examples reveal starkly the problem of selective vision in acknowledging workplace health hazards: hazards are often ignored or minimized, except when they involve female aspects of reproduction. Then, employers act quickly to exclude fertile or pregnant women workers on the basis of preliminary, inconclusive, and sometimes speculative information. The prevalent assumption is that an exposure is unsafe for the fetus unless and until its safety is conclusively proved. In most other contexts, chemicals are presumed safe until proven otherwise. Once the proof is in, chemicals that are particularly harmful to men (i.e.,

DBCP, kepone) have been banned, while women have been barred from working around chemicals suspected to cause fetal harm. These patterns once again reinforce the notion that the workplace must accommodate the needs of men, but not women, and that society requires and benefits from the paid labor of men, but not that of women. Men, however, are wrongly presumed invulnerable to the effects of chemical exposures until conclusive and undeniable evidence of hazard has been amassed.

III. Federal Laws Implicated by Fetal Protection Policies[23]

Two federal laws are directly implicated by policies that single out women for disadvantageous treatment. The Occupational Safety and Health Act, 29 U.S.C. § 651, *et seq.* ("OSH Act") requires employers to maintain a workplace "free from recognized hazards that are causing or are likely to cause death or serious physical harm." 29 U.S.C. § 654 (a)(1). It also requires the Secretary of Labor to promulgate health and safety standards to assure, to the extent feasible, "that no employee will suffer material impairment of health or functional capacity. . . . " 29 U.S.C. 655 (b)(5).[24] Title VII of the Civil Rights Act of 1964, as amended, 42 U.S.C. § 2000e, *et seq.*, prohibits sex discrimination in employment, including discrimination on the basis of "pregnancy, childbirth, and related medical conditions." 42 U.S.C. § 2000e (k). Discrimination is defined simply as different treatment, regardless of motive. Thus, a practice or policy that openly treats women differently, even if for purportedly benign purposes, would be apparently discriminatory under Title VII.

Nothing in either Title VII or the OSH Act indicates any Congressional intent to offer pregnant employees less or different legal protection. Both statutes are comprehensive in their design, one promising a "clean and healthful workplace" to "every working man *and woman*" (emphasis added), the other prohibiting discrimination in employment on the basis of sex, race, national origin, and religion, unless justified by specific defenses. These statutes would appear to disallow exclusionary policies, not only because of their facially discriminatory aspect, but also because of their inherent admission that a hazardous condition exists in the work environment. Read together, these laws appear to require that employers protect the health of *all* employees in a nondiscriminatory fashion, i.e., to provide whatever degree of workplace safety and health protection is necessary to safeguard fully the employment rights of pregnant women or any other allegedly hypersusceptible group protected by Title VII.[25] The courts to date, however, have declined to enforce these statutes as written, and thus no clear message has been sent to employers about the full extent of their apparent statutory obligations.

Three federal Courts of Appeals have examined the legality of exclusionary employment policies under Title VII.[26] Although each approached the issues somewhat differently, all agree that policies that target pregnant or fertile women are discriminatory, even if the purported justification is a benign one related to health protection. Two courts accept the notion that some defense should be available,[27] but in both cases the burden on the defense appears to be substantial. For example, employers would have to show that the chemical or condition poses a significant or unreasonable risk, not just a speculative or hypothetical risk; that exposure at hazardous levels is likely to occur; that the risk is confined to the fetus *in utero*, i.e., the male reproductive system is not similarly susceptible to injury;[28] and that such a substantial body of expert opinion in relevant fields supports this conclusion that "an informed employer could not responsibly fail to act." Even an otherwise justifiable policy would be unlawful if there were an alternative that would provide protection with a less discriminatory effect.[29]

One court has held that the OSH Act's "general duty clause" does not bar all exclusionary policies. In *OCAW v. American Cyanamid*[30] the Court of Appeals for the District of Columbia held that the OSH Act's requirement that employers furnish workplaces "free from recognized hazards . . . likely to cause . . . serious physical harm" was not clearly intended to prohibit the use of exclusionary policies, but was aimed instead at chemical or other exposures posing direct health threats. This decision did not address the ability of OSHA to promulgate occupational exposure standards that would preclude use of exclusionary policies or fatally undermine the rationale for such policies. OSHA's authority in this regard has been upheld in another decision of the District of Columbia Court of Appeals. The court sustained OSHA's determination that lead adversely affects both male and female aspects of reproduction and held that OSHA "has statutory authority to protect the fetuses of lead-exposed working mothers. Harm to fetuses, as OSHA contends, is a material impairment of the reproductive systems of the parents."[31] This court went further in affirming the degree of protection potentially afforded by the statute: "fertile women can find statutory protection . . . discrimination [resulting from exclusionary policies] in the OSH Act's own requirement that

OSHA standards ensure that '*no* employee will suffer material impairment of health. . . .'"[32] This statement reflects this court's clear understanding that, even if hypersusceptible, fertile and pregnant women are entitled to the full protection of the OSH Act, at least so long as OSHA sets comprehensive standards.

IV. Legislative Proposals

A. Long-Range Legislative Goals

Any attempt to amend existing legislation requires important tactical decisions and judgments about the political composition of the particular legislative body, the likelihood of success, the possibility that success in one area will require compromise in another, and so forth. These considerations may be especially relevant in this context since, no matter how framed or characterized, any proposal will necessarily implicate existing antidiscrimination and occupational health and safety law. Advocates for change may well be unwilling to achieve progress in this area at the cost of other statutory protection or benefits. Timing and tactics may thus be critical, and it may well be necessary to wait for an auspicious political moment to make any of the following proposals, which are included on the assumption that such a moment will ultimately arrive.

Because of the inherent connection between regulation of exclusionary policies and existing federal/state laws, the first two proposals are framed in terms of amendments to these laws. Several considerations recommend this approach. First, this statutory framework already exists, and as a matter of logic, exclusionary policies raise questions under both types of laws. It is unlikely and undesirable that the existing framework will be dismantled, and, so long as it exists, questions about the legality of exclusionary policies must be resolved consistently with the interpretations and applications of these statutes. Second, health and safety standards that apply to reproductive risks should mirror those that apply to other health risks. The adult worker is entitled to the same degree of protection for all other body systems as is accorded the reproductive system. Granting preferred status to reproductive or procreative health concerns would suggest that these functions and concerns have higher priority than others, and would also grant more rights to the workers' prospective or actual fetus than the workers themselves enjoy. Finally, a body of generally favorable caselaw relating to burdens of proof, defenses, and presumptions has already developed under existing federal and state anti-discrimination laws. Any proposal ought to take advantage of this circumstance, without the necessity of recreating it in a separate statutory scheme, which, in any event, would not address the range of issues reached in the caselaw.

Some of the following proposals would merely clarify current statutory language to provide explicitly protection that is arguably already included within these statutes. Other proposals would expand the statutory entitlement in potentially far-reaching ways. In either case, the effort to achieve amendments in important federal statutes must be undertaken as a long-term project, since it inevitably involves significant planning and coordination, requires careful timing, and depends in part on both politics and serendipity.

A more limited alternative, available at either the federal or state level, might be appropriate in certain circumstances. One model, included here, would prohibit any form of exclusionary policy based on any genetic, physical, or biological characteristic. This approach is the simplest way to address the issue, but would not achieve overall improvement in workplace safety and health conditions, an objective that can only be met through laws or regulations to ensure occupational safety and health or to control toxic substances. This approach is thus less than optimal to achieve the desired results, and should only be undertaken if comprehensive reform is not attainable, or if health and safety concerns have been otherwise addressed.

In each instance where actual language is proposed, the language reflects the optimum goals sought, not what may result after the bargaining that inevitably accompanies the legislative process. What compromise, if any, might be acceptable will depend heavily on the particular context and cannot be proposed or predicted in the abstract. These proposals thus represent a starting point for individuals and groups advocating the kinds of changes suggested here.

1. Proposed Amendments to Occupational Safety and Health Laws

The most obvious and direct way to address exclusionary policies is to require employers to provide a safe workplace, and to prohibit use of exclusionary policies or sterility requirements to achieve that end. In so doing, legislation should affirm its intent to protect against injury to the reproductive or sexual system, including impairment of the ability to produce healthy children, to the same extent that it protects against all other injuries or physical or functional impairments.

First, a definitional amendment would clarify that the statutory mandate is comprehensive and encompasses protection against injury to reproductive health, broadly defined to cover men, women, and any effects upon the offspring of either. Arguably, this would not expand the scope of current law, but would clarify existing language, which contains neither exceptions to coverage where there are risks to reproductive health, nor limitations on what kinds of injuries or risks are included. This kind of amendment would highlight the past failure to fulfill the statutory mandate to protect against injury to reproductive health, and it would reinforce the proposition that protection of reproductive health must focus on both sexes.

The second aspect of the proposal would directly prohibit the use of exclusionary policies and any other employer practices that might induce workers to alter their bodies in order to qualify for employment. This amendment would legislatively reverse the decision in *OCAW v. American Cyanamid Co.*[33] and make clear the legislative intent that employers bear the primary responsibility for carrying out the purposes of the law by improving workplace conditions, not by excluding workers. Notwithstanding the single opinion to the contrary, this too is arguably only a clarification of existing statutory language.

The final element of this part of the proposal is the creation in individual employees of a private right of action to enforce the provisions of the statute of which they are the intended beneficiaries. The OSH Act can be enforced only by the Secretary of Labor and provides no remedies for aggrieved individuals, a characteristic that undermines the efficacy of the Act generally. The extent to which its provisions are enforced is thus variable but, in any event, is often outside the control of workers. Private enforcement (optimally providing individuals with an optional inexpensive administrative forum, in addition to court remedies), which could be undertaken by any aggrieved person, would expand tremendously the impact and efficacy of the act. Such a proposal to increase the overall effectiveness of the statute would therefore likely be most controversial and would require a major legislative effort, in which the support of organized labor would be critical.

A potential impediment to amendment of state OSH acts is the possible preemptive effect of federal law and regulation. As a general proposition, the states are precluded from acting when the federal government demonstrates its intent to regulate the field, and the OSH Act has generally been held to express such an intent. However, some limitations exist. The OSH Act expressly permits states to "assert [ ] jurisdiction over any occupational safety and health issue with respect to which no standard is in effect." 29 U.S.C. §667(a). When a standard is in effect, the states can still act pursuant to an OSHA-approved "state plan." 29 U.S.C. § 677(b). If the primary purpose of the state law or regulation differs significantly from the purposes reflected in federal law or regulation, preemption doctrine should not apply.[34] Finally, there may be an exception from preemption rules for political subdivisions of states.[35]

Although the following proposals pertain specifically to occupational safety and health legislation, they would also apply, with some modifications, to environmental legislation. Environmental legislation emphasizes regulation of chemical substances themselves, not regulation of the workplace environment.[36] The major elements necessary to adapt these statutes would be similar to those outlined above: protecting against reproductive injury, broadly defined to include injuries to adult males and females and to future offspring of either; prohibiting any kind of exclusionary or other policy that shifts the burden for health protection to the individual from the maker or user of the chemical; and vesting in the statute's intended beneficiaries the right to enforce its terms.

Amendment of environmental legislation, if successful, could provide extensive protection, at least in theory.[37] In practice, problems might persist. EPA, like OSHA, has been criticized for regulatory delays and inaction. Of more significance is EPA's own limited interpretation of its statutory obligation to protect fertile or pregnant women. For example, EPA's proposals have frequently incorporated discriminatory elements, either barring women from hazardous exposures or simply requiring labels to warn of the risk--rather than banning the chemical or seriously restricting its use, as with DBCP.[38] There is legitimate concern that efforts to amend these laws would enjoy neither institutional support from the relevant regulatory agency nor a consistent history of efforts to provide equal protection to both sexes.[39]

Proposed Language to Amend Occupational Safety and Health Statutes

1. Definitions

The terms "serious physical harm" and "material impairment of health or functional capacity" shall include, without limitation, harm to or impairment of the reproductive, procreative, or sexual function of any employee; and any workplace condition, exposure, or activity that is causing or is likely to cause any of the following effects shall be deemed included:

a. Sexual dysfunction (libido, potency)
b. Sperm abnormalities (number, mobility, shape)
c. Subfecundity or infertility of male or female origin
d. Illness during pregnancy or parturition
e. Embryonic or fetal loss
f. Perinatal death
g. Abnormal gestational age (prematurity, postmaturity)
h. Birth defects, decreased birth weight, or multiple births
i. Chromosomal or genetic abnormalities
j. Childhood illness or death
k. Premature menopause.[40]

2. Duties of Employers[41]

a. Each employer shall comply with the provisions of this chapter and shall fulfill all the duties and obligations created thereunder in a manner that does not discriminate or condition employment in any way on the basis of sex (including pregnancy or the capacity to become pregnant, childbirth, and related conditions), race, national origin, ethnicity, or disability.

b. No employer may fulfill the obligations imposed by this chapter by any practice, rule, or policy that has the intent or effect of inducing or requiring any employee to alter his or her physical or biological characteristics in order to avoid risk of harm or material impairment of health or functional capacity, including without limitation any policy permitting only sterile workers to perform specified jobs or tasks.

3. Private Right of Action

Any employer who violates the provisions of this Act [or specified subsections] shall be liable to any person(s) aggrieved thereby.

a. A complaint may be filed with the Secretary within two years of an alleged violation. The Secretary shall conduct an investigation into the charges, recommend a disposition of the charges prior to holding a hearing, hold a fact-finding hearing if the proposed disposition is not accepted by the parties, and render a determination, which shall be enforceable by the Secretary or the complainant in the United States district court, as provided in Section (b), below.

b. In addition to, or in lieu of, the remedy provided in Section (a) above, an action may be brought in any court of competent jurisdiction, including any United States District Court in the state in which the alleged violation occurred or in which the employer has its principal place of business, within two years of the alleged violation, and may be maintained by one or more aggrieved individuals on his or her own behalf and on behalf of others similarly situated.

> The court in such action, upon a finding of a violation, shall award compensatory damages, costs, and attorneys' fees, and may award declaratory and injunctive relief and exemplary damages.

2. Proposed Amendments to Antidiscrimination Laws

Even if the previous proposals are adopted, amendments of fair employment laws would still be appropriate, both to clarify the dual nature of the problem created by exclusionary policies and to provide alternative administrative and court remedies. In addition, the private right of action already available under these laws might be a necessary enforcement tool if one cannot be constructed in the health and safety context.

The following proposal would unequivocally define exclusionary policies, or policies that have similar effects, as discriminatory and therefore unlawful employment practices. They would thus require all health-related employment needs to be resolved on a nondiscriminatory basis.[42] Emergencies may be handled either by suspending production until the hazard can be controlled, or if complete control is impossible, by offering all employees paid disability leave or transfers with no loss of pay or benefits. The proposal corresponds with accepted principles of nondiscrimination law and would clarify that there is no defense for intentional discrimination in this area, as is the case in analogous situations. Alleged hypersusceptibility provides no basis for a defense because to allow such a defense would undermine the express purposes of existing laws to protect workers and to prevent discrimination[43] and would in addition suggest wrongly that the worker is responsible for his/her physiological makeup and can be penalized for it. Nor would the cost of abating a hazard excuse noncompliance, a principle incorporated in present law.[44]

This proposal would remove any incentive under law to identify workers as hypersusceptible. However, at the same time, it recognizes the possibility of hypersusceptibility in fact and protects individuals by assuring that no loss of rights or benefits can flow from their biological characteristics. It is assumed that any disincentive to employers to undertake research about hypersusceptibility will be negated by the fact that most research is done under academic or government auspices or sponsorship, and that research on many substances is required by law and the protocols prescribed by regulation; such research is therefore likely to proceed as before.

As noted previously, the feasibility of any proposal to amend federal civil rights legislation depends heavily on the political environment and the willingness of the civil rights community to undertake such a project. Even though the substantive elements of this proposal should be less controversial than, for example, creation of a private right of action under the OSH Act, the attempt to amend existing law may threaten other aspects of the law and must be approached with suitable care.

Proposed Language to Amend Fair Employment Laws

Unlawful Employment Practices

> It shall be an unlawful employment practice for an employer to adopt or implement any health, safety, medical, or industrial hygiene plan or program that has the purpose or effect of discriminating against applicants for employment or employees on the basis of sex (including pregnancy or the capacity to become pregnant, childbirth, and related medical conditions), race, national origin, ethnicity, or disability; except (i) no employer shall be liable under the provisions of this title for suspending production or operations in order to prevent or abate an unpredictable emergency health hazard, provided that layoffs and recalls occur in an otherwise lawful, nondiscriminatory fashion; (ii) if technology does not exist to abate a hazard, an employer who provides complete information regarding the nature of the hazard and offers paid disability leave or voluntary transfers without loss of pay, fringe benefits, or promotional opportunities to all affected employees on a nondiscriminatory basis shall not be liable under the provisions of this title, provided however that nothing herein shall be deemed to excuse noncompliance with any applicable state or federal law regulating occupational safety and health.
>
> Nothing contained in this title shall be construed to permit an employer or labor organization to avoid liability for discrimination (i) because the health hazard is allegedly or apparently confined to one sex or racial,

ethnic, or other group; or (ii) because of the cost of abating a health hazard.

3. Freestanding Legislation to Prohibit Exclusionary Policies

The following proposal may be feasible and appealing because of its simplicity. This proposal merely prohibits the use of any kind of exclusionary policy or practice. As freestanding legislation, it would not fit into a preexisting statutory scheme, complete with judicial interpretations. A further danger of this limited approach is that it does not reinforce the affirmative obligation to provide a safe workplace for all workers, whether pregnant and/or hypersusceptible. If existing occupational safety and health or environmental protection laws were vigorously enforced, as written, this would not be a problem. Because they are not, the risk remains that all workers will face workplace hazards, including reproductive hazards, without effective recourse. For this reason, the following proposal is offered only with the substantial caveats stated earlier.

Discrimination on the Basis of Physical Characteristics Forbidden

> No employer subject to the provisions of this Act shall deny, limit, or condition employment in any way on the basis of the following, nor shall any of the terms, conditions, rights, or privileges of employment be affected in any way, either directly or indirectly as a result of the operation of any employment-related policy or practice, on the basis of any of the following:
>
> A. reproductive capacity, including the ability or inability to bear children
>
> B. genetic characteristics, either individually determined or presumed based on an individual's race, sex, religion, national origin, or disability
>
> C. any other physical or biological characteristic that does not impede an individual's ability to work.[45]

Commentary

The purpose of the foregoing proposals is to safeguard the right of women workers to remain fertile (if they wish to do so) without negative economic consequences, and to ensure that workplace health protection is afforded to all in a nondiscriminatory fashion. The present proposals do not address issues relating to monetary recovery for workplace-related injury to reproductive health or function, or to certain deficiencies in the law that extend beyond the problem addressed by this position paper.

Monetary recovery for work-related injury is governed by the workers' compensation system and the civil (tort) law generally.[46] These systems have been criticized for various deficiencies that are not unique to occupationally induced reproductive injury, in particular the inadequacy of the remedy, the difficulty of proving causation, and the operation of statutes of limitations that may prevent recovery altogether for injuries that are not immediately apparent.[47] However, since one purpose of the present proposals is to attain consistency with regard to treatment of workers' health needs--not to prefer (or disadvantage) some needs (or some workers) over others, but to provide adequate protection for all--the deficiencies in these remedies must be addressed comprehensively to eliminate the unfairness of the system to all workers.

For similar reasons, problems regarding regulation and testing of toxic substances or new chemicals are not addressed here, nor is the need to eliminate gaps in coverage under the OSH Act. Many workers, such as household and agricultural workers, disabled workers in sheltered workshops and institutions, and others, are excluded from coverage despite the fact that they may encounter substantial occupational health risks. It is critical that occupational and environmental legislation protect the most vulnerable, least powerful segment of society, which often bears a disproportionate share of the burdens inflicted by toxic exposures. This is a generic issue that must ultimately be addressed, although a solution is beyond the scope of this position paper.[48]

Because various genetically identifiable groups might be hypersusceptible to the effects of certain chemicals, and because of the need not to imply that other forms of discrimination are permissible, the proposals comprehensively bar all forms of discrimination that might arise in this context, including discrimination against disabled workers. Employers may develop and enforce medical programs that sometimes affect job placement, but they must do so in a way that does not unfairly disadvantage workers because of membership in designated groups and in a way that reasonably accommodates workers who are disabled within the meaning of the law. Thus, an employer may adopt a neutral policy that provides transfer opportunities for workers whose job placement threatens their physical well-being,[49] so long as the policy is voluntary and fairly administered and does not have a disproportionately adverse effect on a particular group of employees.

These proposals would not bear significantly on the question of employer liability for injuries to workers' future children resulting from negligence in failing to discover or abate a workplace reproductive hazard. The effect of these proposals would be to eliminate exclusionary policies as a technique to avoid such potential liability, but that would likely have a negligible impact on employer activities, because this is a negligible source of potential liability in industries where exclusionary policies have been commonplace. In chemical, petrochemical, and other industries, hazardous chemicals abound. The production process, the use of products by consumers, and the disposal of waste byproducts all involve significant risks intrinsic to the nature of these businesses. The recent discovery of dioxin contamination from disposal of industrial waste underscores this point. The Agent Orange claims, the Love Canal episode, the DES lawsuits, and the problems resulting from asbestos exposure, kepone contamination, and the discharge of polychlorinated biphenyls (PCBs) all reveal the multiplicity and magnitude of risks that chemical and other manufacturing entails.[50]

Although one stated rationale for exclusionary policies has been the necessity of avoiding liability for injury to future children of women workers, companies that have these policies are frequently unable to cite a single instance in which such a lawsuit has been brought.[51] In contrast, numerous widely reported charges have been pressed by male workers that *their* workplace exposures have resulted in injury to their children.[52] This is not surprising, given the fact that more men than women are employed in many of these industries and that men experience much greater cumulative occupational exposure per child born than do women.[53]

The rationale for removing potential liability as a justification for exclusionary policies has been stated recently by the Eleventh Circuit Court of Appeals:

> In today's litigious society, the potential for litigation rests in almost every human activity. For example, every employer faces the risk that a pregnant employee will

> encounter a workplace activity that would not normally be hazardous to a nonpregnant employee, but which could prove injurious to a developing fetus. These hazards range from slippery floors to uncontrolled cigarette smoke, asbestos, known and unknown carcinogenic materials used in the workplace and a plethora of other hazards of modern society. The employer is, of course, free to protect itself from financially ruinous lawsuits by purchasing insurance and by maintaining the degree of care required by law.[54]

This accords with the rule that discrimination may not be excused because of costs associated with nondiscrimination. E.g., *Arizona Governing Committee v. Norris*, 103 S.Ct. 3492 (1983) and *Los Angeles Dep't of Water and Power v. Manhart*, 435 U.S. 702 (1978) (discrimination in pension and insurance benefits may not be excused because it will cost the employer more to provide equal benefits). Further, and more to the point, an employer's best protection is to prevent foreseeable injury. The employer who conscientiously monitors the workplace environment, informs employees fully, abates known hazards, and when necessary offers voluntary transfer without loss of pay or benefits will be best insured against an adverse judgment. The law does not require an employer to have a crystal ball--only to be prudent and diligent. No less should be expected from those whose profit-making activities[55] expose others to potential hazards.

Exclusionary policies exist because of cost consideration: because women are viewed both as marginal workers and as more physically vulnerable, and thus more expensive employees. The costs may be those associated with cleanup, with voluntary temporary transfer systems, with compound substitution of nontoxic materials, with more sophisticated and wearable personal protective equipment protection, or with insurance. The money might be spent in various ways, but necessarily directed at the goal of affording full employment rights without sacrificing safety or bodily integrity. Arguably Congress has already mandated this approach: "even a very high cost could not justify continuation of the policy of discrimination against pregnant women that has played such a major part in the pattern of sex discrimination in this country."[56] The foregoing proposals are intended to help fulfill this commitment.

B. Interim Measures

It is reasonable to view the preceding legislative proposals as long-range goals, at least at the federal level. The likelihood of their ultimate enactment will be enhanced if intermediate steps are taken to heighten public awareness of the problem and proposed solutions, to educate lawmakers, to achieve reform and create workable "models" on the state and local level, and to stimulate regulatory reform.

Some of the legislative proposals could be accomplished through regulation, since they are consistent with existing statutory language. Regulatory reform is less secure and more subject to political influence than is legislation, but it nonetheless could provide a significant remedy in the absence of legislative reform. Other activities would also create a favorable climate for comprehensive reform: legislative hearings would serve an important informational and educational purpose even if not attached to proposed legislation; testimony in support of rulemaking and before state and local legislative bodies; comments to regulatory agencies on relevant proposals; pressure on agencies and academic bodies to conduct or sponsor appropriate research and studies; and general monitoring of governmental and employer conduct.

Persistence and diligence will be required in this effort. No exclusionary policy or proposal should go unquestioned. Scientific and medical data must be provided in each instance to document precisely what is, and is not, known about the effects of a particular exposure. Accurate data and professional risk assessment are central to any opposition effort. Articles in local or national press will alert politicians and others of the significance of the issue, and the need for regulation, but systematic and coordinated efforts to get lawmakers to hold hearings will still be necessary. A coalition effort will be most effective. A natural constituency exists among labor, health, environmental, and women's rights activists. If the efforts of these individuals and groups can be directed, coordinated, and mobilized, a powerful force for change can be created.

V. Additional Areas for Education and Policymaking

A. Sexism in the Design and Evaluation of Scientific Research

1. Research has focused disproportionately on female aspects of reproduction, creating an illusion that women (and fetuses) are at greater risk. However, the greater number of studies on female reproduction is not evidence of hypersusceptibility, but simply reflects the fact that women have been studied more intensively. In these circumstances, studies of males showing adverse effects, even if fewer in number, cannot be discounted, nor can studies demonstrating a mechanism of toxic action that could occur in either sex. Likewise, the absence of data on the effects of a substance on one sex (usually male) does not signify the absence of risk. It only means the question has not been examined. These problems will be ameliorated if research examines as fully as possible the effects of chemical and other agents on both sexes and on the whole body, and if those who rely on this information openly recognize and acknowledge these limitations.

2. Epidemiological studies frequently fail to control for paternal exposures, but assume that the only possible route of injury to a future child is through maternal, *in utero*, exposure. However, in environmental studies, both parents must be assumed to be similarly exposed. In studies of populations living near smelters, it may be reasonable to assume that paternal exposure may be significantly greater, since more men are likely to be exposed occupationally as well. For purposes of documenting a sex-specific effect, such studies are critically flawed, although they may be useful indicators of toxic exposures with undetermined mechanisms of action.

3. Inconsistent attitudes towards experimental or epidemiological data are manifested in a highly critical approach to studies demonstrating adverse male effects, resulting in a rejection of the findings or implications of such studies, although similar data demonstrating adverse effects of chemicals on the fetus are accepted, since such an effect is expected.

4. It is assumed that any substance may be hazardous to a pregnant woman unless its safety has been proved. In contrast, it is commonly assumed that substances are safe for males unless they have been proved unsafe. These diametrically opposing presumptions are based on unexamined assumptions of fetal hypersusceptibility, and ignore elements of male reproductive physiology, most

particularly the vulnerability of sperm during cell division, called meiosis. The potential of toxic chemicals to influence delicate biological arrangements, which exist in both sexes, should ordinarily create a presumption of hazard for both, absent evidence to the contrary.

B. Assistance to Individuals Confronting Occupational Reproductive Health Hazard

Little information is available to prospective parents to help them evaluate the presence of possible reproductive hazards in work environments. Employers may be less than fully informative, and may try either to minimize concern or to induce the concerned individual to quit. Obstetricians may advise a pregnant patient to stop working if she is worried about work-related exposures, without realizing either that unemployment may have adverse nutritional and other health consequences or that it may be impossible or undesirable. Obstetricians and other physicians may also fail to counsel female patients about hazards their mates may encounter, unless infertility is involved. The employer, union, or physician who tries to provide useful information may also be stymied by the absence of meaningful data correlating exposure levels, timing of exposures, and effects.

More information addressing these practical concerns is needed. Even more important is the fact that the information already available is not always communicated to those most concerned--prospective parents. Medical schools, research facilities, unions, and government agencies can and do provide this service, but more specific targeted information is needed. Further, outreach is necessary, since information must be communicated in some cases before individuals realize they need it. For example, workers need to know about exposure that can cause preconception reproductive injury before they attempt to conceive.

C. Workers' Bill of Rights[57]

The optimal goal in dealing with occupational reproductive health hazards, and other occupational hazards, would be the creation of an enforceable comprehensive "workers' bill of rights." Though it is improbable that such a goal is achievable legislatively in the near future (at least until the groundwork is laid by the acceptance of the modifications of existing law set forth above), it provides both a longer-term objective and an educational and organizing tool for individuals and groups counseling workers. The following are proposed as elements of such a bill of rights for workers.

- Each employer should be required to furnish a workplace, and working conditions, that are safe for every worker.

- Employers should be precluded from any activity, or any personnel practice, that burdens reproductive choice. (Exclusionary policies, exposure to reproductive toxins, and provision of inadequate medical insurance would be included among prohibited acts.)

- Government should set permissible exposure limits for hazardous substances at levels that protect against all known or reasonably suspected adverse effects.

- Government should require comprehensive testing of workplace hazards or potential hazards, and testing should always include both male and female reproductive effects.

- Government policies and programs should always emphasize source reduction as the solution to occupational health hazards.

- Workers should have an integral role in the design, development, and implementation of workplace health and safety programs, and should be responsible for the administration of all worker education programs.

- Workers should have the right to refuse to work under hazardous conditions.

- Employers may offer, on a voluntary basis only, pregnancy testing, genetic testing, and participation in medical research projects, but mandatory testing or participation should not be permitted. Any information obtained by employers should be kept strictly confidential, and should not be used for any employment-related purposes other than voluntary medical counseling.

- The health care profession should be encouraged to provide extensive training, on a continuing basis, in the areas of occupational health hazards and reproductive hazards to both sexes, and should require certifications of competence for all appropriate health care providers, e.g., internists, obstetricians and gynecologists, midwives, urologists, nurse practitioners, etc.

- All employees should receive benefits to ensure that their work-related obligations do not impair their ability to meet essential personal or home-related obligations. Included would be paid medical or disability leave, and reasonable leave to care for dependent family members temporarily incapacitated by illness or injury.

Notes and References

1. These facts appear as stated in the Second Amended Complaint, Christman v. American Cyanamid Co., Civ. Action No. 80-0024(P) (N.D.W.Va.).

2. This citation was ultimately sustained with regard to the employer's failure to provide adequate training on use of respirators. The company contended that the government had failed to prove exposures above the permissible levels because it had not taken readings inside respirators, and prevailed on that argument, notwithstanding the ironic result of the two diverse holdings. An appeal was taken and the case was ultimately settled. Secretary of Labor v. American Cyanamid Co., OSHRC Docket No. 79-2438.

3. Regarding risks of low-level lead exposure, see 43 Fed. Reg. 52,952-53,014 (1978) and 50 Fed. Reg. 9386-9408 (1985).

4. See 45 Fed. Reg. 7514 (1980) and 46 Fed. Reg. 3916 (1981).

5. American Cyanamid Co., Olin Corp., General Motors, Gulf Oil, B.F. Goodrich, and Globe Union have been the subject of court or administrative proceedings challenging their policies. Dow, DuPont, and BASF Wyandotte have publicly described theirs. Allied Chemical, Bunker Hill Smelting, St. Joseph Zinc, Eastman Kodak, and Firestone Tire and Rubber have been identified by press and commentators as maintaining such policies. Documents produced by the American Cyanamid Co. in litigation also identify Union Carbide and Monsanto as having maintained exclusionary policies.

6. Minutes of the Environmental Health Committee of the Lead Industries Ass'n, Inc., Sept. 9, 1974, cited in Stellman, J., Women's Work, Women's Health, p. 178 (1977).

7. In 1977, only one percent of married blue-collar working women aged 30 or over expected to bear a child within the next year. U.S. Bureau of the Census, Dep't of Commerce, Current Population Reports, Series P-20, No. 325, "Fertility of American Women: June 1977," Table B, at 3 (1978). The average woman in this country will bear only two children, and working women have fewer children than nonworking women. U.S. Bureau of the Census, Dep't of Commerce, American Women: Three Decades of Change, pp. 6-8 (1983).

8. This does not imply that the fetus is never at increased risk, but only that an unsupported assumption to that effect is invalid.

9. Toxic substances can affect the normal development of the fetus at three stages of the reproductive process. Gametotoxins are substances that cause malformations of the egg or sperm prior to conception thereby impairing the exposed individual's ability to produce a healthy fetus. Mutagens are chemicals that cause alterations in the chromosomal structure of the DNA molecule in the male and female reproductive cells that can be manifested by abnormal fetal development and genetic defects in later generations. Teratogens are substances that operate directly on the fetus and impair normal growth after conception. Stellman, The Effects of Toxic Agents on Reproduction, Occupational Health and Safety 36, 38 (April 1979); Strobino, Kline, and Stein, Chemical and Physical Exposures of Parents: Effects on Human Reproduction and Offspring, Early Human Development 371, 378 (Jan. 4, 1978).

10. U.S. Dep't of Labor, Protecting People at Work: A Reader in Occupational Safety and Health, pp. 305-307 (1980).

11. 50 Fed. Reg. 1123 (1985). The male spermatozoa may be particularly sensitive to chemical and other injury. The testes, vulnerably located in the scrotal sack outside the body, are the "sperm factories," where sperm are constantly being produced. It takes about 72 days for each sperm to mature through the type of cell division called meiosis, known to be one of the body processes most susceptible to chemical toxicity. Exposure of men to chemicals or physical hazards such as radiation may result in mutations in sperm (changes in the hereditary information they carry); may cause sperm deformities, slow movement of sperm, or reduction in sperm numbers; may affect hormones essential to reproduction; and may alter sexual behavior.

12. 43 Fed. Reg. 52959-52960 (1978) (emphasis added). Recent evidence indicates that adult males may be at particular risk for cardiovascular disease from exposure to low levels of lead. See 50 Fed. Reg. 9386-9408 (1985).

13. United Nations Scientific Committee, Sources and Effects of Ionizing Radiation, p. 9 (1977).

14. EPA, Report of the Teratology Policy Workgroup, May 1, 1985, p. 25; Inside EPA Weekly Report, Vol. 6, No. 27, pp. 11-13 (July 5, 1985). See also Council on Environmental Quality, Chemical Hazards to Human Reproduction (1981) and Office of Technology Assessment, Reproductive Health Hazards in the Workplace (1985) for examples of other workplace hazards to male reproductive and sexual health.

15. NIOSH Current Intelligence Bulletin 39 (May 2, 1983), Glycol Ethers, 2-Methoxyethanol and 2-Ethoxyethanol.

16. J. La Dou, Ed., State of the Art Reviews: Occupational Medicine: The Microelectronics Industry (1986).

17. See, e.g., Butterfield, B., "Experts Say Research Raises Questions," Austin American Statesman, Feb. 22, 1987, p. H1, col. 1; Schmitt, C. H., "Group Won't Conduct Chip-Miscarriage Study," San Jose Mercury News, Mar. 6, 1987, p. 13E, col. 1.

18. U.S. Dept. of Labor, Protecting People at Work: A Reader in Occupational Safety and Health, p. 273 (1980).

19. E.g., Doerr v. B.F. Goodrich Co., (N.D. Ohio, Civ. Action No. 81-1745).

20. 43 Fed. Reg. 54415 (1978).

21. Id. at 52954, 54416.

22. Id. at 54435.

23. Many states have laws modeled after the federal laws described here, and in those states many of the following comments apply directly.

24. The OSH Act contemplates that workplace protection from toxic exposures will be achieved primarily through standard setting. OSHA is empowered to set permissible exposure limits, to require monitoring of the workplace and the individual, and to prescribe medical surveillance and medical removal protection (with rate retention), among other things, in fulfilling the standard-setting obligation. However, the standard-setting process has proved to be slow and laborious: in the first 13 years of its existence, OSHA developed standards for only 23 chemicals. (The Registry of Toxic Effects of Chemical Substances compiled by the National Institute of Occupational Safety and Health (NIOSH) lists some 59,000 chemicals.) Office of Technology Assessment, Preventing Illness and Injury In the Workplace, pp. 226-227 (1984). Some standards have been invalidated as a result of industry challenge, and others are still in litigation. Workers also gain protection under the OSH Act from the Hazard Communication Standard, 29 C.F.R. § 1910.1200, promulgated in 1983, under which chemical manufacturers and importers must provide workers with information about chemical hazards and distributors must do the same for customers.

The Environmental Protection Agency also affects the health of workers through its responsibilities for the protection of farm workers under the Federal Insecticide, Fungicide and Rodenticide Act, its responsibility for the regulation of occupational exposure to ionizing radiation under the Atomic Energy Act, and its general responsibility for the control of pollutants and the transport and disposal of toxic substances. EPA has broad regulatory authority over toxic substances pursuant to the Toxic Substances Control Act (TSCA), which is not specifically designed to regulate occupational exposures but may have implications in the employment context. See n. 36, infra, and accompanying text, for further discussion of the relevance of environmental legislation to exclusionary policies.

25. While women (or their fetuses) have most commonly been so classified, blacks have also been targeted by some employer medical policies. See, e.g., Severo, R., "Genetic Tests by Industry Raise Questions on Rights of Workers," New York Times, Feb. 3, 1980 and Office of Technology Assessment, The Role of Genetic Testing in the Prevention of Occupational Disease (1983).

26. Hayes v. Shelby Mem. Hosp., 726 F.2d 1543 (11th Cir. 1984), Wright v. Olin Corp., 697 F.2d 1172 (4th Cir. 1982), Zuniga v. Kleberg Co. Hosp., 692 F.2d 986 (5th Cir. 1982).

27. In Zuniga v. Kleberg Co. Hosp., supra, the court found an alternate basis to rule for the plaintiff and thus did not rule on the question of defense.

28. Neither of these courts explicitly describes the other kinds of risks to male reproductive health which would serve to undermine the validity of a female-specific policy. Heritable injuries clearly qualify, and other injuries such as infertility, damage to sperm, and sterility might also. Hayes indicates that a failure of consistency in overall health protection suggests that a fetal protection policy is a pretext for discrimination.

29. Although cost is not a defense to overt discrimination, Los Angeles Dep't of Water and Power v. Manhart, 435 U.S. 702 (1978), these courts did not consider the extent to which costs could be imposed on employers to provide safer workplace conditions as an alternative to exclusion of workers.

30. 741 F.2d 444 (D.C. Cir. 1984).

31. United Steelworkers of America v. Marshall, 647 F.2d 1189, 1256 n. 96 (D.C. Cir. 1980), cert. denied sub nom. Lead Industries Ass'n, Inc. v. Donovan, 453 U.S. 913 (1981).

32. Id. at 1238, n. 74 (emphasis in original; citation omitted).

33. See supra for a discussion of this case.

34. For example, that portion of a state law providing workers with information about chemical hazards may be preempted while the portion providing information about environmental hazards will survive. See, New Jersey State Chamber of Commerce v. Hughey, 774 F.2d 587 (3d Cir. 1985).

35. Ohio Manufs. Ass'n v. City of Akron, 801 F.2d 824 (6th Cir. 1986), cert. filed, 55 U.S.L.W. 3544 (U.S. No. 86-1242) (Jan. 28, 1987).

36. In particular, the Toxic Substances Control Act ("TSCA"), 15 U.S.C. § 2601, et seq. grants the Environmental Protection Agency broad regulatory authority over chemical substances or mixtures; registration of pesticides comes under EPA jurisdiction pursuant to the Federal Insecticide, Fungicide and Rodenticide Act ("FIFRA"), 7 U.S.C. § 136, et seq.

37. Some of the limitations in OSH Act coverage, noted in the Commentary, infra, would be avoided by this approach.

38. For example, an exemption for use of ferriamicide, a pesticide, was granted on the condition, inter alia, that "[w]omen of childbearing age are prohibited from loading/applying ferriamicide." 47 Fed. Reg. 46884-46885 (1982). A similar requirement was imposed on users of lindane, 45 Fed. Reg. 83668-83669 (1980); further testing revealed that the fetus was only affected at exposure levels which cause "general toxic effects in the mother." 48 Fed. Reg. 48514 (1983). During the special review process, EPA required warning labels regarding the possible teratogenic effects of triphenyltin hydroxide (TPTH), even though the pesticide is a potential reproductive hazard for both sexes. See 50 Fed. Reg. 1107 (1985). An exemption was recently granted for certain uses of dinoseb on the condition that "[w]omen of child-bearing age, i.e., under the age of 45, may not be involved in mixing, loading, or any aspect of dinoseb application." Decision and Final Order Modifying Final Suspension of Pesticide Products Which Contain Dinoseb, FIFRA Docket No. 612, U.S. Environmental Protection Agency, March 30, 1987, p. 4. This occurred notwithstanding EPA's own findings of multiple toxic effects of exposure, including adverse effects on "the reproductive system of male laboratory animals" and "acute toxicity [at] relatively low doses . . . when compared with other pesticides." 51 Fed. Reg. 36634, 36637 (1986). See also 52 Fed. Reg. 11121 (1987).

39. One possible exception is EPA's Radiation Protection Guidance, which recommends certain precautions to prevent radiation exposure to the fetus in utero (protection which should arguably be available to all workers) but cautions specifically against discrimination to achieve protection. 52 Fed. Reg. 2822, 2828-2829 (1987).

40. Adapted from Bloom, Ed., Guidelines for Studies of Human Populations Exposed to Mutagenic and Reproductive Hazards, p. 44 (1981).

41. For application to environmental laws, this section should be modified to apply to manufacturers, users, distributors, processors, or other commercial actors.

42. While the main target of these proposals is the employer, most nondiscrimination statutes also make a union liable if it acts "to cause or attempt to cause an employer to discriminate." See, e.g., 42 U.S.C. § 2000e-2(c)(3).

43. The present proposal is not confined to a proscription on sex discrimination, since racial and ethnic groups have also been identified as hypersusceptible, and the nondiscrimination principle extends to all such groups.

44. See Newport News Shipbuilding and Drydock Co. v. EEOC, 462 U.S. 669 (1983).

45. The private right of action proposal suggested to amend occupational safety and health statutes, supra, would apply to this provision as well.

46. For a review of the operation of these systems in the area of occupational reproductive health hazards, see Office of Technology Assessment Reproductive Health Hazards in the Workplace (1985). Nancy Gertner's position paper, "Interference With Reproductive Choice," in this book, addresses some of the problems in this area.

47. See Bertin and Henifin, "Legal Issues in Women's Occupational Health," in Stromberg, Larwood, and Gutek, Eds., Women and Work: An Annual Review, Vol. 2, pp. 93-115 (1987).

48. However, many of these problems would be resolved by adoption of a comprehensive Workers' Bill of Rights. See the end of this position paper.

49. For example, a worker with a back condition might require a job that does not entail heavy lifting, a worker with a skin sensitivity to a particular substance ought not to have to work with the substance, and a worker with arthritis might need a light duty assignment. In some states, like New York, future risk and safety to self are not defenses to charges of handicap discrimination. Some conditions, such as obesity, may constitute a disability for employment discrimination purposes, even without an underlying medical cause or even though not a disability for purposes of receipt of rehabilitation services under the Rehabilitation Act of 1973.

50. Similarly, the semiconductor industry faces claims by residents of Santa Clara County, where water supplies have been contaminated by underground waste storage tanks.

51. This kind of evidence is notably lacking from the trial record in the reported cases on this issue which have gone to trial: EEOC v. Olin Corp. 24 FEP Cases 1615 (W.D.N.C. 1980), affirmed in part, reversed in part, and remanded sub nom. Wright v. Olin Corp., 697 F.2d 1172 (4th Cir. 1982), and Zuniga v. Kleberg County Hosp., (S.D. Tex., Civ. Action No. 77-C-62, Jan. 26, 1981) reversed, 692 F.2d 920 (5th Cir. 1982). See also deposition of Robert M. Clyne, M.D., May 16, 1983, p. 191, in Christman v. American Cyanamid Co., supra n. 1.

52. Lawsuits brought by male veterans who served in Vietnam and were exposed to the herbicide "agent orange" exemplify this point. Some actions allege birth defects in subsequently born children, miscarriages, and "serious maladies of servicemen." See e.g., "Five Makers of Agent Orange Charge U.S. Misused Chemical in Vietnam: Companies Replying to Suit, Say Federal Negligence Is Responsible for Any Harm to Veterans and Kin," New York Times, Jan. 7, 1980, p. A14, col. 1. Male workers at a plant in Renssalaer, New York also allege that their exposure to the herbicide oryzalin has resulted in birth defects in their children. They have filed complaints with OSHA and the Environmental Protection Agency. "Union, Citing Birth Defects, Asks Ban On a Herbicide," New York Times, Nov. 9, 1979, p. 16, col. 1.

53. A comparison of male and female occupational exposures preceding live births in 1976 reveals that the occupational exposure per conception is at least twice as great for males as females. The calculation is based on the assumption that the working population procreates at the same rate as the nonworking population, an

assumption that is probably far more true for men than for women. Thus, this figure represents a conservative estimate of the differences between male and female occupational exposure per birth. See V.R. Hunt, Work and the Health of Women, pp. 23-24 (1979).

54. Hayes v. Shelby Mem. Hosp., 726 F.2d at 1553, n. 15.

55. Placing the burden on employers results in cost shifting, as the increased costs associated with the requirement of a safe workplace ultimately get passed on to consumers. In this way, the product or service more accurately reflects the true cost of production.

56. Senate Report, No. 95-331, Senate Subcommittee on Labor, Human Resources Committee, p. 11.

57. I am indebted to Maureen Paul, M.D., and Cynthia Daniels, Ph. D. for their substantial assistance in drafting this section.

INTERFERENCE WITH REPRODUCTIVE CHOICE

Nancy Gertner

Position Paper

This position paper begins with an analysis of all types of reproductive choice, and of the circumstances under which those choices are maximized or undermined. Although we have spoken of reproductive choice most often in terms of the right to choose abortion, in fact, the concept is far broader. The decision whether to bear a child, the timing of childbearing, the means of avoiding or promoting pregnancy, the decision to give birth in a safe and supportive environment, and the techniques for childbirth--all of the varieties of reproductive decisions--are pivotal life decisions. Since they are decisions that enable a woman to determine the quality of her life, they are central to personal autonomy. Indeed, maximizing control over reproductive decisions is a prerequisite to full equality for women.

Moreover, as technology purports to expand reproductive choice, it becomes all the more important to examine the circumstances under which that choice is exercised and burdened. For women who cannot conceive, or for gay women who seek in vitro fertilization, the biases of the medical establishment or economic limitations may determine their exercise of reproductive choice.

A Many-Sided Problem

The assault on reproductive choice has taken many forms, in many settings. Reproductive choices, like all other personal choices, are made in a social, economic, and legal context. In some instances, the major obstacle to reproductive choice is economic; poverty necessarily affects access to health care and information, and thus access to the right to choose. At times, as with the right to abortion, the principal danger is from state action that directly manipulates a woman's choice. More recently, private actors have attempted to interfere affirmatively with choice through social pressure, deception, or overt criminal acts.

The reproductive choice most overtly interfered with is the choice of abortion. From 1973 until the present, the constitutional right to choose abortion, announced by the Supreme Court in *Roe v. Wade*, 410 U.S. 113 (1973), has been severely constricted. Although the state may not make the performance of abortion the subject of criminal penalties, more and more, the Court has been willing, within some limits, to burden the right to choose with increased regulation[1] and by permitting the state to defund abortion. Access to federal funds for abortions under Medicaid was denied by the Hyde Amendment in 1977,[2] and has been limited further by state legislatures. More recent proposals include measures that would put a permanent ban on any federal assistance to programs that provide abortions or counsel clients about abortion services.[3] The net effect has been that this right to abortion, so-called, has become the entitlement of a few relatively privileged women.

Added to the burdens imposed by a state indifferent to creating equal access to a constitutional right is the very affirmative threat offered by private anti-choice groups that have resorted to violence, deception, or threats to deter women from choosing abortion. In Philadelphia, the Northeast Women's Center has been successful in bringing criminal charges of racketeering against 13 anti-choice protesters for their unrelenting pattern of criminal acts against the clinic: destruction of property, harassment and intimidation, and trespass and seizure of the premises. The National Abortion Federation reported 51 bomb threats in 1986 alone, and six facilities were either destroyed or heavily damaged in that same year. Seventy-three percent of facilities were the target of at least one of these illegal activities, including invasions, intimidation, vandalism, arson, kidnapping, bombing, and bomb threats.[4]

"Bogus clinics" designed to intimidate and misinform women under the guise of pregnancy counseling accomplish indirectly what the bombs and arsons do directly. Staff persons at these centers are instructed to lie about the results of pregnancy tests, thereby endangering the lives of women who are pregnant under hazardous circumstances; disseminate false information about contraceptives and abortion; and otherwise threaten or coerce the individual's decision of whether to carry a pregnancy to term. In New York, the Attorney General has announced specific guidelines that cover the advertisement of such facilities so that they may not be confused with medically licensed family planning centers.[5]

Interference with reproductive choice also results from harm done to a pregnant woman that causes fetal death. The harm may take the form of intentional injuries, such as spousal abuse or accidental injuries such as might occur from an automobile accident. A number of well-publicized cases of shocking and brutal assaults on pregnant women by their male partners have come to trial. "Feticide" laws, which have now been adopted by 16 states, are designed to combat this problem by labeling the resulting fetal death "feticide" and punishing it as a form of homicide.[6] However, this type of legislation remains an unsatisfactory solution because, by calling the crime "feticide," it elevates the status of the fetus to personhood and may have a long-range chilling effect on abortion rights.

The reproductive option of sterilization has already long had a pattern of abuse associated with it. Women of color and third world women have been sterilized in disproportionate numbers. Twenty-five percent of women of color have been sterilized as of January, 1983, as opposed to 16 percent of white women.[7] They have been coerced into permanently losing their reproductive capacity in lieu of abortion by state action and inaction, and private pressures. The same government that callously cuts off federal and state funds for abortion assumes full responsibility for coverage of sterilization procedures under Medicaid. Some of the same actors who would deter women from seeking abortion by flooding them with inflammatory religious and moral appeals would take advantage of their ignorance about the irreversibility of sterilization. The issue here is less providing access to sterilization, and blocking threats to its performance, than maximizing information about sterilization, and ensuring informed and free choice.

The corollary of forced sterilization is the effort to deny access to new fertilization technology to disabled women, poor women, or women of color. Health clinics and hospitals ostensibly making choices about the allocation of scarce resources have reportedly opted to choose who may have access to these techniques and who may not according to their own views of social engineering.

New attempts have been made to control a woman's behavior during pregnancy. Obstacles to her freedom may be created to force her to undergo genetic testing, to forego it, or to have a particular form of birth, i.e., cesarean section. More and more frequently, the fetus is being placed in an adversarial relationship with the pregnant woman bearing it, to the point where the woman is valued merely as a fetal container and forced to surrender her rights as a person to a medical/legal community that values her fetus more than it values her.

Recently, the *New England Journal of Medicine* published a special report on the incidence of court-ordered obstetrical interventions.[8] The authors found that court orders had been obtained in 86 percent of the cases in which they were sought, for procedures deemed necessary for the life of the fetus, including cesarean section (15 orders sought in 11 states), intrauterine transfusions (2 orders granted in Colorado), and hospital detention of the mother (2 orders granted out of 3 sought). In 18 states, 36 attempts to override maternal refusal of therapy were reported in the past five years. Out of 21 court orders sought, 18 were granted. Doctors, under the protection of the courts, have usurped the decision-making rights of the patient for the sake of the fetus. As medical technology enables doctors to perform previously impossible procedures on the fetus *in utero*, these new techniques will come to be considered the standard of care.

Forty-six percent of the heads of fellowship programs in maternal/fetal medicine surveyed replied that mothers who refuse medical advice should be detained in hospitals to ensure their compliance. Twenty-six percent advocated state surveillance of women in the third trimester who elect to remain outside of the hospital system, putting the future legality of home birth in question. Only 24 percent consistently upheld the right of the mother to refuse medical advice. Court orders force women to forfeit their legal rights in a manner not required of nonpregnant women or "competent" males.[9] It is not difficult to envision the length to which precedents like these may be stretched. What began allegedly in the best interest of the mother and fetus may end in the tightening of restrictions around the pregnant mother, including demands for prenatal genetic screening and fetal surgery, and restrictions on the diet, employment, and activity of pregnant women.

The Existing Legal and Political Challenges

Over time, it has become more apparent that existing statutes and constitutional theories are less and less adequate to protect reproductive choice. This is so even in the clearest cases--where abortion clinics are bombed, and women seeking abortions are assaulted. Criminal prosecution has a limited efficacy, since it depends upon the cooperation and approval of the state prosecutor. In some jurisdictions, political and religious pressures may keep a prosecutor from seeking to apprehend an individual accused of clinic violence. In some instances, courts appear to confront threshold issues involving reproductive technology with archaic and simplistic legal concepts. The new technology raises questions that have not yet been sorted out in the courts.

Even the centerpiece of litigation concerned with abortion and procreative rights--the constitutional right to privacy--is under considerable attack. Should an anti-choice jurist fill the current vacancy on the United States Supreme Court, it is reasonable to predict that state legislation that goes so far as to criminalize abortion, and perhaps even contraception, will be sustained. Although some state courts may interpret their own constitutions to provide for a right to privacy,[10] there is no guarantee that such a strategy will work nationwide.

It may be that the only way to protect reproductive choice will be through the state legislatures. Whatever the risks of such a course, pro-choice advocates may have no alternative. Below, I propose a statute that creates a tort cause of action for all types of interference with reproductive choice. Such a statute could be introduced in state or local lawmaking bodies as a defensive measure to counter anti-abortion or "feticide" bills or as an affirmative measure. Whether it should be used at all needs to be carefully considered in terms of the particular circumstances in a given jurisdiction at a given time.

By broadening the debate, the proposed statute offers the possibility of creating a coalition that has never before existed--of individuals concerned with incursions on the right to abortion together with those concerned with growing threats to a woman's right to make decisions about her own body during pregnancy.[11] It does so by valuing a woman's reproductive choices equally, no matter whether the decision is abortion or birth. On the one hand, the proposal would give greater protection against abortion clinic violence than currently exists by characterizing such acts as something more than simple assault and battery. At the same time, by diminishing the focus on the fetus and reshaping the unseemly adversary relationship in which the law has placed woman and fetus, it reaffirms the right of the pregnant woman to make decisions during pregnancy. The statute, for example, characterizes situations in which a pregnant woman is assaulted and a fetus dies as "interference with reproductive choice," rather than "feticide," "murder," or "wrongful death."

The proposed statute seeks to protect all kinds of reproductive choice, including not only the choice of abortion, but also the decision to become pregnant. It seeks to respond to abuses on a number of different fronts, not previously addressed in a single statute. At the same time, it seeks to counter interference with reproductive choice in all of the settings in which it occurs, public and private, and as it occurs through both direct and indirect means. However, the statute stops short of calling lack of financial resources "interference with reproductive choice."

It is precisely the breadth of the problem that makes me offer this proposal with some reservations. There is always considerable risk in attempting to codify and concretize the law's response to a complicated and delicate problem. There is the risk of treating uniformly situations that may well call for different approaches--e.g., addressing together abortion clinic violence and the crime of "feticide." It may be that there is less need for legal intervention in some areas of reproductive choice than in others. A statute that appears to weigh all reproductive choices equally in a society that pressures women to have children may dilute the rights under the most fire, like the right to choose abortion. Indeed, it may be argued that the law is best used to protect the most vulnerable choices.

Yet, allowing the law to respond to the problems of reproductive choice in an ad hoc, patchwork-quilt fashion may in fact exacerbate the problems. Permitting a legislature to target particular reproductive choices for special protection will result not in special protection for abortion, but rather in special protections for pregnancy (as we have already seen with Medicaid funding of birth but not abortion). Although no piece of legislation could possibly address all of the issues and concerns just outlined, I offer the statute below as a step in the direction of addressing some of them.

The Statute

I. Findings and Declaration of Policy

The decision whether to bear or beget a child is a personal decision, and is central to personal autonomy. In addition, it is a choice that is crucial for women's equality. All of the protections against discrimination in the workplace will be undermined if women do not have reproductive choice.

Reproductive choice is necessarily part of the sphere of individual privacy protected against state interference. However, this statute also seeks to prevent interference in that personal decision by private actors.

In the past, the phrase "reproductive choice" has been used to refer to the choice to terminate a pregnancy. This statute recognizes that reproductive choice means more than the choice to abort. It seeks to provide protection for all kinds of reproductive choices including, but not limited to, the choice to avoid pregnancy or to terminate an existing pregnancy; the choice to become pregnant or to carry a pregnancy to term; and the choice to be sterilized. This statute also recognizes that reproductive choice may be burdened or interfered with in a wide variety of settings, by a wide variety of actors. It provides a nonexclusive list of some of the ways in which such interference occurs.

It is hereby declared to be the public policy of this state to protect the reproductive choice of all individuals, regardless of their race, color, religious creed, age, national origin, marital status, sexual preference, genotype, or disability and to protect that choice from interference by others.

II. Definitions

A. Reproductive choice shall be defined as:

a. an individual's choice to exercise her constitutional right to the performance of an abortion to the extent protected by state and federal constitutional law

b. an individual's choice to exercise her/his constitutional right to be sterilized or to refuse sterilization to the extent protected by state and federal constitutional law

c. an individual's choice to carry a pregnancy to term

d. an individual's choice to obtain and to use any lawful prescription for drugs or other substances designed to avoid pregnancy, whether by preventing implantation of a fertilized ovum or by any other method that operates before, at, or immediately after fertilization

e. an individual's choice to become pregnant through in vitro fertilization, artificial insemination, or any other procedure.

B. Interference with reproductive choice shall be defined as:

a. using force or coercion that interferes with an individual's exercise of reproductive choice

b. using threats of force or intimidation that interferes with an individual's exercise of reproductive choice

c. engaging in activities that otherwise burden the exercise of the right to reproductive choice, except where those activities are constitutionally protected. Such activities shall include, but shall not be limited to: imposing conditions on an individual seeking to exercise reproductive choice that are not imposed on individuals seeking other types of medical procedures; limiting the exercise of reproductive choice to individuals of a certain age or marital status or with a certain number of children, or otherwise discriminating on the basis of race, color, religious creed, age, national origin, marital status, sexual preference, genotype, or disability; imposing conditions on the exercise of certain reproductive choices that are not imposed on other reproductive choices

d. retaliating or discriminating against an individual because that individual has exercised his/her reproductive choice

e. aiding, abetting, inciting, compelling, or coercing an individual to interfere with the reproductive choice of another

f. interfering with the performance of a duty or the exercise of a function by an employee of a health care facility where reproductive choice is exercised, as interference is defined in sections (a) through (e) above.

C. The term "person" means one or more individuals, partnerships, associations, corporations, legal representatives, or any organized group of persons.

III. Prohibited Acts

A. Access to health care facilities: If a person interferes with an individual's access to, or ability to obtain, health care sought in connection with the exercise of reproductive choice as permitted by law, or with the ability of a health care provider to provide such legal health care, then such individual or such health care providers shall have a cause of action pursuant to this section against the person, provided that the act or acts of interference were undertaken knowingly and either with the intent or purpose of interfering with access to the provision of that type of legal health care, or with reckless disregard of the consequences of his/her acts to an individual's exercise of reproductive choice.

B. General interference with reproductive choice: If an individual otherwise interferes with a person's exercise of reproductive choice as defined above, then such person shall have a cause of action pursuant to this section against the individual, provided that the act or acts of interference were undertaken knowingly, and with the intent or purpose of interfering with the exercise of reproductive choice.

IV. Defenses

A. It shall not be a defense to a cause of action pursuant to this section that a defendant acted in the good faith belief that his/her actions were necessary to prevent another wrong from occurring.

B. It shall be a defense to a cause of action pursuant to this section that a defendant did not know that the individual had exercised, was in the process of exercising, or was attempting to exercise his/her right to reproductive choice.

V. Penalties; Injunctive Relief

A. The plaintiff shall recover the greater of actual damages or liquidated damages as defined herein.

a. actual damages shall be treble the amount of damages proved by the plaintiff. Such proof may include evidence of pain, suffering, and emotional distress;

b. liquidated damages for each act of interference shall be $5000.

B. The plaintiff shall also be entitled to punitive damages.

C. If actual damages equal or exceed liquidated damages, then all individuals engaged in acts of interference at the time said actual damages are suffered shall be jointly and severally liable. If liquidated damages exceed actual damages, then each individual engaged in acts of interference shall be individually liable only for liquidated damages resulting from that individual's acts of interference.

D. Attorneys' fees and costs: A prevailing plaintiff shall be entitled to reasonable attorneys' fees and costs. Liability for the award of attorneys' fees and costs shall be joint and several among all individuals found to have engaged in acts of interference.

VI. Miscellaneous Provisions

A. The filing of an action under this section shall not be deemed a waiver of any recognized privilege, and where requested the court shall permit the plaintiff to proceed pseudonymously.

B. Any interested person may commence an action by mandamus, injunction, or declaratory relief for the purpose of stopping or preventing violations or threatened violations of this article, or to determine the applicability of this article to actions or threatened future actions.

Commentary

Findings and Declaration of Policy

In "Findings and Declaration of Policy," I attempt to provide the public policy underpinnings for the statute. The original draft of this statute was circulated to a wide range of theorists and activists committed to achieving gender equality. Although I have taken some pains to draft the statute in sex-neutral language, I share one reader's concerns that, in so doing, we ignore the particular impact these issues have on women. In addition, I have taken seriously another reader's point that a sex-neutral concept of reproductive choice could be used by men trying to force women they have impregnated to continue their pregnancies--i.e., the men could claim that their reproductive choice is being burdened. Although I have not altered the definition of reproductive choice or the concept of interference, in this section I have described the centrality of reproductive choice issues to the question of discrimination against women. Presumably, any litigation involving this statute would have to be informed by the "Findings" section of the statute. Whether or not that would be sufficient protection to address these concerns is an open question.

Definitions

a. Reproductive Choice: I have defined "reproductive choice" as broadly as I could in II(A), to cover all of the settings in which the choice to bear or beget a child is impinged upon. I have listed the medical procedures that implicate the right to choose, but have tried to broaden my language so as to avoid the symbolic problem of reducing pregnancy solely to a "medical procedure." Although I have included questions of access to medical procedures like in vitro fertilization and artificial insemination in the statute, I, and many of the readers, continue to have concerns about whether or not all of the issues in this area are adequately addressed.

I have adopted the suggestion of one reader by defining the breadth of the right in constitutional terms. The original draft language, "to the extent protected by law," was too vague and might well encourage arbitrary additional laws by state legislators. If federal constitutional law changes dramatically over the next few months or years, this definition may have to be modified.

Section (c) raises several problems. I had in mind situations in which a pregnant woman is assaulted and as a result, loses the pregnancy. I was concerned about whether to penalize a perpetrator under circumstances in which the perpetrator could not have known whether or not a woman was choosing to carry a pregnancy to term. I have resolved the latter issue preliminarily by limiting the tort of interference with reproductive choice to knowing and intentional interference (*see* IIB, and IVB).

It should also be noted that this section is a response to ill-considered legislation making the death of a fetus wrongful death on the civil side, or homicide on the criminal side. It ratchets down the status given to the fetus in these situations--from "feticide," "murder," or "wrongful death" to "interference with reproductive choice." Given that lower status, I am willing to include in the interference tort involving the termination of a pregnancy situations where the mother wanted to carry to term, where she communicated that desire, and even where the fetus was not viable.

Section (e) would penalize clinics that refuse to perform in vitro fertilization, for instance, on lesbian patients, unmarried patients, or those with a disability. There are a number of unresolved medical and legal problems concerning the circumstances under which a medical facility can lawfully refuse to use "high-tech" methods to impregnate a woman. Attempting to deal with such concerns might unnecessarily complicate this statute. Several readers have suggested that one way of resolving the problems is simply to indicate that, if these procedures are done for anyone, they must be done for everyone. I believe that I include this concept in the section in which I define "engaging in activities that otherwise burden . . . choice" (section II(B)(c)). In that section, I have included as interference with choice any efforts to impose conditions on some people not imposed on others. Thus, if a clinic denies access to these techniques because of "medical reasons" (i.e., it would be dangerous for a woman to become pregnant), it must use the same standards across the board.

Bear in mind that the concept of "coercing or threatening" is not so limited; if anyone coerces or threatens a woman not to have these procedures, even if his/her abusive behavior is uniformly administered, it would still constitute improper "interference." Finally, I have decided to continue to list the existing medical procedures as examples of procedures to which access must be guaranteed. It was pointed out by readers that this runs the risk of being outstripped by technology. I believe that, as long as the language is broad enough, there will not be a problem.

b. Interference: I have defined "interference" with reproductive choice in II(B) to cover some of the ways in which choice can be burdened. I have included within the concept of "interference" only overt acts burdening choice. This includes using force or coercion, or threats of force; retaliating or discriminating against an individual because of the choice she has made; aiding and abetting another to interfere with choice; and interfering with a health care facility in which reproductive choice is exercised. In addition, there is a general, residual category of "engaging in activities that otherwise burden the exercise of the right to reproductive choice." This provision would cover employers who affirmatively interfered with a woman's exercise of choice--i.e., making job security contingent on a particular reproductive choice--and would also cover instances in which particular aspects of choice are discriminated against, i.e., insurance coverage that did not pay for abortion. In effect, these affirmative acts form a continuum from criminal activities, which would be offenses under the laws of any state, to activities that are not criminal but that simply burden reproductive choices. I have also included actions that are done with a particular intent; *see* section IV(A).

Not included in this section are situations in which choice is burdened because of inaction, i.e., situations in which the law ought to articulate an affirmative duty to support reproductive choice via funding, provision of facilities, or other types of aid. I have decided that a statute that penalizes actions solely based on the fact that they have the effect of interfering with reproductive choice is too stringent and politically impossible. I believe this to be the case, even if the statute imposes only civil penalties. Therefore, I have limited this draft to actions that have that unlawful effect, but that, in addition, accomplish that result through certain affirmative activities. This is a political judgment. I believe that no legislature will go so far as to require funding to maximize choice. The notion of an affirmative duty to support reproductive choice via funding, provision of facilities, etc., is included in the policy recommendations at the end of this position paper.

The affirmative activities that interfere with reproductive choice are described as inclusively as possible in the statute. In this respect, I have broadened the scope of this statute beyond that which one normally finds in civil rights statutes. For example, in the Massachusetts civil rights statute, interference with civil rights is penalized only where that interference is accomplished through "threats and intimidation," a category that may be broader than the criminal offense category, but not as broad as my concept of "burdening" reproductive choice.

In section (c), I sought language to describe actions that burden choice, beyond that which is accomplished through criminal activities. For example, demonstrators who photograph women on their way to an abortion clinic may not be engaging in criminal activities, but they are nevertheless interfering with reproductive choice by intimidating the women. On the other side would be counselors encouraging minority and/or poor women to be sterilized. I am inclined to leave the word "burden" without further definition, followed by a few examples, leaving the task of clarification to legislative history.

At the same time, I am also concerned with not impinging on First Amendment rights in this regard. Peaceful demonstrations may well burden rights, but do so within the framework of legitimate political activities. Involved here are the same concerns that we find in situations in which access to employment opportunities or public accommodations by minorities or women is blocked by an employer's speech, e.g., the argument that the restaurant name "Sambo's" discouraged black patrons; and the argument that an employer who has pictures of *Playboy* centerfolds in the rest area is creating an environment inimical to women workers. For political purposes, I would include the phrase "except where those activities are constitutionally protected," even though it does not resolve the thornier issues, simply in order to make it clear that we are cognizant of the First Amendment concerns.

A similar problem is presented by doctors who have a conscientious objection to performing abortions. Under current federal constitutional law, such individuals are permitted to decline to perform abortions. Although this is not a problem in the more populous areas, where there are enough physicians, it may be a problem in more remote areas. Presumably, a doctor who declines to perform an abortion, where there are no other physicians available, would be acting with "reckless disregard" of the consequences of his/her actions. I believe that it is an open question as to whether or not such an individual could constitutionally be held liable under this statute.

There is another concern regarding the concept of "burdening" reproductive choice. Where the reproductive choice involved is the choice to continue a pregnancy, it could be argued that it is "burdening" a reproductive choice for an employer to offer inadequate pregnancy leave, even in a situation in which it offers the same inadequate leave provisions for all workers. This situation would be inconsistent with the approach of the statute, although it is one that should be addressed in a separate proposal. (*See* policy recommendations.) The statute, as noted above, deals only with affirmative acts of burdening and not with inaction. Similarly, it could be argued that it is "burdening" a reproductive choice for an employer to require contact with VDTs, or to exclude pregnant women from certain jobs because of occupational health concerns. This statute is not intended to preempt OSHA regulations, but to be an additional cause of action available to plaintiffs.

Finally, there are a number of situations that may not be included within this language and, in fact, that I am reluctant to include. Recently, for example, a New York Court ruled that a man who misrepresents his infertility can be held liable in tort for the consequences. (*Alice D. v. William M.*, 450 N.Y.S. 2d 350, 113 Misc. 2d 940 (N.Y. City Civ. Ct., April 27, 1982)). Would misrepresentation fall within the category of burdening, and should it? Likewise, where do wrongful life actions fit into this category? Although they fit to the extent that they arguably involve interfering with the exercise of reproductive choice, they are generally actions involving negligence and are not knowing and intentional, although they could rise to the level of reckless.

I would include in this section, and particularly in the anti-discrimination clause, one reader's concern that insurance coverage of sterilization but not childbirth or abortion is discrimination. Moreover, given the breadth of the word "burden," I believe this section more generally would include situations in which a woman's job security is made contingent on a particular reproductive choice whether or not one can make out a claim for discrimination. The section, as drafted, describes the "burdening" of reproductive choice, with the discriminatory imposition of standards as one way of "burdening" choice.

In section (d), one reader suggested that the concept of "burdening" should include creating an environment deliberately hostile to reproductive choices. I believe that (d) and (e) would cover that situation.

c. Person: I have defined "person," which is the word I use for the perpetrator of interference with reproductive choice, to include corporate entities and other organizations. There are knotty questions about vicarious liability, that is, holding higher level supervisory personnel and organizations liable for the acts of subordinates. One set of likely defendants in these situations--the protester who blocks the abortion clinic, the individual who dupes the woman into being sterilized, the boyfriend who beats up his pregnant girlfriend--is likely to be judgment-proof, i.e., without assets sufficient to make social control through the use of damage actions meaningful at all. The only potential defendants likely to be controlled by the threat of damage actions are doctors who are likely to have either insurance or assets.

The question is what the model for vicarious liability should be. First, there is the traditional tort model, in which the master is vicariously liable for the torts of a servant acting within the scope of his authority, whether or not the master knew of the servant's specific acts or specifically authorized them. Second is the civil rights model, in which the "masters," like supervisory personnel supervising line police officers or municipalities employing the officers, cannot be held responsible unless some acts or omissions against them are proven, which were the proximate cause of injury to the plaintiff. For example, supervisors of line officers accused of police abuse are liable for performing acts or not performing acts that they ought to have performed, where those acts are the proximate cause of injury to the plaintiff, *see Black v. Stephens* 662 F. 2d 181, 189 (3rd Cir. 1981); *Bowen v. Watkins*, 669 F. 2d 979, 988-89 (5th Cir. 1982). Municipalities employing the offending officers are liable where the unconstitutional actions of the officers implement or execute a municipal policy or where the unconstitutional action is pursuant to a custom. *Monell v. Department of Social Services*, 436 U.S. 658 (1978). In between is the standard articulated by the Supreme Court in *Meritor Savings Bank, FSB v. Vinson*, 106 S.Ct. 2399 (1986), in which the Court held employers responsible under Title VII for the sex harassment of female employees engaged in by managerial personnel, and suggested, in *dicta*, that the employers would only be responsible for the harassment engaged in by other employees where the employer was aware of it.

Reading this definition of "person" as drafted with the "knowing" and "intentional" or "reckless disregard" requirement of section III(A), I believe that this statute would implement the civil rights model. It is difficult to apply traditional master-servant liability to the range of intentional tort situations prescribed here, and this approach is not entirely satisfying. Where an individual announces her intention to exercise a certain reproductive choice to an employer whose policies burden that choice, the employer who continues to implement those policies would be liable. The employer, even if a corporation or other business entity, would be held to have known of the woman's intentions, and acted with the intent of interfering with the choice or in reckless disregard of the consequences. Instances in which the burdening of choice occurs as a result of employee action unknown to the supervisory personnel or to the corporation are lost in this formulation.

An entirely separate problem is posed by political organizations encouraging interference with reproductive choice. Any effort to get at political organizations sponsoring individuals who interfere with reproductive rights will likely run up against First Amendment problems (as in *NAACP v. Claiborne Hardware*, 458 U.S. 886, *reh'g denied*, 459 U.S. 898 (1982)).

Prohibited Acts

a. Access to health care facilities: I have limited the cause of action to actions against perpetrators who interfere with reproductive choice in the ways described above, *and* who do so knowingly and with the intent or purpose of interfering with access to the provision of health care *or* with reckless disregard of the consequences of their acts. Negligent interference with access to health care is not included.

For example, if a pregnant woman slipped and fell on a patch of ice negligently left on a sidewalk in front of a store, and, as a result, aborted, her remedies would continue to be the current common law remedies. Common law remedies would include damages for the woman's pain and suffering; payment of medical bills; and some compensation for the loss of a pregnancy, although nothing approaching a wrongful death award. This is consistent with the way in which I have defined the actions that comprise interference with the exercise of reproductive choice, each of which connotes some measure of knowing behavior. In addition, it suggests that the only interferences with reproductive choice that would be cognizable under this section would be those in which the woman's choice was clear, cf. a woman assaulted a few blocks from an abortion clinic, under circumstances in which the perpetrator could not have known of her intentions, or a woman who is in her first trimester of pregnancy who is assaulted under circumstances in which the perpetrator did not know she was pregnant or what her decision was with respect to that pregnancy.

Although the original draft covered only actions in which the perpetrator specifically intended to interfere with choice, I have broadened the section to include the concept of "reckless disregard." "Reckless disregard" has been defined in *Black's Legal Dictionary* as conduct that "evinces disregard of, or indifference to, consequences, under circumstances involving danger to life or safety to others, although no harm was intended."

b. General interference with reproductive choice: This section is designed to deal with interference with reproductive choice where that choice is exercised somewhere other than in a health care facility. It would involve assaulting a pregnant woman knowingly and with the intent to harm the fetus as well. Given the way I have drafted it thus far, it would not deal with a situation, as in a vehicular homicide, where a driver crashes his car into the car driven by a pregnant woman whom he does not know, and in so doing causes her to abort spontaneously. In such a case, it could not be said that the perpetrator acted in a knowing manner. As the statute is currently drafted then, a question arises as to whether this provision would satisfy the desires of some individuals to enact feticide statutes.

There are also potential problems about the reach of this statute insofar as it appears to permit a child to sue a parent for interference with reproductive choice. That is an issue that must be considered carefully.

Defenses

In section IV, I have outlined the defenses to the action, ruling out the so-called balance of harms defense,[12] but accepting the defense that a defendant did not know that the individual had exercised his/her right to reproductive choice. This is consistent with the concerns raised above. It should also be noted, however, that these defenses pertain only to actions pursuant to this statute. Obviously, if a defendant did not know that the woman was pregnant and attacked her, he may not be liable for interference with her reproductive choice, but plainly would be liable--both civilly and criminally--for assault.

Penalties; Injunctive Relief

The limitation in this section to civil damages reflects my general concern for the use of the criminal law to control conduct. For the most part, I believe that the criminal arena should be reserved for prosecuting the most egregious violent acts, and not used for litigating fundamental social issues. The gatekeeper in the criminal arena--the individual controlling whether or not to bring charges, is a public prosecutor whose views in this area are likely not to be enlightened (witness the treatment of domestic violence). In addition, I am concerned about the risks of error in the criminal arena--the possibility that someone may be wrongfully imprisoned. In any case, given the public pressures on prosecutors, the likely scenario would be one in which this statute would be used to enforce *pregnancy* choices and not *abortion* choices. Finally, there is a danger that, rather than conveying a message about the importance of preventing interference with reproductive choice, precisely the opposite message would be conveyed. In rape prosecutions, for example, it has frequently been easier to secure convictions for the robbery and the assault that accompanied the rape, than for the rape itself. This was so in the past because the penalties for rape were so high that juries would compromise for a sympathetic defendant and convict for the lesser offense. In some instances, where the defense attacked the victim's sexuality, it was easier to secure a conviction for the crimes that had no sexual component. In the abortion context, a prosecutor might charge a defendant who physically blocked a woman's access to an abortion clinic with assault and interference with reproductive choice. If the result of that combination was that the jury wanted to compromise and punish only for the lesser included offense, i.e., assault, that would be one possible resolution of the case. It would not be as satisfying as securing a conviction for violating a constitutional right, but would mean some punishment for the perpetrator.

Another possibility, however, is that the presence of the tort of interference with reproductive choice would somehow divert attention from other crimes that in fact took place. My experience suggests that the members of a predominantly Catholic Massachusetts jury may well be so opposed to abortion that they would balk at convicting someone at all who attempted to "interfere with reproductive choice." If they were presented only with an assault charge--even if they knew the context in which the charge was made--they might be more willing to consider conviction. Moreover, I am concerned about judges who would sentence individuals who interfered with abortion rights to minimal terms and about the message that would convey.

Put in the form of a question: Would the crime of interference with reproductive choice make convictions for acts of clinic violence more likely or less likely, or would it not affect the outcome one way or the other? By creating such an offense, are we politicizing a rather straightforward crime, or is the context already excruciatingly political? To be sure, in the civil setting, juries could unfairly devalue these rights in the damages they award or in their findings with respect to liability. Still, a private party would be able to present a claim to a jury, without relying on a prosecutor as gatekeeper. Moreover, the statute fixes damages, thereby limiting the discretion of the jury to undervalue the plaintiff's claim.

Section A(b)--that there shall be liquidated damages for each act of interference in the amount of $5000--poses some problems. A more precise definition of what constitutes an act of interference is needed. If I had limited the statute to acts that were also criminal offenses, defining interference would be clear-cut. An act of interference would simply be any act that is separately chargeable under the laws of the state or of the United States. Since I have broadened the scope of the acts in question, the problem of definition is more complicated.

Section B--punitive damages--does not include a definition of the circumstances under which an award of punitive damages is appropriate. As the section stands now, a court would likely resort to typical common law principles in which punitive damages would be awarded where the defendant, for instance, acted willfully, maliciously, or fraudulently. As a practical matter, that would mean punitive damages for acts perpetrated with specific intent to interfere with reproductive rights. I believe that this needs to be addressed more carefully.

Miscellaneous Provisions

This section is addressed to situations in which a woman sues her doctor for interference with reproductive choice. Under malpractice law, a plaintiff who sues her physician waives her right to confidentiality with respect to that physician's records. Where the cause of action is considerably more limited as here, it would be unfortunate if the effort to vindicate these rights opened the plaintiff's files in their entirety. Thus, under this provision, the plaintiff would retain her confidentiality rights with respect to portions of her record unrelated to the contention of interference with reproductive choice.

Conclusions

In conclusion, the central problem with this legislation is the same as with any legislation whose goal is to maximize choice. The statute announces: Everyone has a right to do anything he or she wants. "Maximize reproductive choice" is a relatively easy clarion call. The more difficult problems, as always, come with questions of conflicting rights and with provisions for realistic access to rights.

These problems are graphically illustrated in the case of the disabled. I see three levels of concerns. First, there is the question of maximizing the reproductive choices of the disabled to make whatever decisions they wish. That could surely be fit within this statute. Then there are problems about the social context in which these decisions are made--the inchoate and not so inchoate pressures, the social attitudes, and the economic pressures. Such factors are more difficult to address in a statute. Finally, there are problems about individuals' conflicting reproductive rights--a disabled woman decides that she does not want amniocentesis, but her male partner who fertilized her ovum disagrees.

I have opted to address the former concerns, maximizing access to rights and choices, in a separate set of policy recommendations for the reasons I have outlined previously. Moreover, within the space of this position paper, I cannot address the latter concerns, the problems of conflicting rights and choices. Our society is still divided in major ways on these issues. Nor have I found a single philosophical approach that can satisfactorily resolve these questions.

Policy Recommendations

I believe that we should address the issue of requiring the government to maximize reproductive choice in separate legislation. This issue was left out of the statute I drafted for pragmatic political reasons, but nevertheless represents an important goal. Working toward this goal would entail developing laws requiring employers to fund pregnancy leave and parenting leave, regardless of the level of funding offered to other "disabilities"; requiring the government to fund the full panoply of reproductive rights in its welfare programs; and requiring insurance companies likewise to fund the full range of reproductive options and care. It would go beyond the negative discrimination theories to establish an affirmative obligation on the government to maximize choice through funding and through the dissemination of information.

Notes and References

1. See, e.g., Planned Parenthood of Central Mo. v. Danforth, 428 U.S. 52 (1976); but see Thornburgh v. American College of Obstetricians, 106 S.Ct. 2169 (1986).

2. Harris v. McRae, 448 U.S. 297 (1980).

3. "Reagan Said to Back Measure to Bar Any Federal Aid for Abortion," New York Times, February 10, 1987, A:20.

4. National Abortion Federation, "Incidents of Reported Violence Toward Abortion Providers," data sheet, New York, January 1987.

5. "Centers' Abortion Ads Called 'Bogus,'" New York Times, July 16, 1986.

6. "Feticide-Cases and Legislation," American Civil Liberties Union, Reproductive Freedom Project Memorandum, May 5, 1986; "Advances Elevate Status of Fetus," Boston Globe, July 21, 1987, 8.

7. "Caught in the Crossfire: Minority Women and Reproductive Rights," Alliance Against Women's Oppression discussion paper, January 1983.

8. Kolder V.E.B., Gallagher J., Parsons M.T., "Court-ordered Obstetrical Interventions," 316 New England Journal of Medicine, May 7, 1987, 1192-1196.

9. Id. at 1195.

10. See Moe v. Secretary of Administration and Finance, 382 Mass. 629, 417 N.E. 2d 387 (1981).

11. I am reminded of accounts of how the Pregnancy Discrimination Act of 1978, which amended Title VII of the Civil Rights Act of 1964, was passed. Prior to the act, the Supreme Court held that an employer did not have to treat pregnancy and related disabilities like other disabilities under Title VII of the Civil Rights Act of 1964. General Electric Co. v. Gilbert, 429 U.S. 125 (1976) and Nashville Gas Co. v. Satty, 434 U.S. 136 (1977). What has been described as an "unholy alliance" of feminists and "right to lifers" forced the Congress to overrule these cases and provide protection against discrimination because of pregnancy, childbirth, or related medical conditions. To be sure, there are limits beyond which this alliance could or would not go. Coupled with the expansion of the definition of discrimination to include pregnancy discrimination was an explicit exclusion for discrimination on account of the decision to seek an abortion. See infra, for caveats about lawmaking concerning reproduction.

12. The "balance of harms" defense applies to a situation in which a defendant claims to have acted in the good faith belief that his actions were necessary to prevent another wrong from occurring.

Commentary

AN ATTORNEY GENERAL'S OUTLOOK

Robert Abrams

All of the authors of the position papers are grappling with some of the most complex and important questions facing our society. How can the laws governing reproduction be updated in light of technological advances? How can we fashion laws that will be widely perceived as fair and just when there seems to be no clear societal consensus on many of these issues? How can we balance our concerns with maximizing reproductive choice against other legitimate, competing concerns? Human tragedies such as the *Baby M* case in New Jersey teach us that we cannot afford to let the law lag too far behind social and scientific developments. Also, although these issues are enormously complex, we should take heart from the fact that, by grappling with them now, we can help shape laws for the future that will maximize reproductive choice, and respect individual dignity.

As Attorney General of the State of New York, I am proud that our state has one of the best records of any state in terms of supporting reproductive choice. We were one of the first states to legalize abortion before *Roe v. Wade*. We are one of only 14 states that fund Medicaid abortions entirely with our own money. We have resisted the trend toward enactment of state laws mandating parental notification or consent for minors seeking abortions. My office has moved vigorously against any threats to these policies. For instance, in *Donovan v. Cuomo*, we are defending the State Department of Social Services against allegations that it funds too many Medicaid abortions. Several years ago, we were the only state to challenge the Reagan Administration's "squeal rule," which would have imposed parental notification requirements on federally funded family planning clinics, and we won. Just recently, my office signed consent agreements with three phony abortion clinics in New York City, forcing them to stop deceiving women who turned to them for help with their unwanted pregnancies. We will continue to be vigilant against these threats to public health and safety, and to the women of this state.

The proposals presented in the position papers are indeed thought provoking. The paper on fashioning a cause of action of interference with reproductive choice poses several fascinating questions. Is it necessary to create such a new cause of action, or is current tort law adequate? To what extent would such a law tamper with existing First Amendment rights? Should there be corresponding criminal penalties for these acts? Similarly, the paper on reproductive hazards in the workplace raises extremely important policy questions. Exclusion of women from the workplace where hazardous substances exist is often a result of unexamined assumptions based on inadequate data. Research indicates that substances that are hazardous to the reproductive health of women will, in many cases, also be hazardous to men's reproductive health. The law should make clear that an employer's primary response to the discovery of a hazard should be to clean up the workplace, not selectively exclude workers. The laws should be reformed on a federal level, but if Congress fails to act, state legislatures should address the problem. For example, existing remedies need to be strengthened. I have submitted as part of my legislative program this year a bill that would allow workers to sue employers outside of the existing workers' compensation framework for work-related injuries to their reproductive capacity.

Debate on these and other questions will be an invaluable tool for the legislators and other public officials who must make the final policy determinations. I look forward to working with all those interested in formulating a pro-choice agenda.

Commentary

THE VIEW FROM CAPITOL HILL

Edmund D. Cooke, Jr. and Sally J. Kenney[1]

I. Introduction

Thoughtful consideration of complex issues should be a precursor to legislative change. The position papers drafted by Nancy Gertner and Joan Bertin advance public debate. By skillfully analyzing complex and politically sensitive problems, they represent the first step in a long process. To their further credit, they accept the often avoided challenge of offering specific solutions.

Unfortunately, carefully drafted proposals do not enact themselves. As a consequence, in addition to commenting on the papers, it is important to consider the next step. What strategies and tactics must we adopt to ensure implementation of these proposals? Do the realities of the political process mean that some proposals can be ruled out immediately as infeasible? Do others need to be altered? And do the unpredictability of the process and limited support for some of our objectives mean that perhaps some should not be undertaken at all? The goal of the conference and the papers in this volume has been to analyze the problems and develop solutions. The purpose of this commentary is not to dampen our enthusiasm for the proposals, but rather to force us to be realistic, and more important, to answer the question: Where do we go from here?

The perspective advanced in this commentary comes from our experience working for the House Education and Labor Committee.

The Committee has jurisdiction over both discrimination and health and safety issues. Commenting on both papers provides a unique opportunity to draw the connection between current industry practice and the limiting of reproductive choice. Though our Committee does not have jurisdiction over many of the specific issues raised by Nancy Gertner, the broader issue of obstacles to reproductive choice is of concern to us.

In discussing approaches to dealing with reproductive hazards in the workplace, it is essential that we get the message across that, in objecting to policies that exclude women from hazardous workplaces, we are not campaigning for women's equal right to be poisoned and bear malformed children. It must be clear that we are demanding that women have the right to work *and* bear healthy children, and that men's and women's general as well as reproductive health must be of equal importance. In other words, we must seek to achieve a much broader understanding and acceptance of the necessary mandates of the Occupational Safety and Health Act and of Title VII of the Civil Rights Act of 1964, as amended. We must make clear that neither men nor women should have to choose between a job and healthy children or "choose" sterilization in order to keep their jobs.

II. Comments on Changing Tort Law

Though both papers are provocative and important, Gertner's is the more troubling, not because its premises are invalid, but because the legislative solutions it suggests are complex, and skirt or tread directly on extremely controversial issues. Neither the public nor the Congress typically deals with such matters well. This is not to suggest that such issues either ought not or cannot be addressed as legislative initiatives, but rather, that if undertaken, it should be with the clear understanding that the effort will likely be protracted and uniquely difficult, and the ultimate product, once initiated, not entirely predictable.

Without an extensive knowledge of tort law, we offer comments that are necessarily impressionistic, but perhaps a useful predictor of how this proposal might be received on Capitol Hill. The authors do not approach the paper on Interference with Reproductive Choice with all of the insight that Gertner incorporates in her draft. Nevertheless, the proposal raises several questions for which, incidentally, we do not necessarily have answers, but that may have a direct bearing on the likely character of legislative receptivity and response.

These questions include (1) the scope of any necessary relationship of the proposed tort to the issue of abortion; (2) its impact on medical malpractice, including the extent to which it creates an expanded incentive to sue, a disincentive to specialize in obstetrics and gynecology, a diminished availability of related malpractice insurance, or substantial increase in the cost of medical services to the public; (3) the impact of the proposed law on male vs female pregnancy/abortion-related rights; (4) its possible constraints on First Amendment free speech rights; (5) its possible constraints on First Amendment freedom of religion protection; and (6) the sufficiency of existing tort and criminal law, if fully enforced, as a deterrent to the offensive policies and practices with which the law would deal.

Thus, though we can concur that the proposed law is validly based and necessary, we must acknowledge that, given real world constraints, the proposed bill would likely be defined and debated primarily as an abortion bill, irrespective of the drafter's intent or the validity of that assertion. It is also possible that the debate would be focused in terms of increased cost of medical services insofar as physician behavior is called into question. The debate may also raise constitutional issues insofar as the proposed tort may constrain speech or the expression of religious beliefs. Finally, it is vulnerable to challenge as overly duplicative in coverage and as unnecessarily intrusive and costly.

These concerns are not raised in order to curtail, but rather to stimulate debate and discussion. Difficult issues are better confronted and resolved prior to the introduction of legislation. Given the complexity of the issue and the unpredictability of the legislative process, perhaps the most essential ingredient for the success of this proposal will be the extent to which proponents are able to educate both the public and legislators about the need for and the legitimate objectives of the legislation.

III. Reproductive Hazards

The proposal offered by Bertin is a feasible approach to dealing with a very disturbing, yet politically sensitive, convergence of gaps in relatively comprehensive workplace safety and nondiscrimination laws. But the very existence of these gaps, which can facilitate conservative judicial construction and narrower-than-intended scope of prohibition, serves to illustrate the difficulty that attends the enactment of clear, direct, and comprehensive legislative solutions. Compromise rarely clarifies and enhances, but is an essential component of the legislative process. The proposal is solidly based on empirical data, and narrowly focused to deal with demonstrable and easily understood deficiencies in current law. Moreover, it builds, in large measure, on prior legislatively defined areas of workplace regulation.

Regardless of the soundness of the proposal, however, amending the Occupational Safety and Health Act and Title VII of the Civil Rights Act of 1964 (as amended) will not automatically meet with the approval of the interest groups most sympathetic to these critical objectives. That lack of receptivity will be based on an assessment of the likelihood that foreseeable (or unforeseeable) constraints can be imposed on existing law as the new amendments move through the process. The proposal that calls for freestanding legislation, for that reason, seems more desirable. Bertin acknowledges these constraints, and we mention them only to add emphasis. We would also like to reinforce strongly her determination in "Section B. Interim Measures" that the enactment of the legislative proposals "will be enhanced if intermediate steps are taken to heighten public awareness of the problem and proposed solutions, to educate lawmakers, to achieve reform and create workable 'models' on the state and local levels, and to stimulate regulatory reform."[2]

In reference to the proposal to create a private right of action to enforce the OSH Act, suffice it to say that, while that proposal is an excellent idea, devices that involve costly court proceedings are often of more limited utility than one might assume. For that reason, we suggest pursuing a private right of action in the context of a quasi-judicial administrative process incorporating cease and desist authority. From plaintiffs' standpoint, such a process can be faster and more affordable.

If Bertin's proposals are implemented, and even if individuals were to receive a private right of action under the Occupational Health and Safety Act, we would still have to rely to some extent on federal agencies to enforce standards of exposure and to enforce sex discrimination legislation. Thus, legislative oversight of the enforcement agencies will always be an important variable, particularly under administrations that may be hostile to the objectives of the statutes that these agencies enforce. Since the House Education and Labor Committee has jurisdiction over health and safety and discrimination issues, we have responsibility for oversight of both the Equal Employment Opportunities Commission and the Occupational Safety and Health Administration. It may come as no surprise to you that, during the Reagan Administration, protecting workers' reproductive health and ensuring that women are not unfairly excluded from jobs have not been high priorities. In fact, we have been frustrated in our attempts to get either agency to act on these issues.

Last fall, committee staff held a meeting with representatives of the EEOC to ascertain what action they were taking on the potential discrimination resulting from company policies designed to protect workers from reproductive hazards. The EEOC was reluctant to meet, since the agency was doing nothing on this issue. Representatives of the agency seemed unaware that the Office of Technology Assessment had just concluded a lengthy study on this very issue, and it seems, the agency did not feel compelled to respond or implement any of its recommendations.

In the past, however, the EEOC had been very active on the issue of reproductive hazards. In 1980, it proposed guidelines after consulting with the Occupational Safety and Health Administration.[3] Those guidelines set high standards for a company policy to meet in order to be legal under Title VII in the agency's view. For example, it required that the company consider evidence on male reproductive hazards, and that it consider engineering controls and protective equipment before excluding women. These guidelines were withdrawn just before the new administration took office.

The EEOC compliance manual, in section 624 entitled "Reproductive and Fetal Hazards," directs Employment Opportunity Specialists regarding how to investigate charges of discrimination resulting from exclusionary policies. It directs EOSs to look carefully to see if the company is acting consistently--to investigate how the company deals with male reproductive hazards, how it deals with other hazards, and especially, whether there is a scientific basis for the policy. Yet, the compliance manual states, "[s]uch charges are non-C[ommission]D[ecision]P[recedent]. After investigating the charge according to the following subsections, the EOS should contact the Guidance Division of the Office of Legal Counsel for further instructions."[4] This means that the agency can make no determination regarding cause (that is whether in its opinion the policy constitutes prohibited discrimination), because the Commission has not reached a decision to guide the agency. There are currently 18 complaints filed since 1980 awaiting the Commission's decision.

The Committee has been given a copy of a draft Commission decision that is awaiting approval. The events took place in 1980. The charging party is a hospital X-ray technician who was forced to go on maternity leave immediately after discovering that she was pregnant because the hospital feared radiation exposure would damage her fetus. The draft decision states that the evidence on the hazard posed to the fetus by radiation exposure does not warrant removing the woman from the workplace and that the charging party alleged that there were other jobs she could have been moved to rather than being forced to leave. It concludes that "[t]here is reasonable cause to believe that the Respondent engaged in an unlawful employment practice in violation of Title VII of the Civil Rights Act of 1964, as amended, by discriminating against the Charging Party on the basis of her sex."[5]

We are disturbed that the case being held up by the Commission, one on which 18 complaints are pending, is almost identical to cases that have already been decided by the Commission and already ruled on by two Courts of Appeals. In *Hayes v. Shelby Memorial Hospital,*[6] the EEOC advocated as amicus curiae that the hospital's policy was not justified and clearly violated Title VII. Thus, not only is the EEOC on record taking the position that removing pregnant women from their jobs as X-ray technicians is not justified by the scientific evidence, but the 11th Circuit Court of Appeals has ruled that such a requirement violates Title VII.

What is also of major concern is that 20 of the 54 recorded complaints the agency has received on fetal protection policies involve exclusion of women from exposure to lead. EEOC officials suggested that they did not have the scientific expertise to evaluate whether the exclusion of women was appropriate. Yet the hazards of lead exposure have been evaluated by OSHA and a standard set. OSHA's standard, which requires that the standard of exposure for lead be the same for men and women and that any worker planning to become a parent may seek medical removal, was given the stamp of approval by the D.C. Circuit Court of Appeals in *U.S.W.A. v. Marshall.*[7] The court found that the extensive evidence that lead posed a hazard to men's reproductive capacity warranted the setting of a single low standard for all workers rather than the removal of fertile women. Thus, given the extensive evidence of male reproductive hazards caused by lead, and given too, the activity of OSHA on this issue, it is unacceptable that the EEOC is failing to act to find cause in cases where women have been excluded from jobs that expose them to lead.

The Commission should take the lead on this issue. It should send a loud and clear message that industry cannot create the illusion of solving a difficult issue by removing women from workplaces rather than protecting the reproductive health of all workers. The Commission should not be allowed to retract its previous policy and move back from the ground already won through the courts by litigating Title VII cases. The EEOC's failure to act is particularly disappointing in light of the action taken by semiconductor manufacturers to exclude women of childbearing age from certain workplaces. Instead of telling industry that it must carefully draft its policy to deal with reproductive hazards, considering carefully all scientific evidence on men and women as well as legal decisions that set guidelines for evaluating policies, the EEOC has delayed acting on this issue altogether. This is even more disappointing when one considers that, in the past, the agency was in the lead on this issue, actively drafting regulations and litigating cases.

IV. Congress

There are five members of Congress who need to be aware of your concern on this issue. Unless the chairmen of Congressional Committees and Subcommittees are made aware of your concerns, the pressure of other competing interests will push this issue to the background. First, is Chairman Hawkins, who is already pursuing this issue as part of his oversight of the Equal Employment Opportunities Commission. Because of the division of issues within the House Education and Labor Committee, the issue falls under the jurisdiction of two subcommittees, neither of which has addressed it recently. First, is the House Subcommittee on Health and Safety chaired by Joseph Gaydos. Second, is the House Subcommittee on Employment Opportunities chaired by Matthew Martinez. These chairmen need to know why you believe that these issues constitute pressing problems, what solutions you would propose, and who else agrees with you. On the Senate side, which should now be equally responsive to the proposals that we discuss, the Labor and Human Resources Committee is chaired by Edward Kennedy, and its Labor Subcommittee is chaired by Howard Metzenbaum.

We would propose that you apprise the relevant subcommittees on both sides of Congress of your concerns and proposed solutions. In addition, it will be helpful for you to express your concern to your own member of Congress. Advising you to write to your members of Congress may sound like a simple and limited action to take. Yet members need to know that this issue exists, what it involves, and that people feel strongly about it. Contrary to what you might believe, congressional mail is taken seriously.

V. Interest Groups

Groups sympathetic to these objectives should be apprised at the earliest appropriate time of the concerns raised here. Labor unions, women's groups, and environmental groups need to make the issue of reproductive hazards in the workplace and nondiscrimination a high priority. In this regard, we would like to applaud the efforts of Joan Bertin who has carried the torch on this issue for a number of years, and we hope that the coalition of groups formed to protest the recent policies of the semiconductor industry can be continued and strengthened.

The key players in elective office have not always made extending reproductive choice a high priority; in fact, they may frequently not approach the issue from the same perspective that you do. But they do respond to political pressure from their national constituencies and people in their districts. Thus, any effort to get attention drawn to this issue, let alone the extensive legislation that Bertin proposes, will require intense and systematic action on your part. We encourage you to take that action to assist those members of Congress who are trying to take action on this issue.

Notes and References

1. The opinions expressed in this commentary are the opinions of the authors and not those of Chairman Augustus F. Hawkins nor the members of the Education and Labor Committee.
2. "Reproductive Hazards in the Workplace," this volume.
3. 45 Fed. Reg. 7514 (1980).
4. EEOC Compliance Manual 624.1, p. 82 (8/83).
5. Page 7.
6. 727 F.2d 2095 (11th Cir. 1984).
7. 647 F.2d 1189 (D.C. Cir. 1980).

Commentary

PROTECTIVE LEGISLATION AND OCCUPATIONAL HAZARDS: FLAWED SCIENCE AND POOR POLICIES

Jeanne Mager Stellman

The need for each living species to reproduce is incontrovertible. In some species, reproduction seems to be the very *raison d'etre.* However, humans in most societies have other purposes for living and, hence, more exercise of choice in the decision to reproduce. In fact, not all humans, either male or female, do reproduce, and, in modern industrial countries, reproduction is limited and carefully timed by the majority of people.

In general, human reproduction is accompanied by a variety of rituals. In most societies, pregnant women are given "preferential" treatment, from special benefits in social welfare programs to seats in public transportation. Some "preferential" treatment, however, is not considered to be advantageous by feminists, particularly in the area of employment policies and practices. In these areas, "preference" may, in fact, be "discrimination." The extent to which different treatment for pregnant or fertile women is "preferential" is a matter of great debate.

In this commentary, I will discuss one specific aspect of such differential treatment of women on the basis of their reproductive capacity: the development and implementation of employment policies and regulations that treat fertile women differently than men. I will base many of the comments on the proposals for reproductive laws for the 1990s made by Joan Bertin. In her paper, she has described how the large class of "all fertile" women has, by employer policy and/or by government fiat, been excluded from oc-

Table 1

Protective Legislation for Women in the European Community: Proposals for Change

Unlike the United States and Canada, where legislation establishing separate regulations for women workers has long been considered discriminatory and in conflict with equal employment opportunity for women, all member countries in the European Community, the EC, still maintain protective laws, according to a report issued by the Commission of the EC in March, 1987. The Equal Treatment Directive governing employment rights of women, which was passed in the EC in 1976, permits protective legislation, "particularly as regards pregnancy and maternity." As late as 1986 the EC's Court of Justice reaffirmed the need for such special protections based on " . . . a woman's biological condition and the special relationship which exists between a woman and her child." However, the Directive does state that where protective legislation has been based on perceptions of social need, [rather than perceptions of biology], protective legislation may no longer be justified. The Commission's report indicates whether individual protective statutes are discriminatory and should be repealed or be extended to males. The following table summarizes national protective legislation and EC Commission recommendations. (Used with permission of the Women's Occupational Health Resource Center.)

SUMMARY OF REGULATIONS AND EC COMMISSION RECOMMENDATIONS[a,b]

[a]Country codes: B, Belgium; DK, Denmark; D, West Germany; F, France; G, Greece; I, Italy; IR, Ireland; Lux, Luxembourg; NL, The Netherlands; UK, United Kingdom.

[b]Marker codes: (*) regulation in force; no recommendation for change; E, regulation in force; recommendation that it be generalized to all workers; R, regulation in force; recommendation to repeal to conform with EC directives; C, regulation in force; choice of either E or R above; L, limit regulation to pregnant women; S, make sex neutral; N, make "protection" narrower.

BANS ON EMPLOYMENT	B	DK	D	F	G	I	IR	LX	NL	UK
Dangerous & unhealthy work	C				C		C			C
Loaders, blasters,drillers (demolition)					R					
Underground work (mines & quarries)	R		R	R	R		R	R	R	R
Manual digging, earthmoving, excavation	R		R							
Demolition on ovens containing free silica				R						R
Exposure to ionizing radiation	*	*	*	*		*	*	*	*	*
Work involving lead			N		*		N	*		*
Work with storage batteries							*			*
Pottery and ceramics			*				*			*
Glass manufacture (young women)							*			*
Lifting & transport of loads			E	E			E		E	E
Certain construction work			R							R
Work in compressed air caissons	R			R						R
Blast furnaces, iron, steel, molten metals			R							
Manufacture of zinc or zinc exposure			R				R			
Dock work							R		R	
Mercury in the pelt industry				L						N
Prep/packing estherathiophosphomerics				R						
Work with aromatic hydrocarbons				R						
Work with benzene					R					
Internal navigation of the Rhine	R		R	R					R	R
Cargo-stowing activities									R	
Work requiring shotgun licenses					*					

CONDITIONAL TERMS OF EMPLOYMENT	B	DK	D	F	G	I	IR	LX	NL	UK
Night work	R		R	R	R		R	R	R	R
Total length of workday			C		C					C
Weekly work limit									C	C
Part-time work rules				C						
Specific overtime rules			E						E	
Extra leave allowance				E					E	
Family leave allowance				E					E	
Leave to do housework			E							
Ban on Sunday work			C	C			C		C	
Ban on working holidays			C	C			C			
Work rules outside shop				R						
Age limit for some jobs				*					*	
Early retirement for some jobs				E						
No work with running machinery							R			
Controls on labor				R			R			
Compulsory worker information				R						
Seats in shops, etc.				E			E		E	
Length of work break rules			E	E	E					E
Ban on some means of transport				*						
Regulations for shipboard work			R		R					R
Ban on shift work/continuous work			R	R						R
Hygiene facilities (e.g. toilets)	E		E	E			E		E	

COUNTRY CODES

B:Belgium
DK:Denmark
D:West Germany
F:France
G:Greece
I:Italy
I:Ireland
Lux:Luxembourg
NL:The Netherlands
UK:United Kingdom

MARKER CODES

- * Regulation in force; no recommendation for change
- E Regulation in force; recommendation that it be generalized to all workers
- R Regulation in force; recommendation to repeal to conform with EC directives
- C Regulation in force; choice of either E or R above
- L Limit regulation to pregnant women
- S Make sex neutral.
- N Make 'protection' narrower

cupations and industries where women are thought to be at higher risk because of hazards to potential embryos or fetuses they may be carrying. The implication is that they are carrying these conceptuses either unknowingly, secretly, or even perhaps abusively, in flagrant disregard of potential hazards.

These policies, generally called fetal protection policies, are a subset of a larger domain of employment policies that restrict women from certain jobs simply on the basis of their being female. (Here, the implication is that females are ipso facto weaker, more vulnerable, and in need of more "protection" than males.) In general, restrictive legislation and practices have been eliminated in the United States by the passage of the Civil Rights Act. A similar trend has existed in Canada and most Scandinavian countries.

It is, however, sobering to consider the situation in Western Europe, where extensive restrictions on the employment of women are still in effect, and where, as late as 1986, the Court of Justice reaffirmed the need for special protections for women because of " . . . a woman's biological condition and the special relationship which exists between a woman and her child."[1] As can be seen in Table 1, based on a recently released report by the European Community,[2,3] restrictive legislation is prevalent in Western Europe, and only some regulations have been found to be in conflict with the civil rights of women workers.

Clearly, much progress for women has been made in the United States and Canada in comparison to the discriminatory employment conditions for women in Western Europe. We can begin this commentary from the perspective that positive change can occur and has occurred for women workers. In addition, we should note that, at the same time that ineffective and discriminatory "protective" laws for women were lifted, other laws, such as the Occupational Safety and Health Act, the Mine Safety Health Act, and the Toxic Substances Control Act, were enacted, which greatly broadened the health protections for all workers. Thus, we can appreciate the positive strides in worker well-being of the previous decades, but our enthusiasm for progress has to be tempered by the reality that there is much room for improvement in the enforcement and structure of each of these pieces of legislation.

It is also important that our optimism about the future be tempered by the realities of the present. It is particularly important to examine closely some underlying issues raised by Bertin with regard to corporate fetal protection policies. In these comments, I will concentrate on examining underlying medico-scientific justification put forth for restrictive employment practices. Ostensibly, one might presume that the scientific and medical data upon which laws and customs are based would be factual, the sturdy warp, if you will, upon which different legal patterns for employment and safety and health policy could be woven. The patterns would, of course, depend on the moral fashions of the day, and on the relative power of those whose life threads comprise the design. The essential presumption is that the backing is firm, because the data are derived from the impartial pursuit of pure scientific knowledge.

Belief in the impartiality of science as applied to social policy is, however, as Bertin notes, and as so many of us know, not well founded.[4] Rather, to use the words of Henifin, Fried, and Hubbard, "Biological woman [has served as] the convenient myth" in the formulation of rules and practices ostensibly designed to assure reproductive health in the workplace. In this commentary, I shall attempt to shed more light on these myths and on how the science of reproductive health has consistently failed to provide a firm or true basis for the laws and policies currently in effect here and abroad.

Bertin identifies at least two basic premises upon which disparate occupational safety and health policies are based. First is the comparative indifference to hazards other than female reproductive hazards, specifically, hazards to the male reproductive system and systemic hazards to all adult humans. I call this premise the *male invulnerability myth*. The second premise is the assumption that women have not controlled and cannot and will not control their own fertility, leading to what I call the *permanent pregnancy ploy*, or the premise that all women, in the time between their puberty and their menopause, shall be considered pregnant until proven otherwise. Permanent pregnancy, in the eyes of many employers, requires banning women from all potentially toxic exposures because any potential fetus may be most susceptible to toxins during the earliest stages of pregnancy, when a medical diagnosis is not yet available.

Bertin describes one instance in which the male invulnerability and permanent pregnancy premises played themselves out in the now infamous American Cyanamid case in Willow Island, West Virginia, where four women underwent sterilization in order to keep their jobs in operations involving lead pigments. The case generated extensive public and legal discussion. In commenting on this case in the much-respected *New England Journal of Medicine*, medico-legal columnist William Curran wrote that "The dilemma was class: the women had to make a choice between higher-paying work and control of their ability to conceive and bear children. . . . Apparently, the hazard was not determined to be of danger to men of child-producing age. They were not barred from the same work areas. *The women understandably resented that advantage* [emphasis added] to the men."[5]

In fact, the exposures on the job *are* a hazard to male reproduction. However, it appears that the greater vulnerability of women, as represented by the employers, was accepted as fact by the *New England Journal of Medicine.* This acceptance of incorrect data by so eminent a source of authority as the *New England Journal of Medicine* is unfortunately typical of the state of the art of occupational reproductive hazards.

The Semiconductor Industry - A Case Study

In her paper, Bertin cites the semiconductor industry as an example in which selective concern for female reproductive hazards has led to exclusion of fertile females from the manufacturing processes of at least two major companies. This is a striking case because it is one of the only existing examples of such a policy in a female-intensive occupation. Generally, women have been excluded from occupations in which their employment has not been an economic necessity for the employers. As she notes, the policy is also particularly striking because one of the main chemicals in use is ethylene glycol, a widely recognized male reproductive hazard.

I have recently had the opportunity to review the unpublished and not generally available report to the Digital Equipment Corporation (DEC) on workers in their Hudson facility by E. Calabrese and H. Pastides, two researchers at the University of Massachusetts School of Public Health (November 7, 1986). It is now recognized that the "study's methodology and sample size significantly affect its findings," as the Semiconductor Industry Association stated in a letter to Joan Bertin. Yet, the exclusionary policies remain in effect. The methodological errors and assumptions in the report typify the conceptual inaccuracies routinely encountered in studies of this subject matter. They are briefly summarized in the excerpt from the *Women's Occupational Health Resource Center (WOHRC) News* that follows:[6]

The Study Population: A Hodgepodge Group

Authors' Statement: The "primary health outcome of interest was a person's reproductive history in relationship to employment in manufacturing. Glycol ethers was the primary exposure of concern. . . . Studies to date have identified testicular toxicity of acute glycol ether exposure . . . reversible fertility loss has also been identified in these male species . . . as has teratogenicity. . . . "

Actual Study: The study obtained questionnaire data from female production workers and other Digital employees who were not exposed to chemicals and spouses of currently married males. The participation rate of male workers was very low, with only 45 spouse pregnancies reported among 273 males, compared to 433 pregnancies among 471 females. No follow-up efforts to obtain better male participation were described.

Although the aim of the study was to examine health with respect to employment at Digital, a second study group of 306 people was added to the initial group halfway into the study. This additional group consisted primarily of females who had borne their children prior to employment at the company. No explanation was given for this post hoc change in definition of the study population, an improper study technique. However, as will be discussed below, the additional female "controls" created sufficient numbers for results to become "statistically significant" for females, but, since the male population was still small, not for males. Also, there is inadequate description of which former workers were "eligible" for the study, raising the possibility that those who had experienced problems in the past were simply not included. It is customary to address this possibility in studies of this type.

Selective Analyses: Male Problems Misstated

Authors' Statement: " . . . significantly increased risk of miscarriage among female Diffusion employees and a non-significantly higher risk among female Photo employees . . . compared to an internal non-exposed group. . . . "

Actual Study: The primary aim, study of a known experimental male reproductive hazard, was never realized. The investigators failed to enroll sufficient males in the study or to follow-up adequately those enrolled. This . . . [turned] the study into one of females only, for whom, the results are, at best, equivocal since the "non-exposed" group is poorly constructed and not reflective of childbearing and work at Digital. Indeed, if one compares the "non-exposed" who bore children prior to employment at DEC, versus the "non-exposed" who had their children after DEC employment, one finds that mere employment at DEC raised the miscarriage rate, regardless of chemical exposures, assuredly also not a "finding," but a reflection of the poor study design.

Why It Seems That Women Have Poorer Health

Authors' Statement: Women were reported to suffer from a "statistically significantly" elevated array of health problems but males were found only to have more nausea in one department.

Actual Study: The authors studied 17 health problems. In 5 of these, analyses were not carried out for males because of [the] *small number of cases*, while the *same small number* were available for females and were analyzed. The summary table does not indicate, however, that males were not analyzed -- it just shows them reporting no ill health. It appears that the authors used an inappropriate statistical test and that analysis for males could have been carried out using a different statistic. Additionally, statistical significance was achieved for the females because of the padding of the control group discussed above. The study had insufficient statistical power to detect any effects for males.

Finally, although significance was not attained in several of the health symptoms, males were actually found to be at much higher risk than females, sometimes showing 3 times the relative risk for health symptoms like nausea, sore throat and a variety of aches and pains.

> WOHRC suggests that the authors heed their own admonition: "The fact that animal experiments indicate that a biological effect of glycol ether was to cause spontaneous abortion through its action on male gametes should not be viewed as an argument against a potential effect on the human female, since sufficient studies of females have not been conducted."
>
> "Likewise, the fact that this study could only produce data on female workers should not be interpreted to show that only exposed female workers are at risk, if at all."

The *Women's Occupational Health Resource Center News* analysis concluded that the Digital report is so seriously flawed in execution and analysis that it should not serve as the basis for any personnel or public health decisions. Use of these data for establishing landmark policies on the hiring of pregnant women is a disservice to the health and well-being of all workers and the field of occupational health.

Conclusions

Since the beginnings of modern labor legislation, women have had the dubious distinction of breaking new ground in protection for working people, but have often suffered job discrimination as a result. Modern policies in reproductive health protection are no exception. The tension between the right to work and the need to protect pregnancy has not really been eased in most circumstances. However, positive change is possible. Recently in Canada, there has been a change in federal radiation protection regulations in which women are no longer treated differently than men and both are protected by a sufficiently stringent standard.[7]

The Canadian actions are a positive example of appropriate scientific analysis and its application to public policy. The words of William Curran in the *New England Journal of Medicine* show us the opposite approach. He commented that the fetal protection policies decreed by American Cyanamid show that "in terms of a hierarchy of values, avoidance of danger to a developing fetus was placed above the woman's choice to risk exposure and retain a higher-paying job."[8] This statement fails to recognize the risks to fetuses posed by the exposures of their fathers. It fails to differentiate between a working woman and a pregnant woman. It fails to recognize that women can and do exercise control over their pregnancies. It fails to recognize the choice available to the employer of cleaning up the workplace for all workers. Clearly, the choice envisioned by Curran was not based on facts or on true risk appraisal. The challenge for the future is to develop the legal and social system that does present true choices.

Notes and References

1. Court of Justice ruling on pregnancy and protection as cited in note 2.

2. Commission of the European Community, "Protective Legislation for Women in the European Community," Brussels, 1987.

3. Table 1 is taken from Women's Occupational Health Resource Center News 8(3), Spring 1987, p. 8 and is based on information contained in the source in note 2.

4. See Mary Sue Henifin, Barbara Fried, and Ruth Hubbard, "Biological Woman: The Convenient Myth" (1982) for extensive documentation of the misinterpretation and misuse of science with regard to women.

5. Curran, William J., "Law-Medicine Notes: Dangers for Pregnant Women in the Work Place," New England Journal of Medicine 312, January 17, 1985, pp. 164-165.

6. This analysis appeared as a signed editorial in Women's Occupational Health Resource Center News 8(3), Spring 1987.

7. See Stellman, Jeanne, "Protective Legislation, Ionizing Radiation and Health: A New Appraisal and International Survey," Women and Health 12(1), 1987 for a discussion of these changes and the rationale behind them.

8. See note 5.

Commentary

A MODEL FOR ADVOCACY: FROM PROPOSALS TO POLICY

Helen Rodriguez-Trias

My comments on the papers "Reproductive Hazards in the Workplace" by Joan Bertin and "Interference with Reproductive Choice" by Nancy Gertner are based on some of my experiences in the movement for an improved consent process concerning sterilization procedures.[1] The movement arose from a growing outrage around the country following the disclosures of the Relf case in 1973.[2] A notorious case, it involved two sisters, Mary Alice, then 14, and Minnie Lee Relf, who was 12 at the time of their sterilizations in Montgomery, Alabama in June 1973. As the girls' mother described in court, two representatives of the federally financed Montgomery Community Action Agency called on her requesting consent to give the children some birth control shots. Believing that the agency had her daughters' best interest and health in mind, she consented by putting an X on paper.[3] Judge Gerhard Gesell, who heard the case, declared:

> Although Congress has been insistent that all family planning programs function purely on a voluntary basis there is uncontroverted evidence in the record that minors and other incompetents have been sterilized with federal funds and that an indefinite number of poor people have been improperly coerced into accepting a sterilization operation under the threat that various federally supported welfare benefits would be withdrawn unless they submitted to irreversible sterilization.[4]

Other cases followed in the 1970s. In Aiken, South Carolina, a number of women sued Dr. Clovis Pierce, a white former Army physician, for his coercive tactics in obtaining consent, including threats to refuse to deliver their babies. In 1973, black women were subjects of 16 of the 18 sterilizations paid for by Medicaid and performed by that physician.[5] Norma Jean Serena, a Native American mother of three children, was the first woman to raise sterilization abuse as a civil rights issue. She charged that, in 1970, health and welfare officials in Armstrong County, Pennsylvania had conspired to have her sterilized when her youngest child was delivered.[6] Ten Mexican-American women sued Los Angeles County Hospital for obtaining consent in English when they spoke only Spanish. Some were in labor at the time of ostensibly giving consent; others even under anesthesia. A few reported being told such things as "Sign here if you don't want to feel these pains anymore," while a piece of paper was waved before their eyes.[7]

It became apparent to many members of professional and lay organizations that sterilization procedures were being forced upon some women by means ranging from subtle to crassly coercive. More and more organizations began to take up the fight against sterilization abuse. A great deal of evidence accumulated, most of it coming from women's testimony, but some from surveys of hospital practices conducted by such groups as the Health Research Study Group,[8] the American Civil Liberties Union,[9] and the Centers for Disease Control.[10] Although organizations differed as to analyses, priorities, tactics, membership, and many other characteristics, there was a convergence in terms of focusing on regulations, with an eventual goal of state and federal legislation that would ensure an informed consent process, safeguarding the areas in which abuse was most likely.

In New York City, in the mid-1970s, the existence of a quasi-public corporation governing the municipal hospital system, which had incorporated community boards and advocates within its structure, allowed for the creation of a mechanism for new guidelines. Briefly, I shall outline some of the lessons extracted from a decade of experience in establishing guidelines to deter sterilization abuse for the New York City Health and Hospitals Corporation,[11] a public law for New York City in 1975,[12] and guidelines for the Department of Health, Education, and Welfare (now Health and Human Services) in 1978.[13]

The first realization was of a need for an empirical base. It was necessary to know whose rights had been infringed. A profile emerged: women who were low-income, black, Hispanic, Native

American, young, both urban and rural, and on and off Medicaid. How had their rights been infringed? Health care personnel had given inaccurate or misleading information about "bandaid surgery," "tying the tubes," and "undoing operations"; requested consent at the time of abortion, childbirth, or other procedures; performed surgery within hours or days of consent; offered "package deals," i.e., abortion and sterilization for one fee; threatened to withhold services; and performed procedures on incarcerated or otherwise institutionalized women. We drafted the initial guidelines using the information we gathered. We attempted to address those situations that placed women at risk of abuse.

Second was the incorporation of affected individuals and groups into the drafting process. It was not easy, but representatives of patients' rights groups from various ethnic communities were invited into the working groups to develop the guidelines. This aspect of the process was essential in developing a practical approach to drafting regulations and ensuring their later adoption.

Third, there was a process of development that included dialogue with the hospital personnel: social workers, nurses, and above all physicians in obstetrics and gynecology.

Fourth, we mapped a strategy for adoption that included massive public education of the members of community boards, and of any and all organizations that could exert pressure. We worked closely with civil rights and women's organizations.

Finally, we devised mechanisms for continuing education and monitoring that we built into the system. These mechanisms included efforts to update educational materials on procedures and a chart review process with required reporting from the hospitals.

During the decade of work on sterilization, it became clear to many of us that the guidelines and laws were only as strong as the process by which they were brought about and the public's awareness of their existence. Our impact stopped short of obtaining federal legislation and generating intensified efforts at the state level for state legislation. I, therefore, welcome this conference, which attempts to bring together the people interested in legislative approaches, and may help focus the broader array of advocacy activities to safeguard and expand reproductive rights. There is much unfinished work.

Joan Bertin gives us a fine start with her comprehensive and strong proposals for legislation that will prevent reproductive hazards in the workplace in a nonexclusionary and nondiscriminatory way. The notion that legislation, education, and public policy must be completely interrelated is particularly sound. The proposed reforms to the Occupational Safety and Health Act and fair employment laws are sensible and appear feasible.

The questions that arise refer to the basic weaknesses in the occupational safety and health statutes and in their regulatory body, OSHA, itself. It seems evident that, without forceful trade union involvement and pressure, the reforms are likely to follow the wavering pattern of the existing statutes. A stronger empirical basis as well as the incorporation of affected groups in the drafting process are necessary before the likelihood of success becomes real.

Providing a workers' Bill of Rights as "a longer-term objective and an educational and organizing tool for individuals and groups counseling workers" is particularly worthwhile. Although it acknowledges that Americans and perhaps all workers will choose employment over unemployment and often a hazardous workplace over no workplace at all, it is still important that safety be among the issues on every agenda. Perhaps a broader vision may emerge as a result, with reconsideration of our societal priorities in terms of industrial development.

Nancy Gertner focuses our attention on the need to broaden our definition of reproductive choice commensurate with the realities of women's lives. The prospects for a new coalition including those advocating for choice as the right to bear a child are very appealing. Women from groups underrepresented in the pro-choice movement--women on welfare, adolescents, and members of ethnic minorities--often objected to what they saw as the movement's lack of support for their choice to have children. The lack of commitment to women's basic needs such as housing, child care, family support services, and prenatal and postnatal care on the part of most pro-choice groups has indeed limited their constituencies.

Gertner's analysis that "choice" takes place in a defective socioeconomic and cultural setting is to my mind extremely sound. The proposal attempts an omnibus approach of protecting all kinds of reproductive choices. It grows out of, as did the work on sterilization, an empirical base, in this case one documenting the likelihood of interference in different settings.

As Gertner recognizes, defining reproductive choice as "an individual's choice to exercise her constitutional right to the performance of an abortion to the extent protected by state and federal constitutional law" does leave the right vulnerable to limitation by the state. Furthermore, the proposal does not cover certain areas in which legislation is presently lacking, as is true for the case of sterilization abuse, except in New York City, and for the issue of prenatal care as a right. Nevertheless, "as a starting point for discussion" as the author intends, this proposal is an excellent beginning for comprehensive legislation.

Notes and References

1. See generally H. Rodriguez-Trias, "Women and the Health Care System--Sterilization Abuse, Two Lectures," The Women's Center, Barnard College, New York, 1978; H. Rodriguez-Trias, "The Women's Health Movement," a chapter in Reforming Medicine, V. Seidel and R. Seidel, Eds., 1984.

2. Relf vs. Weinberger, 372 Federal Supplement 1196, 1199 (D.D.C. 1974).

3. Jack Slater, "Sterilization: Newest Threat to the Poor," Ebony, October 1973, p. 150.

4. Relf vs. Weinberger, 372 Federal Supplement 1196, 1199 (D.D.C. 1974).

5. Slater, p. 152.

6. Joan Kelly, "Sterilization and Civil Rights," Rights (publication of the National Emergency Civil Liberties Committee), September/October 1977.

7. Claudia Dreifus, "Sterilizing the Poor," The Progressive, December 1975, p. 13.

8. Robert E. McGarraugh, Jr., "Sterilization Without Consent: Teaching Hospital Violations of HEW Regulations: A Report by Public Citizens' Health Research Group," January 1975 (prepared for the Public Citizens' Health Research Group, Washington, D.C.).

9. Elissa Krauss, "Hospital Survey on Sterilization Policies: Reproductive Freedom Project," ACLU Reports, March 1975.

10. Carl W. Tyler, Jr., "An Assessment of Policy Compliance with the Federal Control of Sterilization," June 1975 (available from the Centers for Disease Control, Atlanta, Ga.).

11. Guidelines on Sterilization for the New York City Health and Hospitals Corporation.

12. New York City Council Public Law #37.

13. U.S. Department of Health and Human Services, Guidelines on Sterilization. Federal Register, Volume 43, Number 217.

3. Alternative Forms of Reproduction

ALTERNATIVE MODES OF REPRODUCTION

Lori B. Andrews

Position Paper

The past decade has witnessed the birth of new reproductive technologies (such as in vitro fertilization) and the application of older technologies to new situations (such as the use of artificial insemination to facilitate surrogate motherhood). The new and old technologies[1] have been the subject of a vast discussion in the legal and ethical literature, but little attention has been paid to the feminist perspective in the majority of books and articles.[2] However, there is a growing separate literature addressing alternative reproduction written by feminists themselves.[3]

Though the media have heralded alternative reproduction arrangements, and physicians and lawyers have quickly cashed in on them, feminists have asked penetrating questions about their applications and effects. Questions were raised about whether valid consent had been obtained from the women who served as subjects in the basic scientific studies that led to the development of in vitro fertilization,[4] whether the women who participated in the initial clinical applications of the technology had been told about how experimental and risky it was,[5] and whether women who undertook the risky in vitro fertilization procedure were doing so because of excessive social pressures to be mothers and the dearth of other options for alternative roles for women in society.[6]

Feminists have reacted to surrogacy with even more concern and questions. There is a concern that the process might demean motherhood and minimize the woman's contribution to reproduction by allowing a man to purchase the use of a woman's body for nine months to gestate a fetus created with his sperm. The very use of the term "surrogate" is troubling, since it seemed to be an attempt to disguise the fact that the contracting woman was actually the genetic and gestational mother. There is also a concern that a man's contracting with a woman to gestate his child, outside of a personal, intimate relationship with her would lead him, physicians, and legislators to attempt to control the surrogate's activities and behavior during pregnancy in order to assure a healthier, more perfect "product." There is a further concern that such control in the surrogate context would lead to similar constraints being imposed on all pregnant women.

It would be erroneous to suggest that there is total consensus among feminists about what regulatory policies should be adopted to cover alternative reproduction.[7] However, there is substantial agreement about what *values* should be promoted and protected in trying to design legal rules to address alternative reproduction. Although feminists may differ with respect to whether and how these values are threatened by certain applications of alternative reproduction--and to what extent any potential risks can actually be averted by legal rules--the values themselves ring loud and clear. This position paper describes those values and the policy issues they implicate.

The Infertility Issue Should Be Put into a Larger Social Context

The experience of childbearing and childrearing is of importance in many people's lives. Currently, there are a number of barriers to achieving that experience--such as infertility, deficient or absent prenatal care, or deficient or absent infant care. Any policies for dealing with the treatment of infertility must also consider the prevention of infertility, fetal demise, stillbirth, and infant mortality.

One in six people of childbearing age in the United States cannot achieve a pregnancy. Other individuals, though they can achieve a pregnancy, may not be able to bring it to term. In addition, infant mortality in the United States is high, with 10.9 deaths per 1000 live births per year--higher than the 10.3 in Germany, 10.2 in the United Kingdom, 9.0 in France, 8.5 in Canada, and 6.2 in Japan.[8] Nutritional deficiencies among pregnant women are implicated in problems in pregnancy and breastfeeding.[9] There is thus a need to redefine "infertility" to encompass not only physical barriers to fertility, but also social ones and to extend the term "infertility" to cover women whose children do not live through infancy.

Infertility is not just a biological phenomenon. It is a social phenomenon. Lack of access to health care can turn a temporary medical problem (such as pelvic inflammatory disease) into permanent sterility, causing the rate of involuntary childlessness to be much higher among the groups with the poorest medical care--nonwhites and the economically disadvantaged. Moreover, social structures make work and family incompatible. Men and women who are serious about establishing their careers often have to postpone childbearing. Yet, since peak fertility for both sexes is age 25, the delay may correlate with a diminishing of fertility. The problems caused by a dearth of options (such as child care arrangements) for making parenting and working compatible have a particularly harsh impact on women, who face both a steeper decline in fertility, and also a much shorter range of childbearing years.

The medical profession has been notoriously poor in developing and offering preventive measures in health rather than high-tech solutions.[10] Federal and state statutes have sometimes been adopted to help redress that imbalance by specifically providing for financial and other encouragement of preventive health care services.[11] Similar provisions are appropriate with respect to infertility in its broadest sense.

The prevention of procreative risks entails a combination of strategies. These include the education of the public generally about known and probable risks, actions to eliminate these risks, and access to health care to treat preventable problems. The educational aspects could be accomplished through a combination of state public health programs, community group programs, efforts by women's groups, school curricula changes, and so forth. The types of risks to be addressed include disease, such as pelvic inflammatory disease, as well as environmental and social factors presenting procreative risks, such as certain workplace hazards[12] or chemicals released in the environment. There is a need to enforce and even develop laws to protect against environmental and workplace threats to people's fertility. It will also be necessary to make health care services, research, and other measures available to prevent procreative risk and to treat infertility in its broadest sense.

Women Should Have Control over Their Bodies, Their Gametes, And Their Conceptuses

A woman's right to control her body, particularly her reproductive capacity, has been a prime tenet in feminism. Just as there has been support for autonomy in women's decisions *not* to reproduce (for example, recognition of a right to contraception and abortion), there is support for the right of women *to* reproduce (for example, to refuse and not be coerced into sterilization). The latter right is thought by many to encompass a right to reproduce using alternative reproduction. For example, there is considerable (though not unanimous) agreement among feminists that women should be able to use in vitro fertilization (IVF). With respect to third-party involvement in reproduction, there is widespread agreement that women should be able to reproduce using a gamete from a sperm donor or egg donor or even an embryo from an embryo donor.

Many states, however, have enacted restrictive laws aimed at protecting embryos that limit women's choices to use reproductive technologies. State laws that were adopted to restrict fetal research appear to prohibit physicians and women from using innovative reproductive technologies involving embryos, such as embryo freezing or transfer.[13] Other laws enacted to cover in vitro fertilization similarly restrict procreative options. For example, a new law in Louisiana declares an in vitro embryo a person and specifically provides that an IVF embryo "shall not be intentionally destroyed by any natural or other juridical person or through the actions of any other such person."[14] This provision takes from women a substantial degree of control over the fate of their embryos. If a woman undergoing IVF does not want to have all of her fertilized eggs reimplanted (because of the risks of multiple gestation), she would not have the right to terminate any excess embryos. Her choice would be limited to running the risk of reimplanting all the embryos or having the state take possession of the embryos and give them to a second woman. Laws that interfere with a woman's decision to use reproductive technologies are viewed with suspicion by feminists, many of whom are concerned about the extension of protections for embryos into the context of contraception and abortion.

When a woman, because of infertility or other reasons, decides to become a rearing parent for her male partner's biological child by contracting with a woman to bear his child (known as the biological mother or surrogate mother[15]), the concern with protecting women's bodily integrity switches its focus from the intended rearing mother to the surrogate. Even with the extensive reservations that feminists have about surrogacy,[16] there is considerable agreement that surrogate arrangements (particularly private arrangements in which a woman volunteers to bear a child to be reared by a friend or relative) should not be legally banned. To support such a ban would seem to acknowledge a role for government in dictating the circumstances under which a woman should be allowed to have a child and under which families may be formed.[17]

However, to eschew a legal ban does not necessarily mean that the feminist position would encourage surrogacy. Feminists are more concerned with encouraging alternatives to surrogacy. There is support for efforts to prevent infertility, so that the need to consider contracting with a surrogate will be less. There is a concern that new ways be found for infertile people to relate to children in society at large and that women be given equal opportunity to develop themselves in other ways, so that motherhood is not the only role in which a woman can feel fulfilled. There is also a concern that an infertile woman should not enter into a surrogacy arrangement merely because of pressure from a male partner to have "his" child or from parents or other family members to become a mother.[18] On the other side of the arrangement, there is concern with developing alternative employment opportunities so that, if commercial surrogacy is allowed, women who need money do not become surrogates because other preferable jobs are denied them.

Of all the concerns for bodily integrity raised by alternative reproduction, concerns for the surrogate's autonomy in reproductive decisions loom the largest. When a woman has contracted to bear a child for someone else, and has promised to turn the infant over for rearing by its genetic father and his partner after its birth, the arrangement may be viewed by the intended rearing parents as giving them a right to control the biological mother's activities during pregnancy. Some surrogate contracts claim to give the couple the right to force her to follow doctors' orders, undergo amniocentesis, and have an abortion (or not have an abortion) based on their desires. A law proposed in Michigan a few years ago would have prohibited a surrogate from smoking or drinking during pregnancy.

There is universal agreement among feminists that it is important to reaffirm the right of a pregnant woman, including a woman who has agreed to be a biological mother pursuant to a surrogacy arrangement, to make decisions about herself and the fetus during pregnancy. The surrogate should be able to engage in whatever activities she wishes, to refuse any medical consultations or treatments, and to abort or not abort based on her own decisions.[19] To protect her bodily integrity and self-determination,[20] the contract should not be enforced to control her prenatal behavior, nor should she be liable for damages, nor should she be subject to a subsequent tort suit.

Women Should Not Be Exploited

The idea that women should not be exploited in alternative reproduction arrangements has several components. It includes a notion that women should enter into these arrangements voluntarily and with sufficient information about the potential physical, psychological, social, legal, and financial risks to ensure informed consent, and that they should not be pressured into such arrangements on account of personal or economic needs.

There have been numerous questions raised about the exploitation of women in the research and clinical phases of alternative reproduction. The way in which female patients are "recruited" as subjects in medical research is troublesome. Some physicians do research on women who are their patients for traditional gynecological and obstetrical services, and there is evidence that some of these women are not even aware that they are being used for research purposes.[21] In the area of reproductive technologies, Gena Corea raises serious questions about whether doctors have obtained eggs and embryos from women without their consent. She points out, in an extensive review, that in the published studies on research using women's eggs, "there is, in almost every case, no indication that the women consented to the extraction of their eggs or even knew that their eggs had been taken."[22] In some cases, researchers collected fertilized eggs from women undergoing hysterectomies who did not realize they were pregnant at the time of surgery.[23]

Researchers and their attorneys argue that, because there is no additional physical risk to the patient when an egg or embryo removed in the course of surgery is experimented upon, there is no need to even inform her about the fact that research will be done.[24] However, this argument overlooks the enormous significance to individuals of reproductive materials and their handling.[25]

Women seeking medical services sometimes unwittingly serve as research subjects in experiments on reproductive technologies, and even women who seek out clinical services related to the new reproductive technologies may not realize that such procedures are experimental and not standard medical practice. All new medical technologies, when first applied to humans, are experimental in nature. However, when a new technology is touted as being potentially therapeutic to the recipient, she may not realize how experimental the technique actually is. When new infertility treatments are quickly embraced by the profession and heralded by the press, their experimental nature may not be entirely clear to the people who undergo them. Leon Kass has pointed out that, in the initial in vitro fertilization attempts, the women apparently considered themselves patients, whereas they were actually to be experimental subjects.[26]

In addition to information about whether the proposed procedure is experimental or not, women need to have ready access to information about the potential risks of alternative reproduction as well. Both men and women have infertility problems in equal proportion. In infertile couples, 40 percent of the time the infertility lies in the male, 40 percent of the time it lies in the female, and 20 percent of the time it is a combined problem. Nevertheless, the main physical risks presented by alternative reproduction procedures are risks to women. At most, the man will be asked to give a sperm sample--a simple and painless procedure. With in vitro fertilization, the woman generally needs to undergo hormone treatment to hyperstimulate her ovaries and a laparoscopy to recover the ova, both of which present the risk of death to the woman. When an embryo is transferred to the woman's womb after IVF or egg donation or embryo donation, the embryo donor and the embryo recipient may risk an ectopic pregnancy. As the recipient of donated sperm, eggs, or embryos, the woman may contract an infection such as hepatitis or AIDS.

In addition to the risks to women who use reproductive technologies, there are risks to women who serve as donors and surrogates. Although sperm donors face virtually no physical risks, donors of eggs or embryos and surrogate mothers do face serious physical risks in order to help provide another individual or a couple with a child. Surrogate mothers face all the risks of pregnancy and birth. Egg donors may need to undergo a laparoscopy under anesthesia. Even though a woman who donates an embryo via lavage is not subject to anesthesia, she risks infection and ectopic pregnancy from undergoing the procedure.[27]

Alternative reproduction may pose psychological risks to the donors or surrogates as well. Some surrogate mothers go through a period of depression and grieving after the surrendering of the infant.[28] Likewise, some sperm donors may later regret that they have created children whom they may never see.[29]

Women also need information about the efficacy of the proposed procedures. In vitro fertilization's overall low success rate may be material information for a woman who is deciding whether or not to undergo the procedure. Moreover, the live birth rates vary dramatically from clinic to clinic. In 1985, a survey was sent to the 123 in vitro fertilization clinics in the United States. Of the 53 that responded, only 38 had successfully achieved the birth of a child.[30] When artificial insemination by donor is used, the average length of time from artificial insemination to pregnancy ranges from 2.5 months at some clinics to 9.5 months at others. This points to the need for the clinics and programs to provide information, not about the overall success rate in the field, but about the particular qualifications and track record of that particular physician and clinic.

Traditionally, people have been allowed to participate in risky activities (such as firefighting) based on their voluntary informed consent. The risks of participating in alternative reproduction do not seem to be greater than risks women take in other areas of their lives. (Indeed, many of the risks are similar to the physical and psychological risks of normal reproduction and motherhood itself.) Informed consent of the patient is legally required by case law in all states before a medical procedure is undertaken. The legal doctrine requires that physicians disclose to patients, among other things, the nature of a proposed procedure, its risks and benefits, and the available alternatives.[31] Patients have a right to refuse medical intervention. However, physicians do not have a good track record for obtaining informed consent generally.[32] Moreover, male physicians in the past have been particularly reluctant to give women adequate information and an adequate opportunity to refuse. For that reason, some states have adopted special informed consent statutes applicable to certain women's health issues. One example is a statute that requires that women with breast cancer be told of alternatives to a radical mastectomy.

With respect to alternative reproduction, there may similarly need to be policies to foster the provision of information, such as on the risks of and alternatives to the proposed procedure; on the availability of counseling, mutual aid groups, and other resources for making alternative reproduction a more physically and psychologically satisfying experience; and on the track record of the particular health care provider and clinic with that procedure.[33]

Since some state legislators are opposed to alternative reproduction in any form, care must be taken so that, if informed consent to alternative reproduction is legislatively mandated, legislators do not enact informed consent provisions that unduly pressure women's decisions in this area. It is useful to recall the abortion experience in which lawmakers adopted statutory provisions requiring physicians to provide information that unduly discouraged women from obtaining abortions. Such provisions have been struck down as unconstitutional.[34] In *Akron v. Akron Center for Reproductive Health,* for example, the Supreme Court struck down statutory provisions that required physicians to give speculative information such as the characteristics of the fetus, including ability to feel pain,[35] and provisions that required physicians to present "a 'parade of horribles' intended to suggest that abortion is a particularly dangerous procedure."[36] Similarly, informed consent laws that unduly discouraged individuals from using alternative reproduction (such as a statute that required physicians to make the unsubstantiated statement that embryos feel pain in the embryo transfer process) are inappropriate.

Even if a woman is given sufficient information about alternative reproduction, other policies may be necessary to assure that her decision to participate in alternative reproduction is voluntary. Some feminists argue that societal pressures are so great on women to be mothers that they are not making "voluntary" decisions when they choose to use a reproductive technology.[37] This leads some to a policy recommending banning reproductive technologies such as in vitro fertilization. In my opinion, however, such an approach unduly penalizes infertile women.[38] If it is true that women have been brainwashed into having children, then both decisions to use alternative reproduction and decisions to have children naturally are involuntary. For an equitable policy, then, we should not forbid women to be mothers through alternative reproduction without forbidding them to be mothers through normal reproduction as well.

Gena Corea asks "What is the real meaning of a woman's 'consent' to in vitro fertilization in a society in which men as a social group control not just the choices open to women but also women's *motivation* to choose?"[39] She points out that women may view having a child as a better option than a seemingly dead-end job.[40]

My personal opinion is that it would be a step backward for women to embrace any policy argument based on a presumed incapacity of women to make decisions. That, after all, was the rationale for so many legal principles oppressing women for so long, such as the rationale behind the laws not allowing women to hold property. Clearly, any person's choices are motivated by a range of influences--economic, social, religious. A better approach would be a policy attempting to enhance decision making. Among the factors that are necessary to assure that women can make informed, voluntary choices to use reproductive technologies are enhanced participation of women in the development and implementation of reproductive technologies, greater access to information and resources, and greater control of the use of these techniques.

In order to protect against exploitation, there are two alternatives to prohibiting all forms of alternative reproduction on the presumption that a voluntary choice to undertake them is not possible. One alternative is a policy that would scrutinize each individual agreement to assure that the woman has not been coerced to participate. Some proposed laws would require that mental health workers interview the potential rearing parents and the potential gamete donor, embryo donor, or surrogate to assure that they have an adequate understanding of what they are getting into.[41] With respect to surrogacy arrangements, additional procedures have been proposed, such as a court hearing in which a judge would interview the potential surrogate to assure that her participation was informed and voluntary.[42] Some feminists are uncomfortable about such an approach, since it seems to smack too heavily of government intervention in private reproductive decisions.[43]

Another alternative is to focus on particular types of arrangements that seem conducive to exploitation, and either prohibit or regulate just those types of arrangements. For example, there has been a concern that surrogates who do not have separate legal representation may be likely to be exploited (and consequently some proposed laws provide that a surrogate must have her own separate counsel). There has also been a concern that surrogates are more likely to be exploited by commercial agencies, and so there has been some interest in encouraging nonprofit agencies (such as those handling adoption) to provide oversight for surrogacy arrangements and eliminating for-profit brokers.[44]

For some feminists, surrogacy arrangements in which the surrogate is paid are viewed as inherently exploitative and thus should be impermissible.[45] They point out that in our society's social and economic conditions, some women--such as women on welfare or in dire financial need--will turn to surrogacy out of necessity, rather than true choice. In my view, this is a harsh reality that must be guarded against in part by vigilant efforts to assure that women have equal access to the labor market, and that there are sufficient social services so that poor women with children do not feel that they must contract to create and give up another child just to provide care for their existing children.[46] Feminists also raise concerns about the fact that a surrogate can now be used as the carrier of a couple's embryo, creating the possibility that minority and poor women may be financially induced to serve as a breeder class for the needs of affluent white couples. They point to the history of slavery and wet nurses.

Some feminists believe the solution is to ban paid surrogacy. Other feminists, though, question the propriety of protecting potentially vulnerable women through economic legislation. They point to the history of such legislation in which the laws did not really protect women; instead, they merely closed off certain jobs to women, generally higher paying jobs no more dangerous than the ones women were permitted to take. Janet Gallagher, for example, notes that "Women's unequal treatment before the law has often--much more often than not--been justified by claims that it's necessary to protect women and their special function as childbearers. But defining and protecting women in terms of reproductive capacity has been the basis for women's inequality and lack of economic and political power."[47]

One potential surrogate has asked me, "Why is it exploitation to go through a surrogate pregnancy for someone else if I am paid, but *not* if I am not paid?" If indeed the underlying activity is one that we can countenance or even want to encourage, our focus should not be on banning payment, but on making sure the surrogate gets paid more. The more I have listened to responses to surrogacy, though, the less convinced I am that exploitation is the true issue. There is not just a concern that women will be *coerced* into being surrogates, there is a concern that women will be surrogates at all. That is, it is the underlying activity, not just its potentially exploitative nature that is horrifying to many.[48] And this perspective, I think, is one that should closely be explored for its implications for our views generally on reproduction, motherhood, and family.

In her novel, *A Handmaid's Tale,* Margaret Atwood went to great lengths to create a situation in which it was clear that women were being forced to be surrogates against their will. To do so, she had to go to the following extremes. She postulated a society in which women's jobs were taken away from them, money was taken away from them, their family and support systems were taken away from them, they were forced on the threat of physical violence to obey orders, and they were even denied the means of suicide. We do not have to go that far to postulate exploitation. Instead, we can imagine circumstances in our own society in which a woman would feel compelled to be a surrogate to put food on her table, to pay for health care for a loved one, or to buy some other item or service that we legitimately feel that society has an obligation to provide. Although the facts of the Muñoz case that have been released in the press have been sketchy, that case may very well have been a case of such exploitation.[49] Alejandra Muñoz, a 20 year old, apparently was brought to the United States from Mexico to help create a child for her relatives, Mario and Nattie Haro. She claims that she was told that she would be artificially inseminated with Mr. Haro's sperm, and then the embryo would be flushed out and transferred to his wife.[50] Muñoz claims that, after the insemination, she was told the transfer was not possible and she would have to carry the child to term. During the pregnancy, Muñoz apparently was forced to remain inside the Haros' home. In her seventh month of pregnancy, Muñoz decided to keep the child. After court proceedings, Muñoz and Haro were given joint custody.

The vast majority of women who have been surrogates do not allege that they have been tricked into it, nor have they done it because they needed to obtain a basic of life such as food or health care. Mary Beth Whitehead wanted to pay for her children's education.[51] Kim Cotton wanted money to redecorate her house.[52] Another surrogate wanted money to buy another car. These do not seem to me to be cases of economic exploitation; there is no consensus, for example, that private education, interior decoration, and an automobile are basic needs nor that society has an obligation to provide those items.

If exploitation per se is not the key issue, then what troubles us so about surrogacy? Some are troubled by the idea that women should get paid for the use of their reproductive functions.[53] However, this is a very tricky issue, because it can cut both ways. Allowing paid surrogacy may accentuate a demeaning societal perspective that sees women solely in terms of their reproductive functions. On the other hand, forbidding paid surrogacy, while at the same time working hard to allow women access to other risky forms of employment (such as firefighting), may itself seem to be singling out women's reproductive capacity as their dominant characteristic.

Although the concerns of feminists have often been voiced in terms of exploitation of the surrogate and commercialization of her reproductive functions, many, when pressed, will express that their true concern is with the symbolic effects of what appears to be an arrangement for selling a child. Some believe surrogacy is improper, even if the contract is voluntarily entered into by a nonexploited surrogate, and even if the child goes to a loving home and is in no way harmed by the arrangement, and even if a line is drawn and no existing children are "sold." The logic is that babies should not come into the world as part of a market transaction.[54] However, even among feminists who consider surrogacy to be a form of baby selling, there is no agreement that the solution should be a legal ban on payment to surrogates.

Current laws in at least 23 states prohibit payment to a biological mother, beyond certain enumerated expenses, in connection with her giving up her child for adoption.[55] The Michigan Court of Appeals and the New Jersey Supreme Court have interpreted this type of provision to ban payment to a surrogate.[56] In contrast, decisions by the Kentucky Supreme Court and a Nassau County, New York court held that the public policy against baby selling does not extend to prohibit payment to a surrogate.[57] Along those lines, a recently adopted Nevada statute specifically exempts payment to a surrogate from the ban on payment to a biological mother in connection with an adoption.[58]

The latter decisions have emphasized the constitutional protection of autonomy in childbearing decisions. The courts distinguished paid surrogacy from baby selling by noting that, because the man who contracts with the surrogate is the biological father, he already has a legal relationship with the child and thus any dealing between the parties cannot be characterized as an adoption. The courts also noted that, since the contract is made before conception, the surrogate arrangement is not made to avoid an unwanted pregnancy, but to "assist a person or couple who desperately want a child but are unable to conceive one in the customary manner to achieve a biologically related offspring." Since the decision is made before the pregnancy ensues and the arrangement is entered into with the specific intention of relinquishing the child, the surrogate is less likely than an already-pregnant woman to be coerced into giving up a child she wishes to keep. Moreover, once the child is born, it cannot be treated as a commodity; laws against child abuse and neglect come into play.[59]

There Should Not Be Prior Screening for Fitness of Parenthood

The private lives, health status, and beliefs of people who want to adopt are generally intensely scrutinized. In contrast, people who give birth through normal reproduction are not screened in advance for suitability for parenting. A policy issue is raised about whether or not screening should be required for users of alternative reproduction who will be rearing the resulting child.

Currently, screening of users of alternative reproduction is not mandated by statute, but nevertheless is undertaken by the health practitioners involved. They may reject women or couples on the grounds that they are unmarried, socially undesirable, disabled, or not sufficiently wealthy.

Screening opens up the possibility for an abuse of discretion in which people who would be good parents are nonetheless denied the opportunity because they are not thought to be socially desirable.[60] A Norfolk, Virginia IVF clinic chooses "deserving and appropriate parents,"[61] but it is unclear by what expertise or by what right clinic personnel make that assessment. The director of a British donor insemination clinic wrote that he would deny access to the procedure to a "couple of mixed colour or even a mixed religious denomination."[62] Mandatory psychological screening of couples is done at some AID clinics, despite the fact that the empirical evidence thus far does not indicate that it is useful.[63]

Since objective criteria for determining who would be satisfactory parents are unknown (and perhaps even unknowable), screening may be done in a discriminatory way (for example, under the assumption that poor people or people of color would not be good parents). In particular, there is a bias toward more traditional type women as mothers. The history of the involuntary sterilization laws in this society is rife with examples of the abuse that can occur when fitness for parenthood is assessed.[64] Moreover, it seems inconsistent with our pluralistic society to assume that there is a single standard for assessing who might be a good parent. Since the people who go through alternative reproduction very much want a child, it is likely that the children conceived through alternative reproduction will have at least as healthy and nurturing an environment as children conceived in the traditional manner.

People advocating screening analogize alternative reproduction to adoption.[65] Yet screening for adoption is subject to much criticism for using faulty measures and traditional stereotypes for determining who is fit to be a parent. Even if screening in the adoption situation is justifiable, there are crucial differences between adoption and techniques such as in vitro fertilization, artificial insemination, and surrogate motherhood. In adoption, there is no biological tie between the child and any of the prospective parents. Thus, the screening becomes a substitute for the biological bond in determining who should be allowed to parent a child. In contrast, with alternative reproduction, there is a biological tie between one or both of the prospective parents and the child. Traditionally, society has considered that biological tie to be a sufficient indicator of parental merit to let a person reproduce and rear a child without prior restraint.

Most proposals for screening potential users of reproductive technology recommend that their physical health be taken into account.[66] Yet, physically disabled individuals should be allowed to have children.[67] Although an argument can be made that offspring could be harmed if they were given a genetic impairment from parents who wish to rear them, society does not prevent people with genetic impairments from engaging in natural reproduction.

Legislation has been proposed and court cases have been decided requiring investigations into the parental suitability of users of alternative reproduction. A Washington, D.C. Superior Court refused to grant a couple's adoption application when it was disclosed that a surrogate was used.[68] Instead, the court ordered a full investigation, including an analysis of the suitability of the parents.

In an article in *Detroit College of Law Review*, Sarah Humphreys proposed a statute to require an investigation of the couples who wish to use in vitro fertilization, whether with their own or donor gametes. The point of the investigation would be to "ensure the physical and mental health, emotional stability, and ability of the prospective parent or parents to promote the welfare of the IVF child."[69] A proposed Michigan law does not go quite that far, but it does require that, with respect to all forms of alternative reproduction, a marriage counselor, psychologist, or psychiatrist counsel the couple on the consequences and responsibilities of parenthood, and certify that they understand and are prepared to assume these responsibilities.[70]

Feminists generally feel that it is too risky to allow the government, the medical profession, counselors, or social workers to determine which women and men are worthy of being parents and which children are worthy of being born. There is a widespread consensus among feminists that, because of the importance of procreative decisions, users of alternative reproduction who will be the rearing parents should not be forced to undergo psychological or medical screening.[71] Moreover, people should not be denied access to reproductive technologies merely because they are unmarried.[72]

There is less agreement among feminists about the type of screening, if any, that should be performed on the third parties who aid in reproduction--donors and surrogates. Despite the existence of professional guidelines regarding donor screening,[73] the actual practice of screening of donors and surrogates by health professionals is lax,[74] leading to potential health risks to the female recipients and their children. Some women who have been inseminated with donor sperm have contracted venereal disease from the sperm.[75] In addition, children have been born with genetic impairments that were passed on by the sperm donor.[76] A 1979 national survey of infertility specialists offering artificial insemination found that only 29 percent performed biochemical testing other than blood typing on the donor--and most of this testing was for infectious disease, not genetic defects.[77] In addition, 70 percent of the practitioners kept no permanent records about the donors,[78] making it difficult to trace a donor who has passed on a disease or genetic defect and to cease using him. A 1986 *New England Journal of Medicine* article notes that screening procedures for sperm donors "are usually cursory."[79]

Only a few jurisdictions regulate the screening of donors in alternative reproduction. A New York City ordinance[80] provides that carriers of genetic diseases or defects and men suffering from venereal disease or tuberculosis cannot be sperm donors. It also provides that the sperm donor and recipient must have compatible Rh factors. Laws in Idaho and Oregon provide that a person who knows he has a genetic defect or venereal disease may not be a sperm donor.[81] But these laws provide no requirement that donors be screened. An Idaho law as well as an Illinois law do require screening of sperm donors for AIDS.[82] An Ohio[83] law mandates extensive medical and genetic screening of donors; similar laws have been proposed in Washington, D.C.[84] and Michigan.[85]

Even in the absence of legislation, it would be possible for a woman who contracted an infectious disease through donation of a gamete (or whose resulting child was adversely affected) to sue the health care professional for negligent screening (if the disease had been detectable by screening of the donor as recommended by medical guidelines). She would also be able to sue the donor of the sperm, egg, or embryo if that donor had fraudulently concealed a genetic defect or an infectious disease. In Canada, the Ontario Law Reform Commission recommended adopting a law that would make it a crime for a donor to knowingly conceal a genetic defect or venereal disease.[86]

Current practice and laws often do not require screening of donors for infectious and genetic disorders. Many feminists would advocate infectious disease screening of donors (for example, for AIDS), but have qualms about genetic disease screening, since it seems to be a step toward an unpalatable eugenics. If genetic screening of donors is not mandated, however, the women who use alternative reproduction should be informed of that fact and have the choice of procreating using a donor who is willing to be voluntarily screened. This is in keeping with the opportunity given women who are procreating in the traditional manner to undergo voluntarily genetic screening with their partner.

Reproduction Should Not Be Medicalized

Medical technologies introduced into the childbearing process have traditionally represented a loss of control for women. Technologies originally introduced as voluntary (fetal monitoring, amniocentesis, cesarean sections) have rapidly taken on a mandatory nature. There is a risk that a similar evolution will occur with the new reproductive technologies. For example, women utilizing the new reproductive technologies may be required to undergo costly medical technologies, such as blood hormone levels, to pinpoint their anticipated ovulation even if they feel adept at predicting ovulation through more natural means, such as an assessment of basal body temperature and/or cervical mucus consistency. In addition, efforts are being made to require that the new reproductive technologies be undertaken only by physicians. For example, some state laws make it criminal for persons other than physicians to perform artificial insemination. This is despite the fact that artificial insemination is a simple procedure that can be accomplished without medical intervention through the use of such readily available implements as a turkey baster or a drug store syringe.

Over half of the artificial insemination statutes are premised on the assumption that a physician or someone under a physician's supervision will perform the insemination.[87] Statutes in an additional five states (Arkansas, Connecticut, Georgia, Oklahoma, and Oregon) specify that the process must be done by a physician.[88] In Idaho, artificial insemination must be performed by a physician or someone under his or her supervision.[89] In Georgia, performing artificial insemination without a medical license is a felony punishable by up to five years imprisonment. Such an approach withholds from women the chance to take control over their reproductive futures even by safe means such as artificial insemination.

Another troubling legal aspect of the medicalization of artificial insemination is a recent court decision holding that the involvement of a physician makes a difference regarding who has parental responsibility for the child. The manner in which the artificial insemination statutes are drafted raises questions about whether the involvement of a physician is necessary in order for the paternity provisions of the statute to apply. For example, the California artificial insemination statute states that "[i]f, under the supervision of a licensed physician and with the consent of her husband, a wife is inseminated artificially with semen donated by a man not her husband, the husband is treated in law as if he were the natural father of the child thereby conceived. . . ."[90] Such wording thus raises questions about whether the consenting husband is the legal father when a physician is not involved in the insemination.

A further provision of the California statute also assumes physician involvement. It states that "the donor of semen provided to a licensed physician for use in artificial insemination of a woman other than the donor's wife is treated in law as if he were not the natural father of a child thereby conceived."[91]

It seems farfetched to think that lawmakers meant to provide that the paternity determination hinged on whether a physician was involved in the procedure. Rather, the lawmakers probably assumed that physicians would naturally be involved, since physicians are generally involved in the diagnosis and treatment of infertility. However, the California Court of Appeals, First District held that the statute providing that the sperm donor is not the legal father does not apply when the mother performs the insemination with sperm provided directly to her, rather than to a physician.[92]

The court noted that "nothing inherent in artificial insemination requires the involvement of a physician."[93] However, the court gave two reasons why physician involvement was appropriate. The physician could obtain a medical history of the donor and screen him. Also, the physician "can serve to create a formal, documented structure for the donor-recipient relationship" to avoid misunderstandings between the parties.[94]

The court also highlighted countervailing considerations against physician involvement: it "might offend a woman's sense of privacy and reproductive autonomy, might result in burdensome costs to some women, and might interfere with a woman's desire to conduct the procedure in a comfortable environment such as her own home or to choose the donor herself."[95]

In the California case, Mary decided to conceive a child with artificial insemination and co-parent with a close friend, Victoria. Mary sought a sperm donor by talking to friends and acquaintances and interviewed several potential donors. She chose as a donor a man named Jhordan. Mary, a nurse, did the insemination herself. During pregnancy, Mary and Victoria jointly attended medical appointments and birthing classes and shared childrearing after the birth of the baby, Devin. Although Mary bore the child, the court noted, "Devin and Victoria regarded each other as parent and child. Devin developed a brother-sister relationship with Victoria's 14-year-old daughter, and came to regard Victoria's parents as his grandparents."[96] However, Jhordan was listed as the father on the birth certificate and, after some initial resistance on the part of Mary, was allowed monthly visits to the child. After five visits, Mary terminated his visitation privilege. Nine months after the child's birth, Jhordan filed an action to be recognized as the legal father and to gain visitation rights. The court granted Jhordan's claims, noting that he had been allowed to develop a social relationship with the child.[97]

There is clear agreement among feminists that alternative reproduction should not be medicalized. The term "reproductive technologies" is a misnomer that tends to make it appear that physician involvement is required. In truth, both artificial insemination by donor and surrogate motherhood involve a procedure that is hardly "technological," since it requires no specialized skills or equipment. Embryo transfer by lavage in cattle is done routinely by farmers; as the procedure is applied to women, it will be feasible for lay personnel in women's health centers to learn it and provide it. In vitro fertilization does require specialized skills and equipment, but there are many steps in the procedure (injecting hormones, assessing menstrual cycle timing) that could be done by the woman or her partner in a less medicalized fashion. Policies to avoid medicalization will require a combination of strategies that include a greater role for women in the design of the technologies and the provision of them, education of women regarding how each step in each reproductive technology (e.g., the timing of ovulation) could be undertaken with the maximum amount of control, and the requirement that health care providers follow patient dictates. In addition, legal policies should be adopted so that involvement in reproductive technologies outside of a medical setting is not criminalized and so that issues of legal parenthood do not turn on physician involvement.

Alternative Family Arrangements Should Be Allowed

It is now possible for a child to have three biological parents--the woman providing the egg, the man providing the sperm, and the woman gestating the conceptus--as well as any number of rearing parents. With such possibilities, society is faced with the questions of how many and which participants in alternative reproduction should be identified as the legal parents.

Our culture has developed with the idea that it is important to a child's sense of security to know who his or her parents are. In part, though, the need to ascertain the identity of parents is based on the fact that society as a whole has not had an adequate system for caring for children, and so some private citizens had to be identified and given the responsibility of serving that role.

"Parents," for the most part, have been defined according to their biological tie to the child. Because traditionally two people--a man and a woman--have been biologically necessary to create a child, every child was assumed to have (at minimum and maximum) two parents.[98]

This particular structure of parenting has been viewed as oppressive to women. Society has assumed that children will and should be raised by two-parent dyads, but has generally put no responsibility on men beyond the pressure to marry and to provide financial support for the child. The woman alone in the couple has shouldered most of the actual responsibilities of rearing the child. Thus, she may be unduly burdened by the traditional approach and might fare better if children were raised in more communal settings.

Alternative reproduction, by so clearly separating gestational from genetic parenthood and biological from rearing parenthood, opens up the possibility of creating family forms other than traditional two-parent families. In theory, feminists are supportive of alternative family structures and would probably be comfortable with a policy that made it possible for a child to have any number of legal parents (one, two, three, and so forth), and to have legal parents of the same sex. Feminists also encourage the exploration of ways for people who are not legal parents to relate to children. For example, Barbara Katz Rothman suggests that "[w]e do not have to be 'donors' and 'hosts' and 'surrogates'--we can be mothers and fathers and aunts and uncles."[99] These creative new arrangements must take into consideration not only the needs of the adult participants in alternative reproduction, but also the needs of the resulting children. There is considerable evidence from the adoption context that no matter where legal parenthood resides, many adopted children feel a need for information about (and potential contact with) their biological parents.

Feminists have championed the use of alternative reproduction to create an alternative family structure in at least one situation--they have supported the use by unmarried women of artificial insemination by donor to create a child to be reared by the woman alone or with her female partner without any involvement of the biological father. However, some feminists have qualms about the use of a surrogate by a single man to create a child. It is an unfortunate fact of biological life that single men who want a child of necessity must involve a woman in a more intense way than the way in which sperm donors are involved. Nevertheless, many feminists believe that it is appropriate to recognize enforceable preconception contracts in which donors of sperm, eggs, or embryos give up their parental rights, but feel that a different policy should govern surrogacy. In large measure, this viewpoint is based on the fact that the surrogate has made a greater contribution to the child (by providing the egg and by gestating) than has the man providing the sperm and that, arguably, she has developed a greater emotional relationship with the fetus than has the man.

Under current law, courts have difficulty with the issue of legal parenthood if the single person uses a known donor or surrogate, particularly when a physician is not involved. In one New Jersey case, a woman artificially inseminated herself with her boyfriend's sperm. They broke up during the pregnancy, and, after the baby was born, he sued and won visitation rights to the child. The court pointed out that the man would have been the legal father if the child was conceived through normal intercourse, and held that he was no less a father because he provided semen by a method different from that normally used. Relevant to the court's decision was the observation that "courts have consistently shown a policy favoring the requirement that a child be provided with a father as well as a mother."[100]

In contrast, in the California case discussed earlier, dealing with the artificial insemination of an unmarried woman, the court specifically stated that "[w]e wish to stress that our opinion in this case is not intended to express any judicial preference toward traditional notions of family structures or toward providing a father where a single woman has chosen to bear a child."[101] Nevertheless, since the donor did not provide his sperm to the woman through a physician, he was considered to be the legal father of the child.

With respect to married couples, artificial insemination has been the subject of specific legislation to clarify legal parenthood. Thirty states have adopted laws providing that the resulting child is the legal offspring of the sperm recipient and her consenting husband.[102]

When a surrogate is used, the law often presumes that she is the legal mother of the child. Her husband may be viewed as the father. In addition, the artificial insemination laws of 17 states provide that a man who donates sperm to a woman who is not his wife is not the legal father of the child; in three of these states, however, the man providing the sperm is allowed by statute to make a contract to be the legal father.[103] Although these laws were intended to protect anonymous sperm donors from needing to support the child created with their sperm, the statutes preventing the donor from being recognized as the legal father may create a barrier to surrogacy arrangements in which the sperm donor desires to have legal responsibilities for the child.

A few court cases have dealt with the issues of paternity and transfer of custody in surrogacy situations. For example, in one case, the Michigan Supreme Court held that the Paternity Act does allow a father to establish paternity, even if the drafters of the act had not specifically envisioned the surrogate mother situation.[104]

The common law presumption that the woman who gives birth is the legal mother works to the advantage of people who use a donated egg or embryo to create a pregnancy. (This generally occurs when an infertile woman's husband's sperm is used to fertilize a donor's egg in vitro and that embryo is subsequently transferred to the man's wife, or when a female donor is artificially inseminated with the husband's sperm and than a lavage and embryo transfer are used.) Since the law recognizes the woman who gives birth as the legal mother, it is generally easy for the wife to gain recognition as the legal mother, even though she is not the genetic mother.

In some instances when an embryo is gestated by a second woman, though, the original couple providing the embryo will want a woman to gestate the pregnancy and then give them custody of the resulting child. This may be the case when a genetic mother cannot carry a pregnancy--for example, because of severe hypertension or a uterine malformation. In the case of the first child born to a surrogate carrier, a court granted the genetic parents the right to have their names put on the birth certificate and to be recognized as the legal parents.[105] This is a radical departure from existing law, because it makes the gestating woman totally invisible at birth, giving her no legal recognition as the mother of the child.[106]

Though there is support among feminists for the enforcement of the parties' preconception intent with respect to gamete and embryo donation, there is nevertheless considerable sympathy among feminists for a right of the surrogate to change her mind. Questions arise, however, about what policies should govern the rights and responsibilities of the parties to a surrogate contract when the surrogate changes her mind. Under existing law, there are three potential approaches for deciding such a case: (1) following the adoption model; (2) using a divorce model; and (3) using a contract approach. However, it is possible that none of these precedents are appropriate and that new legal mechanisms are needed to handle surrogacy situations.

In cases of traditional adoption, some states provide that the biological mother can change her mind up until the time the adoption is finalized. In some states, a woman's consent to adoption cannot be given legally until after the child is born. Under state adoption statutes, there is sometimes a certain waiting period after birth (for example, 72 hours in some states) before a woman can consent to give up her child for adoption.

A question arises regarding whether such a waiting period should also apply in the case of a surrogate. Arguably, there is less of a policy need for such a waiting period in the case of a surrogate. A surrogate makes the decision to give up the child in advance of conception at a time in which she can make an informed, unemotional reflection about whether she wants to bear a child for another couple.[107] This is unlike the biological mother in a traditional adoption who may unintentionally become pregnant, may encounter emotional dilemmas and stigmatization during the pregnancy, and may not be able to make an adequate assessment at that time about whether or not she wishes to give the child up. In addition, the surrogate situation is unlike a traditional adoption, because the man wishing to rear the child is the child's biological father. As the Kentucky Supreme Court has pointed out, the man already has a legal relationship to the child. However, it has been pointed out by feminists (as well as male-dominated medical organizations[108]) that the surrogate may not accurately be able to predict in advance how she will feel after going through the birth process.

Giving the surrogate a chance to change her mind does not put an end to the dilemma of who should be recognized as the legal parents of the child, however. Since the man who provided the sperm is the biological father, he can, in most states, claim paternity of the child, which will lead to a court battle over who should rear the child.

Thus, even following the adoption model could lead to the second model, the custody approach. This approach to handling the issue of the surrogate changing her mind entails the court determining custody according to the standards used in a divorce case--i.e., what is in the best interests of the child.[109]

In the absence of legislation, courts may fall back on social stereotypes in determining whether the contracting man and his wife or the surrogate (and her husband, if any) would provide a better home for the child. The fact that the contracting couple is likely to have a higher income than the surrogate and her husband may weigh in the couple's favor. In addition, the contract in which the biological mother agreed to give up the child may be used as evidence of her unfitness as a mother. Similarly, if she had accepted a fee to turn over the child, the court might view that act as reflecting badly upon the surrogate. Moreover, the *Baby M* case has demonstrated the horrors for the surrogate of trying the issue of custody. Feminists have pointed out that the surrogate in that case, Mary Beth Whitehead, was treated like a rape victim in terms of the courtroom examinations and expert testimony impugning her integrity and dredging up details of her private life (such as the fact that she dyes her hair) that have nothing to do with her fitness as a mother.

Under the custody model, a question may arise regarding whether the surrogate and contracting father should have joint custody or whether the party who does not gain custody should gain visitation rights to the child. Since a determination of custody of the child will be made at or soon after birth, there is less reason for joint custody or visitation than when parents who are raising a child together divorce. In the latter case, the child has developed a bond with both biological parents, and that bond should be allowed to continue. In the Muñoz case in California, the surrogate and the biological father have been granted joint custody. However, that is an unusual situation, in which the two parents were related and in which there was evidence that the biological mother had been tricked into serving as a surrogate.

The third approach to handling the issue of a surrogate changing her mind is to consider it a contract matter and to hold the surrogate to her promise to turn over the child. The issue of the enforceability of a surrogate contract has received widespread debate in the legal literature. Specific performance is generally not granted for a breach of contract for personal services. For example, if an opera singer reneges on a contract to sing, she is not forced to sing. She can instead pay monetary damages for the breach. Similarly, it would be unfathomable to enforce a contract to force a surrogate to be inseminated (if she changed her mind before the insemination) or to carry the child (if she decided to exercise her legal right to an abortion).

Once the child is born, however, specific performance is a realistic possibility. The personal services of the surrogate are no longer needed, so she would not be forced to provide a personal service against her will. Outside of the personal services area, specific performance is an appropriate remedy when monetary compensation is inadequate. The biological father could persuasively argue that money will not compensate for being unable to rear his child whom the surrogate agreed to bear for him. The contract approach has another facet to it as well. By recognizing the contract, the intended parents (the couple who contracts with the surrogate) will have to take custody of the child no matter what the child's state of health.

Only one percent of surrogates have changed their minds and tried to keep the children they contracted to bear.[110] There are dilemmas in sorting out what policies should govern these cases. Clearly, litigation over custody benefits neither side, and is particularly harmful to the child.[111] In my opinion, the child should not have to be subject to the trauma of litigation after its birth to determine who his or her legal parents are. Even if the parties who have temporary custody of the child while the litigation is going on gain permanent custody, their initial childrearing months (or years) will have been troubled by the anxiety they have felt during the legal battles. That anxiety cannot help but be conveyed to the child. Thus I believe it is important to choose a primary family for the infant before his or her birth. One approach would be to have a policy that always holds that the intended parents are the legal parents. This seems harsh to many feminists. It is a policy that makes a cold calculation that, rather than subject all children born from surrogacy to the risk of a custody dispute, it is appropriate to put the psychological risk of the contract on the few surrogates who change their minds. However, in my view, holding the surrogate to the original agreement will have a beneficial effect on the child, since he or she will have identifiable legal parents and will not be subject to a lengthy custody battle (with the uncertainties creating potential damage to the bonding process). Moreover, it may discourage women who are not entirely sure that they want to be surrogates from participating in the procedure. Under current law, with the surrogate being recognized as the legal mother, a woman who is uncertain about whether she can give the child up may nonetheless agree to be a surrogate because of the possibility that she will have a second chance at the child after its birth before the adoption procedure. In addition, some surrogate mothers, because of their current status as legal mothers when the child is born, may threaten not to go through the adoption until the couple pays more money.[112]

An alternative approach is to have primary responsibility for the child rest with the surrogate and her family. This strikes some people as odd, since the child was not conceived to be part of the surrogate's family; the child has only come into being because of the intended parents' interest in being rearing parents. Nevertheless, many feminists feel that the surrogate's greater involvement in the creation of the child entitles her to a greater claim to have the child be part of her family.[113]

How do we arbitrate between these two approaches? Some people may wish to try the Solomonic approach of splitting the baby in half and awarding partial custody to each genetic parent.[114] However, although such an arrangement accords with biological reality, it may not fit with social reality in that the parties may not be able to develop a smooth way of co-parenting. When a single woman uses artificial insemination by donor to conceive a child and the sperm donor wants a role in parenting, feminists have pointed out that "it is difficult enough to rear a child with an ex-spouse, let alone a donor with whom you have never had a relationship."[115] Similarly, in surrogacy, the parents who believed they would be rearing the child might find it difficult to share that role with the surrogate (and vice versa). In my view, if joint custody is a possibility (for example, as the default arrangement if the surrogate changes her mind), that possibility should be clearly spelled out in advance.

If the policy chosen to govern surrogacy is that the intended parents be recognized as the legal parents,[116] then it is crucial for that policy to go hand-in-hand with a policy that gives the intended parents absolutely no control over the surrogate's activities during pregnancy. The protection of the surrogate's bodily integrity must be guaranteed.[117] If it cannot be, then it would be the position of many feminists (including myself) that the law should provide that surrogates should have the option of keeping the children they contract to bear (in order to provide the assurance that they will not be required to follow the mandates of the couples during their pregnancies). As Janet Gallagher points out, the surrogate "is in a different relationship to the child than the man. We recognize this legally in allowing women to unilaterally decide for abortion. The father or the fetus can't be the decision-maker. The geography of pregnancy gives a woman a greater degree of say and power."[118]

Consideration Should Be Given to Financial Aid for Reproductive Technologies

There is a consensus among feminists that wealth should not be a criterion for determining who should be allowed to be parents. Currently, access to alternative reproduction is limited by financial barriers. In vitro fertilization is an expensive procedure--costing $1500 to $5000 per attempt--and few insurance companies cover it. For that reason, policymakers are considering laws that might mandate coverage.

The financing of alternative reproduction raises dilemmas regarding the allocation of health care resources in our society. Clearly, society should give top priority to meeting the life-threatening and other serious health needs of the population. However, reproductive choice is a sufficiently important component of society that insurance coverage should be available to further such choices. Thus, health insurance should cover pregnancy services and abortion services and arguably infertility treatments. In Maryland, a law was passed requiring that insurers that cover other pregnancy-related services must pay for IVF as well, if certain conditions are met.[119]

A similar policy question is whether medically assisted conception should be considered to be within the minimum level of health care guaranteed to people, and thus paid for by the government. In countries with national health care systems, artificial insemination by donor is among the services offered, and some expansion to include in vitro fertilization has been made.[120]

In this country, the claim for public subsidy for technologies to further reproductive choice has not fared well. According to United States Supreme Court decisions, even though the right to privacy protects procreative choice, that protection has not been held to extend to requiring governmental funding of the medical technologies to effectuate these choices.[121] Although some states have adopted laws or interpreted their state constitutions[122] to provide for public subsidies for abortion, state legislatures may be reluctant to provide welfare benefits for reproductive technologies which, in addition to potentially involving procedures that are more expensive than abortion (such as IVF), also give rise to the potential for welfare funds to be spent on the raising of a child.

Nevertheless, a strong argument can be made that, "[t]he human birthright includes the right to give birth" and that infertility treatments to further that birthright are "part of the price we pay for our living together in community."[123] Feminists have recognized the importance of addressing the financial aspects of alternative reproduction, and have begun efforts to prioritize various health care needs and assess how prevention and treatment of infertility fit into that schema.

Even if a poor person can afford the costs of a particular technology (such as the cost of AID), some practitioners refuse to provide the service on the assumption that the person will not be able to afford to raise the child.[124] This may represent a bias on the part of the physician that people of lower social and economic classes would not make good parents. Since this is not a valid assumption, physicians should not be allowed to turn down patients based on future financial considerations.

The Future Work for Feminists

The need for a policy assessment of reproductive technologies is pressing. More children are born each year through the new reproductive technologies of in vitro fertilization, artificial insemination, surrogate motherhood, embryo transfer, egg donation, and embryo cryopreservation than are available for traditional adoption. State legislators have begun proposing statutory schemes that will control whether and how these technologies are available.

The policies that states adopt with respect to the new reproductive technologies will affect not only the infertile women, but will influence reproductive rights generally. If certain rationales are used to limit access to new reproductive technologies (rationales such as that the in vitro embryo should be considered a person or that unmarried people should not be allowed to be parents), these same reasons may be used in an attempt to curtail reproductive choices (such as abortion or adoption) that are currently available to women.

An adequate policy assessment should address the role that women currently have in shaping the development and implementation of alternative reproductive strategies and analyze how that role can be expanded. It should also address the amount of control individual women have in the employment of alternative reproduction and how these technologies affect individual women and women as a group.

Based on precedents in other areas and the limited experience with alternative reproduction to date, certain key areas of concern can be identified. Some concerns, such as a need for greater autonomy of patients who participate in alternative reproduction, may require responses that are not particularly legal responses. Such concerns may be more appropriately handled by strategies, such as organizing patient groups to influence how the services are delivered and publishing directories of clinics so that any variation from clinic to clinic can be taken into consideration by women choosing to use the technologies. In dealing with other concerns, such as determining the legal status of the resulting children, statutory guidance will be necessary. In still other instances, a combination of legal and other strategies may be appropriate--for example, to encourage more effort in the prevention of infertility rather than in high-tech means to overcome it.

A policy assessment of the alternative reproduction strategies should concern itself not just with their effects on infertile women, but also with effects on the third parties (such as egg donors or surrogates) who will be called upon to enable infertile women to have children, and with their effects on the resulting children. There is a concern about preventing poor women from being coerced into being egg or embryo donors or surrogate mothers out of financial need. There is concern that poor women and women of color will serve as surrogate carriers to gestate the embryos of wealthier, white couples, producing a modern day parallel to the phenomenon of wet nurses.[125] Another concern is whether donors and surrogates will be harmed psychologically by being entirely cut off from the children they create, and how the children themselves will fare.

As the merits and drawbacks of alternative reproduction are debated, great care needs to be taken not to portray women as incapable of responsible decisions. There may be many reasons to urge caution or even limit some applications of alternative reproduction. However, such restraints should be based on predictable and well-articulated bad effects on women, their partners, their children, or society, rather than on women's incapacity to apply correctly the technologies. The language we use and the rationales we employ for the positions we adopt to control alternative reproduction techniques can have as important an impact on the treatment of women as the techniques themselves.

In the course of discussing the policies that should govern alternative reproduction, we must make clear the link between these technologies and other reproductive choices. For example, the constitutional underpinnings for reproductive choice regarding abortion and contraception also protect autonomy in the use of artificial insemination, embryo donation, surrogacy, and so forth.[126] A recommendation that the government severely restrict or ban alternative reproduction may be taken as an acknowledgment of governmental power to restrict or ban contraception and abortion technologies. Norma Wikler has pointed out that "[t]he danger to the feminist program, of course, is that once the right to privacy in reproductive decision making loses its status as a natural or constitutional right, women risk losing choices that they now have."[127]

Appendix A

The Forms of Alternative Reproduction, Their Uses, and Their Costs

Advancing medical technologies provide individuals and couples with a growing number of options with respect to whether, how, and when to have a child. The new reproductive models are known as "noncoital," since they involve procreation without the sex act. Many are experimental procedures, since they have not been sufficiently tested to allow predictions about all their risks. They are generally employed to overcome an infertility problem, and the nature of the problem dictates the type of procedure used.

Currently, among people of childbearing age, one in six is infertile. About half of the number of infertile people can be helped by traditional medical means (such as drugs and surgery). Others who wish to have a child biologically related to themselves or their partners need to turn to alternative reproduction.

For example, if a woman has blocked or absent fallopian tubes or her partner has a low sperm count, they can use in vitro fertilization (IVF). An egg is removed from the woman's ovary and fertilized with the man's sperm in a petri dish. The resulting embryo is placed in the women's uterus two days later.

In some cases, one or both of the partners are unable to provide the gamete to create a child, or the woman is unable to carry the pregnancy. Consequently, such people will have to call upon the aid of a third party--a donor of sperm, eggs, or embryos, or a surrogate. Sometimes a combination of third parties will be used.

If the man produces no sperm, his female partner can be inseminated with sperm from a donor. Similarly, if the woman cannot provide an egg, she can call on the aid of a female donor to provide an egg. The donation of sperm or eggs or both can be done in conjunction with in vitro fertilization, or can be done by transferring the sperm or egg into the woman's body for fertilization.

Another option for women who cannot produce eggs is to use an egg that has been fertilized in vivo, inside another woman. Sperm from the partner of the first woman is used to inseminate the donor woman. Five days after the donor woman conceives, the embryo is flushed out of her womb in a nonsurgical procedure known as embryo lavage and is then transferred to the womb of the first woman. This technique is known as embryo transfer after in vivo fertilization. It allows the first woman to have a biological relationship with the child via pregnancy, even though she has no genetic relationship with the child. If the woman's partner cannot provide sperm, donor sperm can be used to create the embryo that will be transferred.

Some women can provide the genetic component for reproduction, but not the gestational one. Such a woman may decide to create an embryo with her partner and then transfer the embryo to a surrogate carrier for gestation only. The process can be accomplished in conjunction either with in vitro fertilization and transfer or in vivo fertilization and transfer. After the birth of the child, he or she is returned to the genetic parents. This option is likely to be particularly useful for women who have disabilities (e.g., due to severe diabetes) and are advised not to risk pregnancy, but would like to rear their own biological child.[128]

If a woman can provide neither the genetic nor the gestational component for reproduction, but has a male partner who can provide the genetic component, her partner's sperm can be used to inseminate another woman (known as a surrogate mother) who will carry the child for nine months and then release the infant for rearing by the couple. The man who provided the sperm will generally be considered the legal father, but the woman who will rear the child will generally have to adopt the child.

All of these techniques allow one or both of the rearing parents to have a biological bond to the child (genetic, gestational, or both). Except in the case of standard in vitro fertilization, these methods also involve using the aid of one or more third parties for a biological component of reproduction. Another variation is possible, although it apparently is not yet offered. That is for the rearing parents-to-be to contract for a child with no biological tie to them. They could use the combination of an egg donor, a sperm donor, and a surrogate. However, such an approach raises different legal and ethical concerns than do the techniques in which at least one rearing parent has a biological link to the child. The use of third parties to perform all of the reproductive functions makes the situation more akin to adoption than to technologically assisted reproduction.

Although alternative reproduction is generally used by people with infertility problems, one aspect of it--gamete donation--may be used by people passing on a genetically based impairment to their children. At least 33 percent of artificial insemination practitioners have inseminated women whose husbands did not want to pass on a potential genetic impairment.[129] Egg donation or embryo donation may be used for similar reasons.

Although infertility and the potential of creating a child with a genetically based impairment are considered to be medical reasons for using reproductive technologies, they obviously have strong social components. The infertile person may use a reproductive technology because of social pressure to have a biologically related child. In our society, there are few ways to participate in child-rearing other than in the role of biological parent. A person who risks creating a child with a genetically based disorder may use a reproductive technology because of insufficient understanding of how impairments that are genetically based will actually affect the functioning of the child and because there are few social supports for rearing children that have such impairments. Alternative reproduction may also be undertaken for other reasons, such as when a person wants to rear a child, but does not wish to engage in sexual intercourse with a person of the opposite sex.

The existence of alternative reproduction has influenced what is considered to be appropriate medical practice. When a young woman undergoes a hysterectomy, for example, she should be given the option of not having her ovaries removed, since it is now possible for her to become a genetic mother even without a uterus. An embryo created with her egg could be implanted in a surrogate carrier. Similarly, when a woman is about to receive a cancer treatment that might prove mutagenic to her eggs, she should be told about the possibility of freezing eggs or embryos in advance of the treatment for subsequent use to create a child. It is unclear whether the existence of alternative reproduction is being taken into consideration by physicians in other practice areas. If it is not, they may be unnecessarily dooming women to infertility.

The costs of reproductive technologies are currently sufficiently high that they are generally only within reach of upper-middle-class or wealthier people. As of 1987, in vitro fertilization, for example, costs $2000 to $5000 per attempt, with a maximum 30 percent success rate per attempt. Surrogate motherhood costs approximately $25,000, with an average of $10,000 being paid to the surrogate, $10,000 to the lawyer or other facilitator, and $5000 going toward miscellaneous expenses such as medical expenses. Even artificial insemination, the lowest-tech procedure, can be costly. Insemination under the supervision of a physician costs approximately $100 per insemination, with two or three inseminations done per month. Clinics vary, taking an average of anywhere from 2.5 to 9.5 months for conception (thus ranging from $500 to

$2850). Even if a woman finds her own donor and performs the artificial insemination herself, she may wish to have genetic or medical screening (e.g., for AIDS) done on the donor. This may cost up to $2000.

Alternative reproduction has raised criticism from a variety of camps. The Roman Catholic Church opposes alternative reproduction techniques (in vitro fertilization, gamete donation, and surrogacy) because they allow procreation outside of the sexual act of a married couple.[130] Right-to-life groups oppose technologies such as in vitro fertilization, embryo cryopreservation, and embryo transfer because they are viewed as putting embryos at risk.[131] Because of the diversity of viewpoints about alternative reproduction, the debates regarding what policies should govern in this area will no doubt be conducted vigorously over the next few years.

Notes and References

1. For a description of the technologies, see Appendix A.

2. For example, a recent assessment of the reproductive technologies and proposal of guidelines by the organization of medical practitioners who offer reproductive technologies contained no mention of feminist writings or arguments about these issues. Ethics Committee of the American Fertility Society, "Ethical Considerations of the New Reproductive Technologies," 46 Fertility and Sterility 1S (Supp. Sept. 1986) (hereinafter Ethics Committee of the American Fertility Society).

3. See, e.g., G. Corea, The Mother Machine (1984); Rothman, "How Science is Redefining Parenthood," Ms. 154 (July 1982); R. Arditti, R.D. Klein, and S. Minden, eds., Test-Tube Women: What Future for Motherhood? (1984); Wikler, "Society's Response to the New Reproductive Technologies: The Feminist Perspectives," 59 Southern California Law Review 1043 (1986); Women's Bureau, Grael in the Rainbow, "Documentation of the Feminist Hearing on Genetic Engineering and Reproductive Technologies," Brussels, March 6-7, 1986; H.B. Holmes, B.B. Hoskins, and M. Gross, eds., The Custom-Made Child? (1981).

4. See, e.g., G. Corea, The Mother Machine 101-104 (1984).

5. Id. at 166.

6. Gena Corea, for example, asserts that "Lesley Brown, mother of the first test-tube baby, had worked in a factory wrapping cheese and could not bear the prospect of spending the rest of her days in the factory." Id. at 220.

7. The varying approaches are most prominent with respect to the issue of surrogacy arrangements. Compare, for example, the varied positions of Lori Andrews, Janet Gallagher, and Barbara Katz Rothman at the New York Legislative Hearings on Surrogate Motherhood and New Reproductive Technologies, October 16, 1986.

8. United Nations, Demographic Yearbook (1984). For information on racial differences in infant mortality in the United States, see Binkin, Williams, Hogue, and Chen, "Reducing Black Neonatal Mortality: Will Improvement in Birth Weight Be Enough?" 253 Journal of the American Medical Association 372 (1985).

9. Allen, "Effects of Nutritional Status on Pregnancy and Lactation," 101 in Malnutrition: Determinants and Consequences (1984).

10. See, e.g., L. Andrews, Deregulating Doctoring: Do Medical Licensing Laws Meet Today's Health Needs? 3 (1983).

11. Prime among these have been statutes regarding research and services in connection with the prevention of birth defects. See, L. Andrews, State Laws and Regulations Governing Newborn Screening (1985).

12. See the position paper "Reproductive Hazards in the Workplace" by Joan Bertin in this book.

13. For a discussion of these laws, see, e.g., L. B. Andrews and J. L. Hendricks, "The Legal and Moral Status of In Vitro Fertilization," in C. Fredericks, J. Paulson, and A. DeCherney, eds., Foundations in In Vitro Fertilization 317-318, 322, 323 (1987).

14. La. Rev. Stat. Ann. § 9:129 (West Supp. 1987).

15. As noted earlier, the use of the term surrogate may serve to mask the fact that the woman is the genetic and gestational mother.

16. For a literary parable expressing the major dangers, see M. Atwood, A Handmaid's Tale (1985).

17. Moreover, some feminists, such as Wendy Chavkin, note on a practical level that "it appears unlikely that surrogate arrangements can be legislated out of existence, but that laws can exist which monitor the conditions under which such arrangements take place."

18. This, of course, is a concern with respect to all women, including those who are considering reproducing naturally.

19. The logic of Roe v. Wade, 410 U.S. 113 (1973) protects such activities. In the trial court decision in the Baby M case, the court specifically recognized that a contractual obligation to abort could not be enforced. In re Baby M, 217 N.J. Super.

313, 525 A.2d 1128 (1987), rev'd on other grounds, (A-39-87) slip op. (N.J. February 3, 1988). For an excellent overview of the legal and policy reasons for protecting a pregnant woman's bodily integrity, see Gallagher, "Prenatal Invasions and Interventions: What's Wrong With Fetal Rights," 10 Harvard Women's Law Journal 9 (1987).

20. This approach is supported by the position papers of Janet Gallagher, Joan Bertin, and Mary Sue Henifin, Ruth Hubbard, and Judy Norsigian in this book.

21. In one study, Bradford Gray interviewed women who had been given an experimental drug to induce labor. The consent form was administered at the time women came to the hospital to deliver, sometimes in the labor room itself. Of the women who had signed the consent forms, 20 out of 49 did not realize until after their participation began, usually when Gray interviewed them, that they were subjects of research. Even those who understood that they were part of an experiment did not understand that participation was not required, there were hazards, and there were effective alternatives. B. Gray, Human Subjects in Medical Experimentation 66 (1975).

22. Corea, supra n. 3 at 135 n. 2.

23. Id. at 102.

24. Wagner, "The Legal Impact of Patient Materials Used for Product Development in the Biomedical Industry," in "Public Policy Symposium--The Legal, Ethical and Economic Impact of Patient Material Used for Product Development in the Biomedical Industry," 33 Clinical Research 442, 446 (1985). In addition, the federal law governing federally funded research on body tissue and parts removed in the course of surgery does not appear to provide for consent of the patient if the research will not identify her as the source of the eggs or embryos. 45 C.F.R. § 46.10(b)(5) (1986).

25. See e.g., Del Zio v. Manhattan's Columbia Presbyterian Medical Center, No. 74-3558 (S.D.N.Y. filed April 12, 1978) in which a woman who was an IVF patient was awarded $50,000 for emotional distress when a doctor destroyed the contents of the petri dish in which in vitro fertilization was being attempted with the woman's egg and her husband's sperm.

26. Kass, "Making Babies--The New Biology and the 'Old' Morality," 26 Public Interest 18, 30 (1972).

27. In a study of embryo lavage of donor monkeys, two deaths occurred as a result of ectopic pregnancies. Hodgen, "Surrogate Embryo Transfer Combined with Estrogen-Progesterone Therapy in Monkeys," 250 Journal of the American Medical Association 2167 (1983). When embryo transfer was tried on women, the lavage did not work properly on one donor, and she retained the pregnancy and subsequently spontaneously aborted nine days after her expected period. Bustillo et al., "Nonsurgical Ovum Transfer as a Treatment in Infertile Women: Preliminary Experience," 251 Journal of the American Medical Association 1171 (1984).

28. "Effects of Surrogate Motherhood, Other Child-Bearing Options Need Closer Study, Says Researcher," Psychiatric News 10 (May 18, 1985).

29. Such is apparently the case with some sperm donors. According to psychotherapist Annette Baran and psychologist Aphrodite Clamar, who counsel former sperm donors, some sperm donors feel regret about not having contact with their biological child. See also research of psychotherapist Annette Baran reported in L. Andrews, New Conceptions: A Consumer's Guide to the Newest Infertility Treatments Including In Vitro Fertilization, Artificial Insemination, and Surrogate Motherhood 267 (1985).

30. Marrs, "The American Experience," Fourth World Congress on In Vitro Fertilization, Melbourne, Australia, November 19, 1985.

31. Andrews, "Informed Consent and the Decisionmaking Process," 5 Journal of Legal Medicine 163 (1984).

32. Applebaum and Roth, "Treatment Refusal in Medical Hospitals," in President's Commission for the Study of Ethical Problems in Medicine and Biomedical and Behavioral Research, Making Health Care Decisions: The Ethical and Legal Implications of Informed Consent in the Patient-Practitioner Relationship Vol. II, 67 (1982).

33. For example, reproductive technology providers should be required to maintain annual reports that detail their particular expertise and experience with procedures. Such a requirement is now considered an ethical mandate by the American Fertility Society. Ethics Committee of the American Fertility Society, supra n. 2 at 72S. These reports should be provided to anyone who requests them. See, e.g., Anderson and Shields, "Quality Measurement and Control in Physician Decisionmaking: State of the Art," 17 Health Services Research 125 (1982), which recommends annual reports for all areas of medicine.

34. Akron v. Akron Center for Reproductive Health, 462 U.S. 416, 444 (1983).

35. Id. at 444 n. 34.

36. Id. at 445. See also Thornburgh v. American College of Obstetricians and Gynecologists, 106 S. Ct. 2169, 476 U.S. 747 (1986).

37. These same commentators are in favor of unmarried women using artificial insemination by donor. The message seems to be that the only time the decision to be a mother is voluntary is when it is done without a male partner.

38. Penalizing infertile women is an approach that certain conservatives take. Since some women need in vitro fertilization because of previous use of an IUD or because of pelvic inflammatory disease as a result of multiple sexual partners, the conservatives make the argument that such women should be denied access to reproductive technologies as punishment for their previous sexual behavior.

39. G. Corea, supra n. 3 at 3.

40. Id. at 220.

41. This is the approach taken by Mich. H.B. 4554 (1985).

42. This is the approach recommended in a proposed New York law. N.Y. Sen. B. 1469 (1987).

43. There is also a concern about whether judges will be able to make an adequate assessment of whether the surrogate involvement is voluntary and uncoerced. Judges have a similar responsibility in assessing whether a guilty plea is voluntary, but there are varying viewpoints as to how well they meet that responsibility.

44. The English Parliament enacted a law prohibiting payment to agencies and other arrangers of surrogacy, but not explicitly prohibiting surrogacy itself. Surrogacy Arrangements Act 1985.

45. In contrast, many feminists have no problem with payment to donors of sperm, eggs, and embryos.

46. In this sense, the conservatives who oppose surrogacy may have had a hand in encouraging it by cutting back on welfare programs for dependent children.

47. Van De Velde, "The Future of Surrogacy," Chicago Tribune (January 25, 1987).

48. To illustrate the point that exploitation is not the key issue, let me posit the following hypothetical case. What if I wanted to be a surrogate? Since I have read extensively about the physical, psychological, and social consequences of surrogacy, it would be hard to argue that I will have been exploited in the sense of entering a situation without sufficient information. Since I have other ways to earn money, I will not have been exploited in a sense of being forced to be a surrogate because of economic duress. Nevertheless, many other feminists would still advise me not to do it, some even suggesting I should be prohibited from doing so by law.

49. The case itself does not stand as an argument against legislation regulating surrogacy and allowing a fee, however. The tragedy of the case might have been avoided if a particular procedure was in place to assure that surrogates were not coerced--such as prior court approval of the arrangement and/or representation of Muñoz by her own attorney.

50. The Haros claim that she consented to be a surrogate for a fee of $1500.

51. Dr. Burton Sokoloff, presentation at the Child Welfare League of America Annual Meeting, March 18, 1987. Dr. Sokoloff was an expert witness in In re Baby M.

52. K. Cotton with E. Daly, Baby Cotton: For Love and Money (1985).

53. Some feminists see a parallel between surrogacy (fee for reproductive services) and prostitution (fee for sexual services). However, just as there is disparity among feminists about what the appropriate policies should be regarding prostitution (i.e., some would decriminalize it), there is disparity about what should be done about surrogacy.

54. Babies should probably also not come into the world in an attempt to hold a marriage together, to provide love for a woman who feels unloved, to provide a son for a man who has a family of daughters, and so forth. In fact, if we start scrutinizing the circumstances surrounding natural conceptions, it is hard to imagine many people whose motives would be pure enough to satisfy our close scrutiny.

55. See, e.g., Ala. Code § 26-10-8 (1977); Ariz. Rev. Stat. Ann. § 8-126(c) (1974); Cal. Penal Code § 273(a) (West 1970); Colo. Rev. Stat. § 19-4-115 (1974); Del. Code Ann. tit. 13, § 928 (1981); Fla. Stat. Ann. § 63.212(1)(b) (West Supp. 1983); Ga. Code Ann. § 74-418 (Supp. 1984); Idaho Code § 18-1511 (1979); Ill. Rev. Stat. ch. 40, para. 1526, 1701, 1702 (1981); Ind. Code Ann. § 35-46-1-9 (West Supp. 1984-85); Iowa Code Ann. § 600.9 (West 1981); Ky. Rev. Stat. § 199.590 (2) (1982); Md. Ann. Code § 5-327 (1984); Mass. Ann. Laws ch. 210 § 11A (Law. Co-op. 1981); Mich. Comp. Laws Ann. § 710.54 (West Supp. 1983-84); N.J. Stat. Ann. § 9:3-54 (West Supp. 1984-85); N.Y. Soc. Serv. Law § 374(6) (McKinney 1983); N.C. Gen. Stat. § 48-37 (1984); Ohio Rev. Code Ann. § 3107.10(A) (Baldwin 1983); S.D. Codified Laws Ann. § 25-6-4.2 (Supp. 1983); Tenn. Code Ann. § 36-135 (1984); Utah Code Ann. § 76-7-203 (1978); and Wis. Stat. Ann. § 946.716 (West 1982). However, since the statutes in Arizona, California, Florida, Illinois, Iowa, and New Jersey exempt stepparents from either the payment ban or the statutory reporting requirements for payment in connection with an adoption, it may be possible to circumvent the ban in these states, so long as the man who provided the sperm can prove that he is the biological father and thus that his adopting wife will be a stepparent.

56. Doe v. Kelley, 106 Mich. App. 169, 307 N.W.2d 438 (1981), cert. denied, 459 U.S. 1183 (1983); In re Baby M, (A-39-87) slip op. (N.J. February 3, 1988).

57. Surrogate Parenting Associates, Inc. v. Kentucky, 704 S.W.2d 209 (Ky. 1986); In re Adoption of Baby Girl, L.J., 505 N.Y.S.2d 813 (1986).

58. Nev. Rev. Stat. § 127.287 (5) (1987).

59. There may be a concern that, even if the rearing parents do not treat the child as a commodity, there has been a symbolic commodification of children when there has been payment to a donor or surrogate. Yet in the area of reproductive rights, the possibility of symbolic harms (such as the potential symbolic harms associated with aborting a fetus) has not been felt to be sufficient to outweigh an individual's or couple's autonomy in making procreative choices. I worry about legitimating symbolic concerns in this reproductive context because of the possibility that it might allow symbolic concerns to be used to justify laws against contraception or abortion. The only way to get around this, in my mind, is to recast the original right to privacy in the reproductive context as a right of bodily integrity (rather than a decision-making right), which would protect access to abortion and contraception but which might allow restriction of paid donation or surrogacy on symbolic grounds. I am not advocating such an approach, however.

60. The Baby M case demonstrated how social stereotypes influence judgments of who is fit to be a parent. In particular, the trial judge implied that Mary Beth Whitehead was less worthy of attaining custody because she had less education than did William and Elizabeth Stern. In re Baby M, 217 N.J. Super. 313, 525 A.2d 1128 (1987). On appeal, the New Jersey Supreme Court specifically rejected such an approach, but used its own questionable criteria, suggesting that Whitehead should not get custody because she was overenmeshed in the child and would not allow her sufficient independence. In re Baby M, (A-39-87) slip op. (N.J. February 3, 1988).

61. Corea, supra n. 3 at 145 citing information from clinic director Mason Andrews.

62. Sandler, "Donor Insemination in England," 19 (5) World Medical Journal 88, 89 (1972).

63. Shapiro, "Some Unresolved Questions about Artificial Insemination," 17 Contemporary Ob/Gyn 129 (1981).

64. Beckwith, "Social and Political Uses of Genetics in the United States: Past and Present," 265 Annals N.Y. Academy of Sciences 46 (1976). The same is true of immigration policy. Corea, supra n. 3 at 27 points out:

> at a time when the privileged feared that the unprivileged (including immigrants) would outbreed them and threaten their ascendancy, they administered IQ tests to immigrants arriving on Ellis Island. In 1912, the IQ tests found that 83 percent of the Jews tested, 80 percent of the Hungarians, 79 percent of the Italians and 87 percent of the Russians were "feeble-minded." For this defect, many aliens were shipped back to their homelands.

65. Pierce, "Survey of State Activity Regarding Surrogate Motherhood," 11 Family Law Reporter (BNA) 3001, 3004 (January 29, 1985).

66. See, e.g., Ontario Law Reform Commission, Report on Human Artificial Reproduction and Related Matters Vol. I 158 (1985).

67. See, e.g., R. Arditti, R.D. Klein, and S. Minden, eds., Test-Tube Women: What Future for Motherhood? (1984).

68. In re R.K.S. (Super. Ct. D.C., April 13, 1984), 10 Family Law Reporter (BNA) 1383 (May 15, 1984).

69. Comment, "Lawmaking and Science: A Practical Look at In Vitro Fertilization and Embryo Transfer," III Detroit College of Law Review 429, 446 (1979).

70. Mich. H.B. 4554 § 4(c) (1985).

71. There is more support for a role for medical screening, however, when there is a medical risk to a third party other than the potential child. For example, a man should be screened for venereal disease before his sperm is used to inseminate a surrogate to prevent the transmission of an infection to her.

72. Health care providers raise concerns about whether an unmarried individual's use of reproductive technologies will be harmful to the resulting children. However, studies of children in single-parent female-headed families have found that they have comparable cognitive abilities to those raised in two-parent homes. McGuire and Alexander, "Artificial Insemination of Single Women," 43 Fertility and Sterility 182 (1985). In addition, children in single-parent homes have a level of self-esteem that is at least equal to that in two-parent families. Raschke and Raschke, "Family Conflict and Children's Self Concept: A Comparison of Intact and Single Parent Families," 41 Journal of Marriage and the Family 367 (1979); Weiss, "Growing Up A Little Faster," 35 Journal of Social Issues 97, 110-111 (1979). The children conceived by single individuals using reproductive technologies should be at least as cognitively developed, if not more so, than those children who have been studied who are in single-parent homes that are the result of divorce or the death of a parent. The unmarried men and women who wish to use alternative reproduction will generally have given considerable thought to their ability to provide adequate nurturing and care to the child, possibly more thought than many couples who conceive sexually.

Other concerns raised about the children of single or gay parents seem unfounded. Research indicates that children from single-parent female-headed households, including those from lesbian-headed households, do not differ from other children in their gender-role behavior or in their sexual preference. McGuire and Alexander, supra at 182-184. See especially the following two studies: Hoeffer, "Children's Acquisition of Sex-Role Behavior in Lesbian-Mother Families," 51 American Journal of Orthopsychiatry 536 (1981); Kirkpatrick, Smith, and Roy, "Lesbian Mothers and Their Children: A Comparative Survey," 51 American Journal of Orthopsychiatry 545 (1981). Nor should the possibility of community intolerance be grounds for denial of access to reproductive technology. A judge in a lesbian custody case replied to the argument that children could be harmed by stigma by stating, "It is just as reasonable to expect that they will emerge better equipped to search out their own standards of right and wrong, better able to perceive that the majority is not always correct in its moral judgments, and better able to understand the importance of conforming their beliefs to the requirements of reason and tested knowledge, not the constraints of currently popular sentiment or prejudice." M.P. v. S.P., 169 N.J. Super. 425, 438, 404 A.2d 1256, 1263 (Super. Ct. App. Div. 1979). Two differing perspectives in the gay press on gay parenthood can be found in Siegel, "Gay and Lesbian Parenting: A Complicated Life," Windy City Times 2 (May 1986) (complicated but joyous experience); Saffire, "Babyboom," The Lesbian Inciter 24, 25 (April/May 1986) (a lesbian arguing it is irresponsible for lesbians to have children when they "can mother the children of the world who are suffering").

73. Ethics Committee of the American Fertility Society, Ethical Considerations of the New Reproductive Technologies (1986); American Association of Tissue Banks, "Reproductive Council Addendum," Standards for Tissue Banking 25-32 (1984).

74. For example, as recently as 1982, there have been reports in the literature from programs that admit they do not test donors for venereal disease or hepatitis. See, e.g., Aiman, "Factors Affecting the Success of Donor Insemination," 37 Fertility and Sterility 94, 95 (1982). This is particularly troublesome in view of the fact that other donor studies show a 1.4 percent rate of venereal disease or hepatitis. Some physicians maintain that the reason they do not need to do genetic testing or other detailed physical examinations on donors is that the primary source of donors is medical students. However, for-profit sperm banks are tapping a much larger population of donors than just medical students. In addition, even when the donors are medical students, it is questionable whether it is sufficiently protective of the child to leave it up to the donors to recognize the genetic disorder in their family history and disclose it. A study of 168 donor applicants at the University of North Carolina School of Medicine demonstrated the problem of relying on donors' self-reports. According to the researchers, "a majority of donors having a positive family history [of the genetic disorder] did not recognize the condition as being genetic even if the individual had had medical training." Timmons, Rao, Sloan, Kirkman, and Talbert, "Screening of Donors for Artificial Insemination," 35 Fertility and Sterility 451 (1981).

75. Mascola, Colwell, and Couch, "Should Sperm Donors Be Screened for Sexually Transmitted Diseases?" 309 New England Journal of Medicine 1058 (1983) (letter).

76. Johnson, Schwartz, and Chutorian, "Artificial Insemination by Donors: The Need for Genetic Screening," 304 New England Journal of Medicine 755 (1981); and Shapiro and Hutchinson, "Familial Histiocytosis in Offspring of Two Pregnancies After Artificial Insemination," 304 New England Journal of Medicine 757 (1981).

77. Curie-Cohen, Luttrell, and Shapiro, "Current Practice of Artificial Insemination by Donor in the United States," 300 New England Journal of Medicine 585, 588 (1979).

78. Id. at 588.

79. Mascola and Guinan, "Screening to Reduce Transmission of Sexually Transmitted Diseases in Semen Used for Artificial Insemination," 314 New England Journal of Medicine 1354 (1986).

80. City of N.Y. Health Code §§ 21.03, 21.05 (1973).

81. Idaho Code § 39-5404 (1985); Or. Rev. Stat. § 677.370 (1983).

82. Idaho Code § 39-5408 (Supp. 1986); Ill. Ann. Stat. ch. 127, para. 55.45 (enacted 1987).

83. Ohio Rev. Code Ann. § 3111.33 (Baldwin 1987).

84. Sperm Bank Licensure and Regulation Act of 1985.

85. Mich. H.B. 4554 § 5(d) and 6(2) (1985).

86. Ontario Law Reform Commission, Report on Human Artificial Reproduction and Related Matters Vol. I 190 (1985).

87. Ala. Code § 26-17-21(a) (Supp. 1986); Alaska Stat. § 25.20.045 (1983) (refers to a physician only); Cal. Civ. Code § 7005(a) (West 1983); Colo. Rev. Stat. § 19-6-106 (1986); Idaho Code § 39-5402 (1985); Ill. Rev. Stat. Ann. ch. 40, para. 1453

(Smith-Hurd Supp. 1987); Minn. Stat. Ann. § 257.56 (West 1982); Mont. Code Ann. § 40-6-106 (1981); Nev. Rev. Stat. Ann. § 126.01(1) (Michie 1986); N.J. Stat. Ann. § 9:17-44 (West Supp. 1986); N.M. Stat. Ann. § 40-11-6A (1986); Ohio Rev. Code § 3111.32 (Baldwin 1987); Va. Code Ann. § 64.1-7.1 (1980) (refers to a physician only); Wash. Rev. Code Ann. § 26.26.050(1) (1986); Wis. Stat. Ann. § 891.40(1) (West Supp. 1986); Wyo. Stat. § 14-2-103(a) (1986).

88. Ark. Stat. Ann. § 34-722 (Supp. 1985); Conn. Gen. Stat. Ann. § 45-69g(a) (West 1981); Ga. Code Ann. § 74-101.1, § 74-9904 (Harrison Supp. 1986); Okla. Stat. Ann. tit. 10, § 553 (West Supp. 1987); and Or. Rev. Stat. Ann. § 667.360 (1983).

89. Idaho Code § 39-5402 (1985).

90. Cal. Civ. Code § 7005 (West 1983).

91. Id.

92. Jhordan C. v. Mary K., 179 Cal. App. 3d 386, 224 Cal. Rptr. 530 (1986).

93. Id. at 535.

94. Id.

95. Id. (footnotes omitted).

96. Id. at 533. In the course of the litigation, Victoria (supported by Mary) moved for joint legal custody (with Mary) and visitation rights. Victoria asserted she was a de facto parent of Devin. The court held it was premature to resolve the de facto parent issue since Victoria had been granted visitation rights by court order. Id. at 537.

97. Id. at 536. The court held that the statute requiring physician participation to prevent the sperm donor from asserting parental rights did not violate equal protection, despite the fact that other paternity statutes precluded a sperm donor from asserting paternity claims to a child born to a married woman, even if a physician was not involved.

98. Even when spouses separate and remarry, the law does not view the child as having additional parents. The two-parent approach is reinforced by requiring that, for example, the only way a stepfather can be recognized as a legal parent is for the biological father to give up the child for adoption. Similarly, in the case of adoption by strangers, the state backs the fiction that the adoptive parents are the biological creators of the child, nullifying the original parents' legal claims and responsibilities.

99. Rothman, "How Science is Redefining Parenthood," Ms. 154, 158 (July 1982).

100. C.M. v. C.C., 377 A.2d 821, 152 N.J. Super. 160 (1977).

101. Jhordan C. v. Mary K., 179 Cal. App. 3d 386, 224 Cal. Rptr. 530, 537-38 (1986).

102. Ala. Code § 26-17-21 (Supp. 1986); Alaska Stat. § 25.20.045 (1983); Ark. Stat. Ann. § 61-141 (Supp. 1985); Cal. Civ. Code § 7005 (West 1983); Colo. Rev. Stat. § 19-6-106 (1986); Conn. Gen. Stat. Ann. §§ 45-69f to 69n (West 1981); Fla. Stat. Ann. § 742.11 (West 1986); Ga. Code Ann. § 74-101.1, § 74-9904 (Harrison Supp. 1986); Idaho Code § 39-5401 to 07 (1985); Ill. Ann. Stat. ch. 40, para. 1453 (Smith-Hurd Supp. 1987); Kan. Stat. Ann. §§ 23-128 to -130 (1981); La. Civ. Code Ann. art. 188 (West Supp. 1987); Md. Est. & Trusts Code Ann. § 1-206(b) (1974); and Md. Gen. Prov. Code § 20-214 (1982); Mich. Comp. Laws Ann. § 333.2824 (1980) and § 700.111 (1980); Minn. Stat. Ann. § 257.56 (West 1982); Mont. Code Ann. § 40-6-106 (1985); Nev. Rev. Stat. § 126.061 (1986); N.J. Stat. Ann. § 9:17-44 (West Supp. 1986); N.M. Stat. Ann. § 40-11-6 (1986); N.Y. Dom. Rel. Law § 73 (McKinney 1977); N.C. Gen. Stat. § 49A-1 (1984); Ohio Rev. Code Ann. § 3111.37A (Baldwin 1987); Okla. Stat. Ann. tit. 10, §§ 551-553 (West Supp. 1987); Or. Rev. Stat. §§ 109.239, .243, .247, 677.355, .360, .365, .370 (1983); Tenn. Code Ann. § 53-446 (1983); Tex. Fam. Code Ann. § 12.03 (Vernon 1986); Va. Code Ann. § 64.1-7.1 (1980); Wash. Rev. Code Ann. § 26.26.050 (1986); Wis. Stat. Ann. § 767.47(9) (West 1981), § 891.40 (West Supp. 1986); Wyo. Stat. § 14-2-103 (1986).

103. Ala. Code § 26-17-21(b) (Supp. 1986); Cal. Civ. Code § 7005(b) (West 1983); Colo. Rev. Stat. § 19-6-106 (2) (1986); Conn. Gen. Stat. Ann. § 45-69j (West 1981); Idaho Code § 39-5405 (1985); Ill. Rev. Ann. Stat. ch. 40, para. 1453(b) (Smith-

Hurd Supp. 1987); Minn. Stat. Ann. § 257.56(2) (West 1982); Mont. Code Ann. § 40-6-106(2) (1985); Nev. Rev. Stat. § 126.061(2) (1986); N.J. Stat. Ann. § 9:17-44 (b) (West Supp. 1986) (unless the woman and donor have entered into a written contract to the contrary); N.M. Stat. Ann. § 40-11-6 (B) (1986) (unless the woman and donor have agreed in writing to the contrary); Ohio Rev. Code Ann. § 3111.32B (Baldwin 1987); Or. Rev. Stat. § 109.239(1)(2) (1983); Tex. Fam. Code Ann. § 12.03(b) (Vernon 1986); Wash. Rev. Code Ann. § 26.26.050(2) (1986) (unless the woman and donor have agreed in writing to the contrary); Wis. Stat. Ann. § 891.40(2) (West Supp. 1986); Wyo. Stat. § 14-2-103(b) (1986).

104. Syrkowski v. Appleyard, 420 Mich. 367, 362 N.W.2d 211 (1985).

105. Smith v. Jones, No. 85 532014 02 (Michigan Cir. Court, Wayne County, March 14, 1986).

106. Some feminists, however, point out that this is precisely what happens now with adoption in which the birth mother is rendered invisible after surrendering the child.

107. This may particularly be true when there are safeguards in place, such as the representation of the surrogate by separate counsel or advance judicial scrutiny of the surrogacy arrangement.

108. In England, the Royal College of Obstetricians and Gynaecologists Ethics Committee on In Vitro Fertilisation and Embryo Replacement or Transfer recommended against the use of surrogate carriers and surrogate mothers on the grounds that these women cannot predict beforehand what their attitudes toward the children will be and the relinquishment might cause emotional stress. Royal College of Obstetricians and Gynaecologists (RCOG), Report of the RCOG Ethics Committee on In Vitro Fertilisation and Embryo Replacement or Transfer § 7.3 (1983).

109. Some feminists point out that many court decisions that are ostensibly made in the best interests of the child instead merely enforce social stereotypes and do not reflect the child's true needs.

110. There have been an estimated 500-600 births in the United States to surrogate mothers, with five reported instances of surrogates changing their minds.

111. Numerous press articles have pointed out that the only conceivable "winners" in the Baby M case are the lawyers.

112. This has already happened in one case. See N. Keane and D. Breo, The Surrogate Mother (1982).

113. However, it may be inequitable to require the biological father to support the child in that situation.

114. In the case of a surrogate carrier, the custody would be a three-way split between the genetic father, the genetic mother, and the gestational mother.

115. See, e.g., the statement of Roberta Achtenberg in Andrews, "Yours, Mine, and Theirs," 18 Psychology Today 20, 29 (1984).

116. This is the policy with respect to unmarried surrogates in Arkansas. If a couple uses an unmarried surrogate, the intended mother (not the biological mother) is the legal mother. Ark. Rev. Stat. Ann. § 34-721(B) (Supp. 1985).

117. There is much dispute among feminists over whether it is feasible to think such a guarantee can be made by statute in the surrogate context. I, for one, think it can. Barbara Katz Rothman thinks it cannot.

118. Van De Velde, "The Future of Surrogacy," Chicago Tribune (January 25, 1987).

119. Md. Ann. Code art. 48A, §§ 354 DD, 470 W, 477 EE (Supp. 1985). The conditions include that the patient's egg is fertilized by her husband's sperm and that certain medical indications for the procedure are met.

120. See, e.g., Department of Health and Social Security, Report of the Committee of Inquiry into Human Fertilisation and Embryology (1984).

121. See, e.g., Maher v. Roe, 432 U.S. 464, 474 (1977).

122. See, e.g., Moe v. Secretary of Admin., 417 N.E.2d 387 (Mass. 1981).

123. Siegal, Case Study, "Baby Making and the Public Interest," 6 (4) Hastings Center Report 14 (August 1976).

124. For example, the recent Ontario Law Reform Commission report makes reference to a medical practitioner who admitted to taking the financial background of the couple into consideration when deciding whether or not to provide the couple with artificial insemination. Ontario Law Reform Commission, Report on Human Artificial Reproduction and Related Matters Vol. I 22 (1985).

125. This type of exploitation, however, may stem less from the technology itself than from our economic and social system.

126. Robertson, "Procreative Liberty and the Control of Conception, Pregnancy and Childbirth," 69 Virginia Law Review 405 (1983); Andrews, "The Legal Status of the Embryo," 32 Loyola Law Review 357, 358-361 (1986); In re Baby M, (A-39-87) slip op. (N.J. February 3, 1988).

127. Wikler, "Society's Response to the New Reproductive Technologies: The Feminist Perspectives," supra n. 3 at 1051.

128. Alternative reproduction (specifically artificial insemination with the man's sperm or with a donor's sperm) has already been used extensively by couples in which the man is disabled.

129. Curie-Cohen, Luttrell, and Shapiro, "Current Practice of Artificial Insemination by Donor in the United States," 300 New England Journal of Medicine 585, 585 (1979).

130. See, e.g., Congregation for the Doctrine of the Faith, Instruction on Respect for Human Life in Its Origin and on the Dignity of Procreation: Replies to Certain Questions of the Day (Vatican City 1987).

131. To assess the actual risk to the embryo in reproductive technologies, one must consider the risk to embryos in natural reproduction. For every 100 eggs fertilized inside a woman's body, only 31 result in the birth of infants. Biggers, "In Vitro Fertilization and Embryo Transfer in Human Beings," 304 New England Journal of Medicine 336, 339 (1981). Alternative reproduction likewise does not guarantee every embryo created a chance to survive. The success rate of in vitro fertilization, at the better clinics, approaches that of normal reproduction. However, the transfer of an embryo from one woman to another subsequent to in vitro fertilization and lavage, or the preservation of an embryo by freezing may all present more risks to the embryo than does natural reproduction.

ALTERNATIVE MODES OF REPRODUCTION: OTHER VIEWS AND QUESTIONS

Wendy Chavkin, Barbara Katz Rothman, and Rayna Rapp

Position Paper

The recent national impassioned debate generated by the *Baby M* case has dramatically increased the necessity of recognizing disparate voices in any feminist discussion of "third-party reproduction." Indeed, the very language of "third-party reproduction" underscores tensions over the meanings of pregnancy and parenting when reduced to the realm of contract law. For many of us who participated in the Rutgers University Law School's Project on Reproductive Laws for the 1990s, these tensions have been discussed fruitfully, but are not ultimately represented in the other position papers prepared for this book.

In the shadow of *Baby M*, we express grave concerns about threats posed to diverse women's interests in pregnancies by the new reproductive arrangements. We offer the following questions in the hopes of provoking extensive societal debate.

Q: In a class-stratified society, where surrogate contractors are bound to have more resources and power than those who "freely" contract to become surrogate mothers, is it possible to protect women from economic exploitation in such a surrogacy contract? Is such an attempt at protection paternalistic?

A: We are dubious that adequate protection can ever be established, given the social relations within which such contracts are inscribed. Ten thousand dollars' payment for nine months' work, 24 hours a day, entailing substantial physical and emotional transformations and risks comes out to about $1.50 hourly wages: a figure that would only tempt someone for whom work has always been a dead end, as it is for so many working-class women. Middle-class women do not find their obstetricians through the Yellow Pages, or ask their adversaries to recommend a lawyer, as Mary Beth Whitehead requested of William Stern. They do not engage in indiscriminate telephone contact once involved in a legal suit, without thinking that the telephone might be tapped. Nor do they misunderstand the language of the courts, while trying to speak in a discourse that is clearly foreign to those without professional credentials. These class differences cannot be erased, making it very difficult for a working-class woman to contest any aspect of her contract.

For the moment, surrogacy is a white-on-white practice. When embryo transfer is perfected (that is, when an egg and sperm of two white donors can be securely united, and inserted into the womb of an unrelated woman), cross-racial surrogacy will loom as a strong possibility. What better way to decrease the costs of this new "revolutionary process" than to rent the wombs of our poorest women, disproportionately black and brown?

In light of these pervasive inequalities within which surrogacy contracts are drawn up, we recommend that surrogacy be taken out of the realm of economic transaction, and re-placed in the realm of adoption. Our position is based on strong social policy precedent. Despite a deeply ingrained American belief in laissez-faire, our society already does not permit "free contract" when it comes to the sale of organs or babies.

We do not permit the sale of organs because we recognize the coercion implicit in the marketplace in a society of economic disparity. We do not permit the sale of babies because we do not believe in slavery. To sell "terminable rights in a pregnancy" is a legal fiction that glosses exactly such sales.

To accept such a fiction would underwrite an old, powerful, and destructive ideology that controls most women's behavior: that women are the "vessels of the race" (to use a nineteenth-century phrase): passive recipients of male activity, defined by their ability to produce babies for a paternal line or family.

Pregnant women are people first, and reproducers secondarily and episodically. Our pregnancies grow out of our biological capacities, but are not reducible to them. Any woman who agrees to carry a pregnancy to term enters into a growing social relationship with her fetus. Like any social relationship, it is fraught with intended and unintended consequences, and its outcome cannot be completely predicted from its beginning. In the struggle to find a new legal form (that is, a social fiction) to handle the transfer of a newborn baby, we cannot ignore the agency of women, or the social nature of the relations that grow throughout a pregnancy. Only a "vessel" can be rented, or used for a "service" without the right to express or change her mind.

To consider pregnancy as "labor or service performed" that can be monetarily compensated is to extend mechanistically the alienation involved in commodification to an arena that involves a person's whole being. A more apt analogy is slavery.

Q: Do we need new surrogacy contracts and new laws protecting surrogacy?

A: We do not. Adoption laws have been hammered out in state after state so as to protect the rights of both birth mothers and adopting parents. In all states, birth mothers cannot surrender parental rights prior to the birth of the child, and in all states, there is a grace period in which a birth mother may change her mind. We realize this places the burden of risk on the adopting couple. However, such a burden has always been held to be more appropriate than risking involuntary termination of the parenting rights of a woman who has voluntarily carried her pregnancy to term.

Q: But does such a position in favor of the rights of birth mothers over sperm fathers violate the feminist concern with gender equality, and the encouragement of men to take a serious role in fathering?

A: We do not think so for two reasons:

1. The issue is not the "equal contribution" of egg and sperm in conception. Genetically, the two are of course equal and interdependent. However, the genetic contribution does not equal the gestational. Surrogacy contracts avoid recognition of the physical and emotional work and investments that carrying a pregnancy to term and giving birth entail.

Our argument recognizes the significance of the gestational relationship over genetic relationship in determining parenthood at the time of birth. This position that the gestational mother has the right of motherhood at birth (including, of course, the right to give a child up for adoption) specifically protects the women of color who may be used to bear white embryos for a fee, by recognizing those women too as the legal and social mothers of the babies they bear.

2. During nine months of pregnancy and birth, a social relationship may well develop between a mother and the fetus-becoming-baby she carries. There is no co-parenting during pregnancy. During the course of infancy and childhood, all feminists ardently encourage the development of a similarly deep and complex relationship between fathers and babies (indeed, between long-term caretakers and babies, whatever their genetic or nongenetic relationship). But such a relationship develops through time, and through hard work. It is not inherent in the biology of eggs and sperm. Fathers (and relevant others) may earn equality by behavior, and develop major investments, responsibilities, and aspirations in and for their children. However, their relations cannot be assumed to begin equally, from the moment of birth, when compared to a woman's nine months of biosocial connection. Such parental concerns develop with time and intentional activity.

Q: Do surrogate arrangements serve the needs of infertile women, just as the use of artificial insemination with donor sperm has been used to serve the needs of infertile men?

A: To date, surrogate contracts are actually largely concerned with the transfer of patriarchal authority from the mother's husband to the sperm donor. In the language of the courts, Judge Sorkow told William Stern that he was "simply returning to you that which was yours." We underline that the reason for expressing concern with the sperm-donating father's rights, but not considering them equal to the rights of the birth mother is this: any other argument elevates his genetic contribution at the expense of understanding that parental relations are primarily social arrangements, not reducible to their genetic origins. If we rest our case solely on "biological contribution," then a woman's genetic *and gestational* contributions need recognition, and surely outweigh a man's genetic contribution alone.

Q: What about the pain of the infertile?

A: Acknowledgment of the pain of infertility does not lead to the acceptability of exploiting another person. There is a big leap involved from acknowledging the pain of infertility or the difficulties associated with adoption to showing a willingness to rent another person's body and buy a baby. Rather, this could push us to attack the root causes of infertility (sexually transmitted disease epidemics, certain contraceptives, lack of parental leave and child care options that lead women to defer childbearing until worklife is well established, etc.), and to challenge and scrutinize our practices and prejudices regarding adoption.

Q: Does surrogate motherhood open the prospect of new ways of relating to children and alternative familial arrangements?

A: Our technological abilities to separate genetic, gestational, and nurturant parenting enable us to visualize concretely the alternative arrangements we have already begun to explore. Social arrangements are not inherent in technology; rather they shape the uses to which we put it. This model of surrogacy does not raise anything new, just the old ugly concept of woman as vessel. All that is introduced squarely in the middle here is commerce and contract.

Commentary

REGULATING THE NEW REPRODUCTIVE TECHNOLOGIES

George J. Annas

I am pleased to comment on the paper of Ms. Andrews. Although I will take a significantly different approach than hers, we are all indebted to her for attempting what may be an impossible task, i.e., giving the feminist perspective on the new reproductive technologies. There are a number of "feminist perspectives" on these highly emotional and personal issues. I see my primary task here as presenting an alternative one.

The New Reproductive Technologies

Each of the new reproductive technologies could be viewed individually. It is, however, more analytically fruitful to identify and explore their common characteristics relevant to public policy, since other methods will no doubt be developed, and permutations of existing methods can also be used. Thus, unless we are to have a separate policy for each method (an unlikely and ultimately unproductive response--like having a separate public policy for each form of treatment for cancer), we will have to identify the characteristics that these methods share that make regulation important and useful.

Table 1 summarizes the social policy issues raised by each of the techniques, and impressionistically assigns weights to each issue to give an overall view of the *relative* social utility of regulating each method, and the generic importance of each social

Table 1
Index of Relative Importance of Societal Issues in Noncoital Reproduction*

Issues**	AID	IVF	SET	GIFT	Surrog. mother	Frozen embryo
Potential for noninfertility use	2	2	2	2	2	3
Protection of embryo	0	3	3	0	0	3
Identification of mother	0	0	3	0	3	3
Identification of father	2	0	2	0	2	3
Donor screening	2	0	2	0	2	2
Donor anonymity	2	0	2	0	2	2
Opportunities for commercialization	1	1	3	1	3	3
Total:	9	6	17	3	14	19

* 1, indicates of societal concern, but not sufficient to require uniform guidelines; 2, of sufficient societal concern to require uniform guidelines; and 3, of sufficient societal concern to justify discouraging or perhaps prohibiting the procedure altogether if reasonable uniform guidelines cannot be agreed on and enforced.

** *Definitions: potential for noninfertility use:* Use of the technology to gain access to embryo for research or genetic manipulation; avoidance of pregnancy for "convenience" of genetic mother; use of technique for eugenic purposes. *Protection of embryo:* Exposure of embryo to potentially hostile laboratory environment; research that would not directly benefit that embryo; use that would devalue the embryo and human life. *Identification of mother:* Difficulty in distinguishing between genetic mother and gestational mother and determining who will be legally identified as presumptive rearing mother. *Identification of father:* Difficulty in distinguishing between genetic father and rearing father, and determining who has legal responsibility for rearing child. *Donor screening:* Requirements for gamete donors and method of ensuring compliance. *Donor anonymity:* What records should be kept, by whom, and how access can be gained to them by the child. *Opportunities for commercialization:* Buying and selling gametes, embryos, or children, and the implications for society.

policy issue.[1] As can be seen from this table, GIFT (gamete intrafallopian transfer) and IVF (in vitro fertilization) are the least socially problematic of the procedures; with SET (surrogate embryo transfer) and the use of frozen embryos presenting the most difficult social policy issues.

The first and paramount social policy issue raised by this group of new reproductive technologies is their *medical nature*, and the implied medical indications and "potential for noninfertility use." The issue of protecting the extracorporeal embryo applies only when such an embryo is produced, and this will occur in IVF and SET. Issues of parental identification, donor screening, and donor anonymity arise when more individuals than the married couple are involved in producing gametes for the resulting child: AID (artificial insemination by donor) and SET. Of course, should we employ donor gametes in either GIFT or IVF, the *same* screening, record-keeping, and parental identification issues would be raised by these techniques as well.

When reviewing the legal options for regulation on both the state and federal levels, we will find it useful to keep in mind the "pressure points" at which regulation can be brought to bear. In general, these will be: controlling medical practice; controlling human experimentation; defining the presumptive rearing father and mother; granting legal protection to the extracorporeal human embryo; making legal provisions for donor screening and record confidentiality; regulating commerce in gametes and embryos; and attaching conditions to the delivery of medical services that are paid for by government programs. Protecting the interests of children, for example, will require detailed record keeping concerning their genetic parents, and, I believe, guaranteeing access to this information by the child, at least by the age of 18.

Overview of Regulatory Activity to Date

It is fair to say that the federal government has not engaged in *any* regulatory activity in this area. On the other hand, the federal government has over the last 13 years formed three important commissions that have made recommendations regarding the new reproductive technologies: The National Commission, the Ethics Advisory Board, and the President's Commission on Bioethics. It is also in the process of forming another, the Congressional Biomedical Ethics Board.

States have been a bit more active concerning AID. More than half of the states have laws making the husband of the impregnated woman the child's father for all legal purposes, so long as he has consented to AID. Also, a number of states have regulations related to fetal research. However, no states have specific statutes on IVF, SET, or GIFT. Since the regulation of medical practice is primarily a state function, regulation of the actual delivery of these technologies is almost always primarily a task for the individual states.

States could also regulate the new reproductive technologies indirectly by statutorily defining which woman, as among a gestational, genetic, and planned-rearing mother, would have presumptive rearing rights and obligations with respect to the child. I believe *states should enact statutes that clearly define the gestational mother* (i.e., the woman who gives birth to the child) *as the irrebuttably presumed mother for all legal purposes.* This is because of her gestational contribution to the child, and the fact that she will definitely be present at the birth, easily and certainly identifiable, and available to care for the child. Such a law would have the effect of helping to legitimate and protect children born from SET, but would give the so-called surrogate mother the right to retain her child, even in the face of a prior contractual agreement to give it up for adoption or to relinquish parental rights in the child after birth. She could do either, but only *after* the child was born, and the standard waiting period for adoption or relinquishment of parental rights had expired. This presumption would also operate in the case of ovum donation in a manner analogous to AID (sperm donation): the gestational, not the genetic mother, would be the presumptive rearing mother.

Because of the decision of Judge Sorkow in the *Baby M* case, at least another word about preconception contracts should be added. Ms. Andrews argues that not to require specific enforcement of such a contract would reinforce the traditional stereotype of women as "fickle" and unable to contract.

I am afraid I just do not see it that way. The logic of this argument would compel us to enforce other contracts (against women, but not men) such as the contract to marry (until death do us part), and an agreement not to have an abortion (or an agreement *to* have an abortion) made with the father or another interested party. I do not really believe that Ms. Andrews wants the state to be able to enforce a preconception contract to carry a pregnancy to term, any more than she really wants the state to be able to enforce a contract to remove forcibly a child from its mother after birth. The state should simply have no constitutional ability to enforce either contract, at least in the absence of a "compelling state interest." The only such interest possible in a "surrogacy" case is proof that the mother is unfit to parent her child.

On the other hand, Ms. Andrews is correct in wanting to spare the child a custody battle after birth, but the best way to do this is not for judges to enforce preconception contracts; it is to have a legal presumption as to which parent in a "surrogate mother" agreement will have legal rearing rights.[2] Since the mother has contributed significantly more to the birth than the father, I believe she should be presumed to have such rights, and in the event of a dispute, have the presumptive legal rearing rights.

Overview of Federal Authority

In the area of health care in general, and the new reproductive technologies in particular, Congress can act in areas where the federal government has indirect authority: primarily taxation and spending, and interstate commerce. The most important area in which Congress has used its spending power to adopt regulations related to the new reproductive technologies has been in the field of research on human subjects, and most physicians and institutions engaged in research on these technologies must follow federal requirements for such research. Regulation of interstate commerce can involve a ban on the sale of an article. Congress has indicated its willingness to ban the purchase and sale of human body parts, and could certainly ban the interstate sale of human embryos (and sperm and ova as well). In 1984, for example, Congress passed the National Organ Transplant Act. Although most of the Act is aimed at promoting organ transplantation in the United States, Title III is directed exclusively toward prohibiting organ purchases. Its operative section reads:

> It shall be unlawful for any person to knowingly acquire, receive, or otherwise transfer any human organ for valuable consideration for use in human transplantation if the transfer affects interstate commerce.

For the purpose of this act, the meaning of "human organ" is defined as "the human kidney, liver, heart, lung, pancreas, bone marrow, cornea, eye, bone, and skin. . . . " Violation carries a five-year maximum prison sentence and a $50,000 fine. *Congress should amend this statute to include human embryos among the items it is unlawful to sell.* The purpose would be to protect children by preventing them from being viewed as and treated as commodities.

Constitutional Limits on Regulation

The right to privacy encompasses decisions to use contraceptives, not to be sterilized involuntarily (except, perhaps when it is the least drastic alternative and in a person's best interest), and to obtain an abortion (i.e., a right not to beget or bear a child). The question is whether this right will be expanded to include an affirmative right to *parent* a child, will be stable, or will be contracted.

In June 1986, the United States Supreme Court decided an exceptionally controversial case dealing with the issue of whether or not a state could constitutionally make sodomy committed by two adult males in the bedroom of a private home a crime (*Bowers v. Hardwick*, 106 S.Ct. 2841). The Court concluded that there is "no fundamental right to engage in homosexual sodomy." Fundamental rights not readily identifiable in the Constitution's text would be found only if (1) they were fundamental liberties that are "implicit in the concept of ordered liberty," such that "neither liberty nor justice would exist without them"; or (2) they are "deeply rooted in this Nation's history and tradition." In terms of the "right to privacy," the Court limited its application to a "connection between family, marriage, [and] procreation. . . . "

In addition to upholding laws against sodomy, the Court indicated that laws against "adultery, incest and other sexual crime" would also be constitutional. As to whether or not a legislative finding that certain conduct is immoral is a sufficient basis for outlawing it, the Court concluded that it was, noting that "the law is constantly based on notions or morality . . . and majority sentiments about the morality of homosexuality" are sufficient justification to outlaw this behavior.

This was a 5-4 decision, with a concurring opinion of Justice Powell and a strong dissent by Justice Blackmun. Blackmun argued that the Court had fundamentally misconstrued and defined too narrowly the "right to privacy," which should be seen as embodying "the moral factor that a person belongs to himself and not others nor to society as a whole":

> We protect the decision whether to have a child because *parenthood alters so dramatically an individual's self-definition*, not because of demographic considerations or

the Bible's command to be fruitful and multiply. (emphasis added)

With respect to the new reproductive technologies, we need to examine the underlying values at stake in procreative privacy to delineate the scope of this right. These include self-identity, self-expression, freedom of association, freedom to make decisions that drastically affect one's identity, and rights to have intimate relationships with a view toward producing a child. Although the Court is badly split on the reach of privacy outside of a heterosexual union, there is no such split concerning privacy within a heterosexual union when that union is aimed at procreation.

All members of the Court would thus likely conclude that IVF and GIFT, if conducted within the context of marriage at least (and probably if done in any "stable" heterosexual relationship) are to be viewed as within the ambit of the "right to privacy." Accordingly, only laws similar to those endorsed by the Supreme Court to regulate previable abortions (i.e., those aimed primarily at restricting performance to physicians, monitoring the safety and efficacy of the procedures, and insuring informed consent) could be used to regulate these activities. AID regulations could be stricter, since they involve another participant--the sperm donor--and could include screening rules and procedures as well. Where nonprocreation issues are at stake, or where public participation is sought that might harm others, including the resulting children, banning altogether might be permissible. Examples would include commercial surrogate motherhood, selling human embryos, and experimentation on human embryos. The view of one religion alone (e.g., the Catholic Church) that any or all of these techniques are "illicit" would, in itself, be an insufficient rationale to ban them.

Conclusion

Regulation of the new reproductive technologies is primarily a matter for the individual states. Just as they have regulated adoption, custody, marriage, medical licensing, and medical practice, it seems most reasonable for the states to regulate the practice of new reproductive technologies insofar as they are seen as medical procedures and performed by physicians. Regulations in the areas of quality control and monitoring, safety, record keeping, inspection and licensing, consent, the identification and obligations of mothers and fathers, and requirements for donor screening are all well within the traditional state activities, and regulation in these areas would not raise any major social policy implications. In extreme cases, such as banning the sale of human embryos or regulating experimentation with human embryos, statutes would have to be carefully drawn (so as not to be voided for vagueness), and based on a reasonable state policy designed to protect the common good and prevent children from being treated like commodities.

Federal activity concerning the new reproductive technologies, on the other hand, has been restricted to setting up and financing national commissions and groups of various kinds to study the scientific, legal, and ethical issues involved in these practices, and to make recommendations on what actions various private and governmental organizations should take. The federal government could, however, become involved in its own "traditional" areas, such as regulation of interstate commerce, by forbidding the sale of human tissue, regulating "false and deceptive" advertising, and promulgating rules for human research, without any major implications. Major federal involvement, however, seems reasonable only when related directly or indirectly to federal financing of these technologies.

Government has only the most limited role in preventing contraception and prohibiting abortion (mainly in protecting the health and safety of the adult participants), but has a potentially much greater role in the new reproductive technologies: not only protecting the interests of the adults in quality services and informed consent, but also taking reasonable steps to protect the interests of future children that are "created" by these methods. Regulations that are firmly grounded in reasonable steps to protect these children are legitimate, and should enjoy broad societal support.

Notes and References

1. A more detailed discussion is contained in chapter 9, "Noncoital Reproduction" in Elias, S. and Annas, G. J., Reproductive Genetics and the Law (1987), and Elias and Annas, Social Policy Considerations in Noncoital Reproduction, Journal of the American Medical Association, 1986; 255:61-68.

2. For a more detailed discussion of the Baby M case see, Annas, G. J., The Case of Baby M: Babies (and Justice) for Sale, Hastings Center Report, June, 1987.

Commentary

ALTERNATIVE MODES OF REPRODUCTION: THE LOCUS AND DETERMINANTS OF CHOICE

Peggy C. Davis

Long before the *Baby M* case ignited passions and made debate about surrogacy a national pastime, the participants in the Project on Reproductive Laws for the 1990s saw the need to consider the profoundly difficult questions surrounding legal regulation of new reproductive technologies. Their position papers are reflective of broad concerns, rather than reactive to an immediate controversy. They enable us both to learn from individual cases, like the case of *Baby M*, and to put the lessons of those cases in a larger context. The Andrews paper, which serves to focus and inform our discussion of reproduction involving third parties, is no exception. It is enriched by a broad perspective, a wealth of information, and the judgment of a woman who has given careful and sustained consideration to the intricate set of policy concerns that surround official regulation of sexual and family matters.

I wish to take full advantage of this opportunity to consider the range of implications flowing from the regulation of new reproductive technologies. My objective is to draw attention to two questions that are often ignored, but inevitably implicated, when we choose whether or how to regulate human reproduction. The first concerns the proper relationship between law and morality. The second concerns the inevitable relationships between social policy choices, and the evolution or stagnation of gender role definitions. Neither of these questions is central to the Andrews paper. The questions are nevertheless important to our consideration of the

Andrews proposals, for each question is designed to elicit a principle against which governmental regulation of sex, reproduction, and family life must be evaluated.

The Andrews paper begins with an enumeration of "values" that should be promoted as we design legal rules to address new reproductive technologies.[1] The enumerated values provide convenient contexts for discussion of Andrews' policy recommendations. Although Andrews may, in some cases, overstate the extent of feminist agreement, each of her enumerated values does command support among feminists. These values are, however, inadequate to justify the most significant of Andrews' policy choices.

The value to women of control over reproductive decision making is at odds with the value of protection against exploitation. Juxtaposition of these unweighted values will not instruct us in deciding whether a woman should be free to enter a surrogacy agreement, despite the risk that she is motivated by poverty. It does not tell us whether she should have the liberty to contract concerning her behavior during the pregnancy.[2] It does not tell us whether government has the right to place risk-laden fertility cures outside the law in order to protect women who are influenced by narrow gender role definitions to experience infertility as an unbearable defect.

Moreover, exploration of values important to feminist interests with respect to reproduction will be inconclusive, unless those values are measured against social interests in the conditions of families, children, fetuses, and gametes that are the products of reproductive choices. Toleration of alternative family configurations seems attractive to feminists, because it holds a promise of liberation from constraining gender role definitions. But legitimate social interests in the protection of children, or in the structure of basic units of socialization, may weigh against the value of toleration. Feminists regard the right of choice with respect to contraception and abortion as essential to self-definition and fulfillment. But the government may have legitimate interests in regulating choices with respect to the destruction or experimental use of gametes and fetuses.

Resolution of these value conflicts requires resort to values of a more fundamental character. Each conflict concerns the distinction between individual moral autonomy and collective moral choice. With respect to each, the locus of choice is more important than the choice itself.

These matters have been raised--but inadequately addressed--by a Committee of Inquiry directed by the British government to examine the social, ethical, and legal implications of developments in the field of human fertilization and embryology. The committee was chaired by a philosopher, Lady Mary Warnock. It expressed a reluctance "to appear to dictate on matters of morals to the public at large."[3] Yet, it set for itself a course of "moral reasoning" to be followed as it devised recommendations for the regulation and prohibition of various clinical and research practices:

> [T]hat moral conclusions cannot be separated from moral feelings does not entail that there is no such thing as moral reasoning. Reason and sentiment are not opposed to each other in this field. If, as we believe, it was our task to attempt to discover the public good, in the widest sense, and to make recommendations in the light of that, then we had, in the words of one philosopher, to adopt "a steady and general point of view." So, to this end, we have attempted in what follows to *argue* in favor of those positions which we have adopted, and to give due weight to the counter-arguments, where they exist.[4]

This statement is murky, as the statements of committees often are. It gives us little understanding of the committee's process of "moral reasoning." But one thing about the statement is clear and striking: it acknowledges no restraint upon the right of government to legislate morality.

This failure is puzzling, for the Warnock Committee was not the first to tread the boundary between private morality and public sentiment. Thirty years ago, the Wolfenden Committee, asked to consider the regulation of homosexuality and prostitution, confronted the issue forthrightly:

> [The law's] function, as we see it, is to preserve public order and decency, to protect the citizen from what is offensive or injurious, and to provide sufficient safeguards against exploitation and corruption of others. . . . It is not, in our view, the function of the law to intervene in the private lives of citizens, or to seek to enforce any particular pattern of behavior, further than is necessary to carry out the purposes which we have outlined. . . .

> There must remain a realm of private morality and immorality which is, in brief and crude terms, not the law's business.[5]

The failure of the Warnock Committee to consider whether its moral judgments about medical research and reproductive technology were "the law's business" affected the scope and content of its legislative proposals.[6] Many of the committee's recommendations are based explicitly on moral grounds, and its approach has been criticized--rightly, I think--by an eminent feminist legal scholar for adopting "a patriarchal, authoritarian vision of families and the role of medical expertise."[7] In its review of each controversial clinical or research issue, the committee considers various moral objections. It proceeds to accept some and reject others, with no basis for its distinctions other than the agreement of the committee membership. In this fashion, it decides that reproductive technologies should only be made available to heterosexual couples[8] and that clinicians may refuse treatment even to heterosexual couples on the basis of "social judgments that go beyond the purely medical."[9] The committee concludes that moral objections to artificial insemination and in vitro fertilization are an inappropriate basis for policy choice and approves the use of those techniques.[10] It concludes that egg donation "is ethically acceptable where the donor has been properly counseled and is fully aware of the risks";[11] and that embryo donation is acceptable if achieved by in vitro fertilization, but unacceptably risky to the donor if accomplished by lavage.[12] Moral objections to surrogacy are considered and accepted,[13] with the result that the committee recommends declaring surrogacy agreements unlawful and void, and criminalizing agency participation in arrangements for surrogate births whether or not the participating agency is operated for profit.[14]

We must avoid in this country the confusion of law and morality that has befuddled the British debate concerning reproductive technology. We must acknowledge the need to distinguish those matters as to which conduct should be regulated and those as to which individual moral choices should be respected. Finally, we must consider carefully the implications of the rationales by which we draw these distinctions. These precautions are important for several reasons. Some of the reasons relate to feminist concerns. Others relate to a concern for preservation of the values of mutual respect and protection in a richly, but not always comfortably, pluralistic society.

The question of whether a government regulation inappropriately constrains individual autonomy is, in this country, a question of constitutional law.[15] The black civil rights movement and the women's movement have led the struggle to give meaning to the Fourteenth Amendment's promise of individual liberty.[16] The evolution in the doctrine of privacy that began with *Griswold v. Connecticut*[17] has taken us from the recognition that the police may not "search the sacred precincts of marital bedrooms,"[18] to recognition of a constitutionally protected "interest in independence in making certain kinds of important decisions."[19] *Griswold* can be read to mean that the prohibition of contraception in marriage is impermissible only because the means of detection would be distastefully intrusive. However, as the Supreme Court considered the constitutionality of statutes prohibiting contraceptive use by single people and statutes prohibiting abortion, it moved from recognition of a right to hide to recognition of a right to choose.

Rights of individual autonomy in sexual and family matters have been expanded on other fronts. The civil rights movement has established a limited right of freedom of choice in marriage.[20] And persons subjected by reason of poverty or cultural difference to state interference with family decision making[21] have expanded the right of choice with respect to marriage, and established rights of choice with respect to family living arrangements[22] and parental decision making.[23] Older cases establishing rights of autonomy in parenting[24] and a requirement of strict scrutiny with respect to involuntary sterilization[25] have been questioned,[26] but not repudiated by the Court.

These rights of privacy and autonomy are narrow. It has recently been held that they do not protect individual decisions about the form and object of sexual expression between consenting adults in the privacy of their homes.[27] The strongest expression of rights of parental autonomy came in a case in which a guardian was denied the right to decide, consistently with her religious beliefs and those of the child who was her ward, that the child might stand on a street corner with her and distribute religious literature.[28]

These rights are also fragile. The right of choice with respect to abortion will almost surely be reconsidered within the next few years by a Supreme Court that is skeptical of the legitimacy of affording constitutional protection to *any* rights of choice regarding sexual and family matters.[29] The scope of rights of parental autonomy is largely undetermined and must be defined by a court that held in 1981 that indigent biological families lacked even the right of counsel when faced with state action to sever permanently the legal relationship with their children.[30]

As the Court reexamines and refines constitutional protections of individual choice with respect to sex, reproduction, and family life, it will be required to juxtapose individual interests in privacy and autonomy and governmental interests of various sorts. The scope of individual rights will shrink or grow inversely as the roster of legitimate state concerns shrinks or grows. When matters like contraception, abortion, sexual freedom, parental choice, and new reproductive technologies are before the Court, everything can turn on whether the enforcement of morals is counted among legitimate state interests.

Nothing fully justifies interference with freedom of choice regarding contraception, save the moral judgment that it is wrong to inhibit the biological consequences of intercourse.[31] Nothing justifies interference with freedom of choice regarding abortion, save the moral judgment that the life of the fetus should be protected. Nothing fully justifies interference with freedom of choice regarding homosexuality between consenting adults, save moral approbation.[32]

So, too, in matters concerning new reproductive technologies, moral arguments predominate, and the protective justifications[33] for sweeping, prohibitive regulation are often strained and speculative. Consider the case of surrogacy. In this country, as in Great Britain, moral objections "weigh heavily"[34] in public and legislative debates. Evidence is lacking to support prohibition on nonmoral grounds. Arguments concerning the risk of exploitation of the carrying mother prove too much. As Professor Law has asked, "Why is work as a surrogate inherently more exploitive than . . . working with toxic chemicals?"[35] The emotional and social consequences for the child depend largely upon whether we create a social climate of acceptance or stigmatization with respect to three- and four-party reproductive and parenting arrangements, and whether we accept in this context the responsibility for prompt, reasoned resolution of custody disputes that we bear in every other family context.[36] It is likely, then, that constitutional approval of a decision to place surrogacy outside the law would require the acceptance of theories of governance that undermine rights of family, sexual, and reproductive autonomy in every context.

The choice between libertarian and regulatory approaches to governance is not a simple matter. History teaches that a laissez-faire approach to the marketplace compromises legitimate efforts to achieve economic justice and a reasonable measure of equality of opportunity. In the early part of this century, the Court consistently invalidated labor and price regulations as violative of principles of liberty of contract. There is a contemporary consensus that that deference to liberty of contract was excessive. "[A]s a picture of freedom in industrial society, the one painted by the Justices [before the mid-1930s] badly distorted the character and needs of the human condition and the reality of the economic situation."[37] Since the New Deal, the Court has been less deferential to the principle of private contractual autonomy, and legislatures have been freer to address power imbalances within the marketplace. In a context of closer market regulation, individuals have been less vulnerable to economic domination--ironically, more autonomous as a result of restraints upon certain forms of contractual liberty in the commercial sphere.

In the private sphere, there are "forms of expression, action, or opportunity . . . [that] touch [ ] deeply and permanently on human personality" and comprise a terrain of constitutionally protected "liberty beyond contract."[38] Behaviors and opportunities related to sex, reproduction, and family life figure prominently among them. The burning constitutional issue of our day is clarification of the limits upon government restriction of these fundamental personal liberties. As we consider the extent to which reproductive choice should be constrained by regulation of reproductive technology, we must remember that prohibition compromises benefits associated with personal liberty and choice. The most important of those values are the personal fulfillment associated with self-definition and moral autonomy, and the social growth and enrichment associated with diversity in life-style choices.

Consider the play of these values in the context of the surrogacy controversy. There are those who see surrogacy as a baby-selling arrangement that demeans the child; exploits and commercializes the reproductive capacity of the carrying mother; and manifests narcissism on the part of prospective custodian(s) who are genetically related to the child, and either selfishness or subordination to a partner's narcissism on the part of the adoptive parent.[39] There are others who see it as a humane arrangement in which the carrying mother and the prospective custodians can participate with self-respect, and moral and psychic comfort, and the child can flourish. These are subjective judgments with a variety of elusive determinants. Receptivity to an arrangement by which the contracting father has, and the carrying mother lacks, rights of custody may depend upon whether life experiences have led one to regard male parents as nurturing or aloof. Confidence that atypical combinations of biological and social parenting are consistent with healthy child development may depend upon experiences with adoption, step-parenting, extended families, and fictive kinship relationships. Tolerance of these arrangements may be higher in communities, like the black community, in which parenting responsibilities are more likely to be shared and to extend widely within family and friendship groups.[40] Different ways of experiencing pregnancy will produce different assessments of the sentiments and self-images of carrying mothers.

If surrogacy is outlawed, we will have reinforced the sense that the right, the moral, and the socially acceptable family pattern is one in which the nurturing obligation of the birth mother is primary (if not virtually exclusive) and inviolable. If surrogacy is permitted and wisely regulated,[41] social attitudes and mores will be influenced by a wider variety of approaches to responsible procreation and childrearing. For people committed to liberation from role stereotypes based upon sex, the second scenario is surely more promising.

Notes and References

1. "The values are:

The Infertility Issue Should Be Put into a Larger Social Context . . .
Women Should Have Control over Their Bodies, Their Gametes, and Their Conceptuses . . .
Women Should Not Be Exploited . . .
There Should Not Be Prior Screening for Fitness of Parenthood . . .
Reproduction Should Not Be Medicalized . . .
Alternative Family Arrangements Should Be Allowed."

See Andrews, "Alternative Modes of Reproduction," this volume.

2. For an especially thoughtful analysis of this issue, see Note, "Rumpelstiltskin Revisited: The Inalienable Rights of Surrogate Mothers," 99 Harvard Law Review 1936 (1986).

3. M. Warnock, A Question of Life 1 (1985).

4. Id. (emphasis in original).

5. Report of the Committee on Homosexual Offenses and Prostitution 23-24 (1963).

6. On the occasion of commercial publication of the committee's report, Lady Warnock added two chapters that more fully addressed questions of law and morality. She acknowledged the Wolfenden position, and the debate it triggered between Lord Patrick Devlin and H.L.A. Hart. She continued, however, to regard the mission of her committee as the giving of "advice to Ministers, based on moral judgments." Warnock, supra note 3, at 95. Moreover, although she acknowledged [in the tradition of Griswold v. Connecticut, 381 U.S. 479 (1965)] a need to avoid legislation that requires intrusive police practices, id. at xii, she again failed to acknowledge the need for a sphere of individual moral autonomy with respect to matters of family, sexuality, and reproduction.

7. Law, "Embryos and Ethics," 17 Family Planning Perspectives 140 (1985).

8. Warnock, supra note 3 at 11. The moral argument related by the committee is as follows: "many believe that the interests of the child dictate that it should be born into a home where there is a loving, stable, heterosexual relationship and that, therefore, the deliberate creation of a child for a woman who is not a partner in such a relationship is morally wrong." The committee also reports having been "told of a group of single, mainly homosexual, men who were campaigning for the right to bring up a child . . . [and were] well aware of the potential of surrogacy for providing a single man with a child that is genetically his." Id.

9. Id. at 12.

10. Id. at 18, 20-23, 31-32.

11. Id. at 39.

12. Id. at 39-40.

13. The committee stated:

> The moral and social objections to surrogacy have weighed heavily with us. In the first place, we are all agreed that surrogacy for convenience alone, that is, where a woman is physically capable of bearing a child but does not wish to undergo pregnancy, is totally ethically unacceptable. Even in compelling medical circumstances the danger of exploitation of one human being by another appears to the majority of us far to outweigh the potential benefits, in almost every case. That people should treat others as a means to their own ends, however desirable the consequences, must always be liable to moral objection.

Id. at 46.

14. Id. at 47. On the basis of its exploration of the question "how it is right to treat the human embryo," id. at 60 (emphasis deleted), the committee also concludes that "the embryo of the human species should be afforded some protection in law," id. at 63, and recommends that unauthorized experimentation with any embryo, and preservation of an embryo in vitro beyond the fourteenth day following fertilization, be made criminal offenses, id. at 64-66. In considering the implications of this recommendation for questions concerning abortion, it is important to bear in mind that the fourteenth day following fertilization is the first day upon which the average woman would, in the usual course, discover a pregnancy--the day that menstruation would otherwise have begun.

15. Similarly, the need to fashion government regulations to protect individual autonomy was urged upon the Warnock Committee as a requirement of the European Convention on Human Rights. See Warnock, supra note 3 at 10.

16. See D. Bell, And We Are Not Saved: The Elusive Quest for Racial Justice 50-73 (1986); Freund, "The Civil Rights Movement and the Frontiers of Law," in T. Paresos and K. Clark, eds., The American Negro 363 (1967).

17. 381 U.S. 479 (1965) (invalidating a Connecticut statute to the extent that it prohibited use of contraceptives by married couples).

18. Id. at 479.

19. Carey v. Population Services International, 431 U.S. 678 (1977).

20. Loving v. Virginia, 388 U.S. 1 (1967).

21. For a description of the dynamic that leads to interferences of this sort, see Davis and Dudley, "The Black Family in Modern Slavery," 4 Harvard Blackletter Journal (Spring 1987).

22. Zablocki v. Redhail, 434 U.S. 374 (1978); Moore v. City of East Cleveland, 431 U.S. 494 (1977). See also Santosky v. Kramer, 455 U.S. 745 (1982); Lassiter v. Department of Social Services of Durham Co., 452 U.S. 18 (1981); Stanley v. Illinois, 405 U.S. 645 (1972).

23. Wisconsin v. Yoder, 406 U.S. 205 (1972).

24. Pierce v. Society of Sisters, 268 U.S. 510 (1925); Meyer v. Nebraska, 262 U.S. 390 (1923).

25. Skinner v. Oklahoma, 316 U.S. 535 (1942).

26. See, e.g., Ely, "The Wages of Crying Wolf: A Comment on Roe v. Wade," 82 Yale Law Journal 920 (1973).

27. Bowers v. Hardwick, 54 L.W. 4919 (1986).

28. Prince v. Massachusetts, 321 U.S. 158 (1944).

29. See, e.g., Thornburgh v. American College of Obstetricians and Gynecologists, 54 L.W. 4618, 4628-31 (dissenting opinion of Justice White).

30. Lassiter v. Department of Social Services of Durham Co., 452 U.S. 18 (1981).

31. For an analysis of whether proscriptions upon the use of contraceptives further the goal of discouraging premarital and extramarital sex, see Griswold v. Connecticut, supra note 17 at 505-507 (concurring opinion of Justice White).

32. Diminishment of rights of individual and family autonomy is not always the result of government efforts to legislate insular moral choices. It may also occur when government is permitted to legislate too broadly in the unquestionably legitimate quest to prohibit choices and actions that result in tangible harm to others. For example, the Supreme Court's failure to uphold the right of Sarah Prince to distribute religious literature with her niece and ward resulted from the Court's trivialization of the principles of moral and religious autonomy and its acceptance of unjustifiably sweeping child labor regulations. Prince v. Massachusetts, supra note 28. As Justice Murphy noted in dissent:

> It is claimed . . . that such activity was likely to affect adversely the health, morals and welfare of the child. Reference is made in the majority opinion to "the crippling effects of child employment, more especially in public places, and the possible harms arising from other activities subject to all the diverse influences of the street." To the extent that

they flow from participation in ordinary commercial activities, these harms are irrelevant to this case. And the bare possibility that such harms might emanate from distribution of religious literature is not, standing alone, sufficient justification for restricting freedom of conscience and religion.

Id. at 174-175.

33. See note 32, supra.

34. Warnock, supra note 3 at 46.

35. Law, supra note 7 at 144.

36. In this connection, the Andrews report that only one percent of surrogate mothers have attempted to alter the custodial terms of their agreements is significant. See Andrews, "Alternative Modes of Reproduction," section entitled "Alternative Family Arrangements Should Be Allowed," this volume.

37. L. Tribe, American Constitutional Law 455 n. 37 (1978).

38. L. Tribe, supra note 37 at 565. Tribe uses the phrase "liberty beyond contract" to describe the zone of individual freedom defined by "preferred rights" that are constitutionally protected "from all but the most compellingly justified instances of governmental intrusion." Id. Infringement of these rights triggers special judicial scrutiny, despite the Court's repudiation of the theories of contractual liberty by which economic regulations were similarly scrutinized, and regularly invalidated, before 1937. Tribe observes that the effort to define the legitimate scope of judicial review of governmental encroachments upon these liberty interests "has preoccupied (one could say obsessed) constitutional scholarship for the last forty years." Id. at 453.

39. The phrase "adoptive parent" is used here to refer to the prospective custodial parent who has no genetic relationship to the child. Adoption may, however, be legally necessary or prudent in the case of prospective custodial parents who are genetically related to the child, for the law may define the birth mother and/or her spouse as the child's legal parents. Participation of an adopting parent is not an essential ingredient of surrogacy arrangements. See note 8, supra regarding surrogacy arrangements involving homosexual men.

40. See C. Stack, All Our Kin (1975).

41. The tragedy of the Baby M case is, of course, the uncertainty and prolonged deliberation caused by a legislative failure to prohibit or regulate an increasingly frequent practice.

Commentary

THE KEY SOCIAL ISSUES POSED BY THE NEW REPRODUCTIVE TECHNOLOGIES: A PHYSICIAN'S PERSPECTIVE

Luigi Mastroianni, Jr.

New technologies in the treatment of reproductive failure have, quite appropriately, stimulated a great deal of public interest and concern. This concern is shared by health professionals in the reproductive field as they too grapple with the medical and social implications of the numerous approaches that have become available as a result of advances in biotechnology. Lori Andrews has focused on a number of important issues regarding reproductive technologies, and, although she says that she presents "feminist perspectives" on the technologies, it is clear that the concerns that arise out of her discussion are not solely feminist concerns. They are societal concerns. Such issues as informed consent, exploitation, and potential for bodily harm, among others, should be the concern of all of us. The new reproductive technologies conjure up visions of a "brave new world," the potential for wrenching reproductive capabilities and choice from individuals for wider and possibly inappropriate purposes, and unauthorized and inappropriate experimentation with human eggs and sperm that could result in the production of humans beings who, if given a choice, might not have wished to be born. The potential for misuse of this new technology is enormous. Our ability to modulate the application of science to human reproduction through law is sadly limited.

It might be useful to review some of the issues regarding the use of reproductive technologies from the point of view of one who is called upon to treat individual patients and, in the case of infertility, couples. As aptly pointed out by Lori Andrews, the right not to reproduce is now well established. I recall as a physician in New Haven, Connecticut in the 1950s, that legislators, acting in good faith within a particular social context, had years earlier passed a law prohibiting the use of contraception. This law was challenged by C. Lee Buxton, the Professor and Chairman of Obstetrics/Gynecology at Yale. Professor Buxton was jailed for his stand, but the concept that we could write into law how people were to behave in a sexual sphere was successfully challenged. This example leaves me with less than complete confidence in the law. Hence, my unease as legislators with good intent now grapple with issues such as the use of surrogacy arrangements and, in our own state of Pennsylvania, in vitro fertilization (IVF). The slippery slope concept engendered by scientific advance is matched by the slippery slope of legal encounter. In the state of Pennsylvania, we are required to report at quarterly intervals on the number of oocytes recovered, the number of oocytes fertilized, and the number of embryos returned to patients. This was incorporated in the state's Abortion Control Act. The intent of the regulation remains unclear, and it is not known how the information that is being gathered will eventually be used, but I can assure you that we are complying.

Some of the new reproductive technologies under consideration are not so new. The issues concerning the *Baby M* case, every detail of which has been dissected in the press, are not so clear as to justify but one conclusion. A woman, labeled a "surrogate," agreed to bear a child conceived from artificial insemination and, at birth, to turn that child over to the father (the inseminator) and his wife. This does not constitute new reproductive technology. Clearly, the method utilized to obtain the specimen for insemination is not new, nor is it technical. The insemination could easily have been carried out by someone with no special training or, for that matter, by the recipient herself. What is new is the concept that this approach can be formalized; what is new is that the legal profession has incorporated a contract system into the process; and what is new is that the state now feels obliged to explore issues concerning the use of surrogacy arrangements and to devise laws that presumably will protect the various parties. Given the track record of the state, and given the changing complexity of attitudes, one can, with some justification, question the wisdom of placing into law requirements that would be, at the very least, unenforceable and, at the very worst, would constitute an interference in private decision making.

Carefully thought-out guidelines are important. The British have considered such guidelines. A government-sponsored commission was established to review issues regarding human fertilization and embryology, and out of it came the Warnock Report entitled, "A Question of Life." Such issues as artificial insemination, in vitro fertilization, egg donation, embryo donation, and surrogacy were reviewed in detail. The wider use of these techniques, including the freezing and storage of human semen, eggs, and embryos; scientific issues; and possible future developments in research were considered. Guidelines were suggested on the regulation of infertility services and research in the British National Health Service. Not unexpectedly, the report included some expressions of dissent, especially on matters of surrogacy and the use of human embryos in research.

Our system is quite different in a number of respects. Our health care delivery programs are not centrally coordinated. Nevertheless, constraining factors are operative. For example, the United States Public Health Service, to this day, does not support research on human in vitro fertilization. Such research was interdicted in the early 1970s when our own program at the University of Pennsylvania, designed to explore the physiology of human fertilization, was discontinued. An ethics advisory committee did meet to explore the issues regarding in vitro fertilization and did present a report to the Secretary of Health and Human Services. The report was pigeonholed, and no official action was ever taken. Thus, it was left to the medical profession itself to review the issues regarding the use of technologies in infertility. In November 1984, the Committee on Ethics of the American Fertility Society was charged with exploring in vitro fertilization (surrogacy, gamete donation, cryopreservation, genetic manipulation, cloning, etc.), and donor insemination including cryopreservation. This committee's report was issued in September 1986. The committee included representatives from a wide variety of disciplines. Thus, in the United States, the medical profession, so often maligned for its lack of concern with the implications of new technology arising out of biomedical research, has taken the initiative and has provided a working document that serves as a basis for further discussion on a number of important issues.

Somehow systems must be structured to avoid exploitation. When the potential risks of any new procedure are reviewed and consent is given, it must be understood that we do not always have the vision to predict untoward complications. Human experimentation committees in universities and research centers are generally careful to review in meticulous detail and discard any proposal that would even remotely subject a patient to harm, unless the benefit to that patient significantly and clearly outweighs the risk, but the unexpected can occur. There are risks that go beyond bodily harm. As the early experiments on human fertilization were carried out, this new technology was described in the press as a potential answer to infertility in patients with hopelessly damaged or absent fallopian tubes. That reporting engendered a great deal of mental anguish. The harm in that case resulted from the false hope that this new technology would provide a solution to infertility problems in the immediate future. There is emotional harm when a procedure with a very low success rate is offered and when the treatment is followed by failure. There is emotional harm when procedures are available but, for financial reasons, must be withheld.

In the early phases of in vitro fertilization research, the nagging question--the most critical one in my set of values--was the issue concerning the normalcy of the embryo after transfer to the uterus. How could we be sure that the babies resulting from in vitro fertilization would be normal? Extensive laboratory experimentation had been carried out in mice and other laboratory animals, but the actual experiment had never been done in the human. In fact, there was no way to have known that the systems were failsafe in the human without taking the first step. Fortunately, the accumulation of information now supports the conclusion that there is no significant increase in fetal abnormalities. What if the first IVF baby, simply as a matter of chance, had been abnormal? Might this have eliminated, once and for all, a treatment that has been so successful in providing much-wanted children to infertile couples? What about the rights of the fetus itself? Would that newborn eventually have the right to claim damages for "wrongful birth"? A considered decision was made based on scientific information indicating that it would be reasonable to proceed. Informed consent was especially complicated in that there was no way to ask the fetus whether the experiment should be carried out. Rules should be made that address values and be designed to avoid exploitation, but potential risks cannot be eliminated entirely.

Lori Andrews points out that the medical profession has been notoriously remiss in developing and offering preventive measures in health; rather, it focuses on solutions. This may very well be true, but throughout medical school and beyond, the dictum "the best treatment is prevention" is emphasized. Many candidates for in vitro fertilization-embryo replacement have suffered from prior pelvic infection--pelvic inflammatory disease. Let us consider the methods that are available to prevent such tragedies. They include abstinence before marriage and a monogamous relationship. Pelvic infection may also be prevented to some degree by the use of condoms. I would question whether it is the role of the physician to espouse abstinence before marriage and monogamy. Here, strong social forces come into play. The economically disadvantaged patient with pelvic inflammatory disease is often treated late. Her counterpart who has ready access to medical care is usually diagnosed and treated early. Incorporated in her treatment is a review as to how the infection was contracted and advice as to how it can be avoided. The incidence of a repeat pelvic inflammatory disease is surely significantly lower among advantaged patients than among the economically disadvantaged for reasons that have very little to do with abstinence and monogamy, but have a great deal to do with educational level and availability of health care. All of us must fight for equal access to health care. All of us should emphasize prevention.

The impact of social circumstance on fertility and on the availability of health care applies also to the availability of infertility treatment. For the less advantaged person who has suffered repeated bouts of pelvic inflammatory disease resulting in hopelessly damaged fallopian tubes, and who is now in a position to have a much-desired child, in vitro fertilization may be the only medical option. For financial reasons, it is often a nonoption. Not that financial constraints do not also apply to people who would not be classified as economically disadvantaged. The cost of the procedure is high, and insurance companies have discriminated in favor of microsurgical repair of damaged fallopian tubes and, by and large, have not incorporated in vitro fertilization into their coverage. More and more, medical decisions must be based not on what is best for the patient, but rather on what is available to the patient.

Reasonable medical precautions can be taken to ensure the safety of reproductive technologies. In programs involving donor insemination, elaborate screening systems should be mandated to ensure insofar as possible the genetic health of the donor and to avoid the transmission of the sexually transmitted diseases--gonorrhea, chlamydia, and AIDS. Such preventive measures have encouraged the additional application of high-tech systems, specifically the use of frozen semen. With frozen semen, one can be assured that the donor is disease-free by testing the donor some weeks after the specimen has been received. The downside is that frozen semen is associated with a lower pregnancy rate. In its guidelines, the American Fertility Society has addressed the issue of quality control. What happens, however, when medical technologies are demedicalized? Do women who elect self-insemination with specimens obtained from presumably appropriate donors accept an additional risk of a sexually transmitted disease? Clearly they do. Laws that have been suggested stipulating that only a physician, or one supervised by a physician, carry out donor insemination are designed to address this problem. It is likely that such laws, which are inappropriate, potentially unconstitutional, and certainly unenforceable, will be ignored.

APPENDIX

PROJECT ON REPRODUCTIVE LAWS FOR THE 1990s

Working Group

Vicki Alexander, M.D.
Alliance Against Women's Oppression

Lori B. Andrews, Esq.
American Bar Foundation

Adrienne Asch
New Jersey Commission on Legal and Ethical Problems in the Delivery of Health Care

Joan Bertin, Esq.
American Civil Liberties Union

Wendy Chavkin, M.D.
Bureau of Maternity Services and Family Planning of the New York City Department of Health

Irene Crowe
The Pettus Crowe Foundation

Dana Gallagher
Formerly at Vermont Women's Health Center

Janet Gallagher, Esq.
Formerly at Civil Liberties and Public Policy Program, Hampshire College

Nancy Gertner, Esq.
Silverglate, Gertner, Fine, Good, and Mizner

Beverly Harrison
Union Theological Seminary

Mary Sue Henifin, Esq.
Debevoise and Plimpton

Linda Janet Holmes
The Traditional Midwife Center International

Ruth Hubbard
Biological Laboratories, Harvard University

Nan Hunter, Esq.
American Civil Liberties Union

Dara Klassel, Esq.
Planned Parenthood Federation of America

Iris Lopez
Department of Puerto Rican Studies, City College

Isabel Marcus, Esq.
SUNY at Buffalo Law School

Judy Norsigian
Boston Women's Health Book Collective

Laurie Nsiah-Jefferson
Bureau of Maternity Services and Family Planning of the New York City Department of Health

Lynn Paltrow, Esq.
American Civil Liberties Union

Rosalind Petchesky
Women's Studies, Hunter College

Rayna Rapp
Department of Anthropology, The New School for Social Research

Barbara Katz Rothman
Department of Sociology, Baruch College, CUNY

Nadine Taub, Esq.
Rutgers University Law School-Newark

Norma Wikler
Department of Sociology, University of California at Santa Cruz

PROJECT ON REPRODUCTIVE LAWS FOR THE 1990s

Advisory Committee

NOTES ON CONTRIBUTORS

ROBERT ABRAMS, J.D.

Robert Abrams is Attorney General of the state of New York. Previously, he was a New York State Assemblyman for three terms and Bronx Borough President in New York City. As Attorney General, Mr. Abrams has initiated effective legal and administrative actions to combat discrimination; protect the interests of workers, consumers, and tenants; guarantee reproductive rights; and prosecute environmental polluters, white-collar criminals, and organized crime figures. In recognition of his efforts, he has received the Margaret Sanger Award from the New York State Family Planning Association, the "Champion of Women" award from the Professional Women in Construction, and the Man of the Year award from the National Association for the Advancement of Colored People.

LORI B. ANDREWS, J.D.

Lori B. Andrews is a Research Attorney with the American Bar Foundation. An authority on medical law, she has testified on alternative reproduction before a subcommittee of the United States House of Representatives Committee on Law, Science, and Technology, and has served on advisory bodies to the World Health Organization, Office of Technology Assessment, and Department of Health and Human Services. Ms. Andrews was on the committee that produced the American Fertility Society's 1986 report on ethical guidelines. Her most recent book is *New Conceptions: A Consumer's Guide to the Newest Infertility Treatments, Including In Vitro Fertilization, Artificial Insemination, and Surrogate Motherhood*.

GEORGE J. ANNAS, J.D., M.P.H.

George J. Annas is Edward R. Utley Professor of Health Law at the Boston University School of Public Health and Edward R. Utley Professor of Law and Medicine at the Boston University School of Medicine. He is the author of *The Rights of Hospital Patients* and coeditor of three sequential anthologies entitled *Genetics and the Law*. He also writes regular columns in the *Hastings Center Report* and *American Journal of Public Health*. Professor Annas has chaired the Massachusetts Task Force on Organ Transplantation and the American Bar Association committees on "Family and Science" and "Legal Problems in Medicine."

ADRIENNE ASCH, C.S.W.

Adrienne Asch is Associate in Social Science and Policy with the New Jersey Commission on Legal and Ethical Problems in the Delivery of Health Care. She has taught at City College and Barnard College. Ms. Asch has authored many articles and book chapters on feminism, disability studies, and ethical issues. She is coeditor of *Women with Disabilities: Essays in Psychology, Culture, and Politics* and a special issue of the journal *Social Issues* entitled *Moving Disability Beyond "Stigma."*

JOAN E. BERTIN, J.D.

Joan E. Bertin is Associate Director of the Women's Rights Project of the American Civil Liberties Union. She was a lawyer with the National Employment Law Project, and has lectured and taught at several law schools. A specialist in women's health issues, employment discrimination, and constitutional rights, she has testified before Congress on genetic screening in the workplace and has advised the United States Congress Office of Technology Assessment on reproductive hazards in the workplace.

WENDY CHAVKIN, M.D., M.P.H.

Wendy Chavkin is Director of the Bureau of Maternity Services and Family Planning of the New York City Department of Health. She has been involved in establishing programs to ensure reproductive health care for low-income women, conducting research on the reproductive health outcomes of New Yorkers, and formulating policy regarding many reproductive health issues. She is the editor of *Double Exposure: Women's Health Hazards on the Job and at Home*, winner of the American Health Magazine Book Award in 1984.

SHERRILL COHEN, PH.D.

Sherrill Cohen is Research and Conference Coordinator for the Project on Reproductive Laws for the 1990s at Rutgers University School of Law-Newark. Dr. Cohen is a historian and analyst of contemporary social policy. She was formerly the founding director of a women's rights project at the American Civil Liberties Union of New Jersey and advocated for reproductive rights as a staff member of Planned Parenthood affiliates in California and Rhode Island.

EDMUND D. COOKE, JR., J.D.

Edmund D. Cooke, Jr. is Counsel to the United States House of Representatives Committee on Education and Labor, chaired by Representative Augustus Hawkins. Prior to taking this post, Mr. Cooke served as Chief Legislative Assistant to Representative Harold Washington. An expert on economic opportunity and civil rights, Mr. Cooke is a former Deputy Director of the Offices of Field Services at the Equal Employment Opportunity Commission. He has also been a staff attorney with the Appellate Court Division of the National Labor Relations Board.

PEGGY C. DAVIS, J.D.

Peggy C. Davis is a Professor at New York University School of Law. She was formerly a Judge in Family Court in the state of New York and has written, among other publications, "Use and Abuse of the Power to Sever Family Bonds," an article appearing in the *New York University Review of Law and Social Change*. Professor Davis underwent a year of training with the New York Society of Freudian Psychologists and also taught at Rutgers University School of Law. She is on the Board of Directors of Lawyers for Children and on the Board of The Vera Institute.

ALAN R. FLEISCHMAN, M.D.

Alan R. Fleischman is Director of the Division of Neonatology and Professor of Pediatrics at Albert Einstein College of Medicine and Montefiore Medical Center. A member of the Governor's Commission on Life and the Law in the state of New York, he has also twice been a Senior Adjunct Associate of the Hastings Center. He currently serves on the editorial board of the *American Journal of Perinatology* and regularly advises a variety of bioethical journals.

JANET GALLAGHER, J.D.

Janet Gallagher is the former Director of the Civil Liberties and Public Policy Program at Hampshire College. During earlier work with the Center for Constitutional Rights, she was involved in preparing important reproductive rights litigation in the cases *Harris v. McRae* and *McRae v. Califano.* Her articles concerning the fetus as patient appeared recently in the *Harvard Women's Law Journal* and in the *New England Journal of Medicine.* She is on the national Board of Directors of Catholics for a Free Choice and the Bioethics Advisory Committee of the Planned Parenthood Federation of America.

NANCY GERTNER, J.D.

Nancy Gertner is an attorney with the Boston law firm of Silverglate, Gertner, Fine, Good, and Mizner. Ms. Gertner has been a Visiting Professor at Harvard University School of Law, and an adjunct teacher at Boston University School of Law. Among the many cases she has litigated have been *Preterm v. Dukakis* and *Doe v. Hale Hospital,* both challenging restrictions on the funding of abortions. Ms. Gertner is on the Board of Directors of the Massachusetts Civil Liberties Union.

MARY SUE HENIFIN, J.D., M.P.H.

Mary Sue Henifin is an attorney with the law firm of Debevoise and Plimpton. Formerly coordinator of the Women's Occupational Health Resource Center at Columbia University and Assistant Professor of Biology at Hampshire College, Ms. Henifin brings public health expertise to her legal work on women's issues. With Ruth Hubbard and Barbara Fried, Ms. Henifin edited *Biological Woman: The Convenient Myth,* and she is now legal editor for the journal *Women and Health.* She frequently writes and speaks on women's health issues.

RUTH HUBBARD, PH.D.

Ruth Hubbard is a Professor of Biology at Harvard University, where she teaches courses dealing with the interactions of science and society, particularly as they affect women. She has written numerous articles and coedited several books, the most recent of which is *Woman's Nature: Rationalizations of Inequality*, with Marian Lowe. She is a member of the National Women's Health Network and serves on the Executive Council of the Committee for Responsible Genetics.

NAN D. HUNTER, J.D.

Nan D. Hunter was formerly Staff Counsel for the Reproductive Freedom Project of the American Civil Liberties Union. At present, she is Director of the new Lesbian and Gay Rights Project of the American Civil Liberties Union. Ms. Hunter has participated in the litigation of landmark cases in reproductive issues, including *City of Akron v. Akron Center for Reproductive Health*, in which the United States Supreme Court reaffirmed the principles of *Roe v. Wade*, the 1973 decision legalizing abortion. She has also taught at George Washington University School of Law and Catholic University School of Law.

DEBORAH KAPLAN, J.D.

Deborah Kaplan is an attorney who has a private practice as a legal consultant and public policy specialist on the civil and legal rights of people with disabilities. She is the founder and former Executive Director of the Disability Rights Center, a Washington, D.C. organization affiliated with and funded by Ralph Nader. She has participated in two important disability rights cases argued before the United States Supreme Court, *Pennhurst State School and Hospital v. Terri Lee Halderman* and *Southeastern Community College v. Frances B. Davis*. Ms. Kaplan is a former member of the Board of Directors of the National Association of Women and the Law.

SALLY J. KENNEY

Sally J. Kenney is an Assistant Professor of Political Science and Women's Studies at the University of Iowa. Previously, she taught at the University of Illinois. Her research focuses on sex discrimination law and health and safety issues in the United States and Britain. In 1986, she was awarded the Women's Research and Education Institute Congressional Fellowship on Women and Public Policy and served as a legislative assistant for the Education and Labor Committee in the United States House of Representatives.

LUIGI MASTROIANNI, JR., M.D.

Luigi Mastroianni, Jr. is William Goodell Professor and Director of the Division of Human Reproduction at the Hospital of the University of Pennsylvania. He is the recipient of numerous awards, including the Barren Foundation's Gold Medal for Contributions to the Field of Reproductive Endocrinology. Dr. Mastroianni is a former President of the American Fertility Society and a past board member of the Sex Information and Education Council of the United States (SIECUS). He is past editor of the journal *Fertility and Sterility,* is an advisor to a number of professional journals, and presently sits on the editorial board of the *International Journal of Fertility.*

THE HONORABLE GEORGE MILLER

Representative George Miller is serving his seventh term in Congress representing the seventh district of California (Contra Costa County). In 1983, he became the first Chair of the House Select Committee on Children, Youth, and Families, which he continues to chair. Congress has enacted many significant pieces of legislation that Representative Miller sponsored or coauthored, including the Adoption Assistance and Child Welfare Act, the Women, Infants, and Children's Act, the Family Violence Prevention and Services Act, the Asbestos School Hazard Detection and Control Act, and the Education for All Handicapped Children's Act. Representative Miller is also a member of the House's Committee on the Budget, Education and Labor Committee, and Interior and Insular Affairs Committee.

JUDY NORSIGIAN

Judy Norsigian is a member and staff member of the Boston Women's Health Book Collective. She coauthored *The New Our Bodies, Ourselves* and its earlier versions and coedited the *International Women and Health Resource Guide.* She speaks widely to the media and medical and public health organizations and has testified on contraceptive research priorities before the United States House of Representatives Select Committee on Population. Ms. Norsigian is a member of the Board of Directors of the National Women's Health Network and a past member of the Board of the Health Planning Council of Greater Boston.

LAURIE NSIAH-JEFFERSON, M.P.H.

Laurie Nsiah-Jefferson is a consultant for the Bureau of Maternity Services and Family Planning of the New York City Department of Health. She was formerly Director of Reproductive Health Programs for the Massachusetts Department of Public Health. Ms. Jefferson has served widely as a consultant and speaker on women's health concerns in the third world, as well as on domestic health issues. In the past, she has been active with the National Black Women's Health Project and other organizations dealing with community health care.

RAYNA RAPP, PH.D.

Rayna Rapp is an Associate Professor and Chair of the Department of Anthropology, Graduate Faculty, at The New School for Social Research. She is writing a book on the social impact and cultural meaning of amniocentesis, based on three years of field research with women of diverse class, racial, ethnic, and religious backgrounds as they encounter this reproductive technology.

JOHN A. ROBERTSON, J.D.

John A. Robertson is Baker and Botts Professor of Law at the University of Texas at Austin and a Fellow of the Hastings Center. Previously, he taught at the University of Wisconsin School of Law and School of Medicine. In addition to authoring recent articles on the concept of procreative liberty, Professor Robertson has written numerous articles on medical-legal topics. He is the author of *The Rights of the Critically Ill* and is currently working on a book entitled *Gender Roles and Noncoital Reproductive Technologies*.

HELEN RODRIGUEZ-TRIAS, M.D.

Helen Rodriguez-Trias is Medical Director of the AIDS Institute in the New York State Department of Health. She was formerly Associate Director of Primary Care at Newark Beth Israel Medical Center. She is an Executive Board Member of the American Public Health Association and has been on the Hospital Review and Planning Council for the state of New York. She has taught at Columbia University's College of Physicians and Surgeons, Albert Einstein College of Medicine, and the University of Puerto Rico School of Medicine. The National Women's Health Network in 1977 named Dr. Rodriguez-Trias among a select group of Distinguished Physicians.

JEANNIE I. ROSOFF

Jeannie I. Rosoff is the President of The Alan Guttmacher Institute. She has worked as a community organizer in East Harlem, New York; Associate Director of the New York Committee for Democratic Voters; and Director of the Washington office and Vice President for Governmental Affairs of Planned Parenthood. Author of many articles on family planning, women's rights, population, and the functioning and financing of government programs, she was a member of the National Advisory Council of the National Institute of Child Health and Human Development (of the National Institutes of Health) from 1980 to 1985. Ms. Rosoff has received numerous awards, among them, in 1980 the Carl S. Schultz award of the American Public Health Association and the Margaret Sanger award of the Planned Parenthood Federation of America in 1986.

BARBARA KATZ ROTHMAN, PH.D.

Barbara Katz Rothman is a Professor of Sociology at Baruch College and the Graduate Center of the City University of New York. Her work has focused on women's experiences in maternity care in America. She is the author of *In Labor: Women and Power in the Birthplace*, published in paperback as *Giving Birth*, and of *The Tentative Pregnancy: Prenatal Diagnosis and the Future of Motherhood*. Her work has also appeared in *The Hastings Center Bulletin*, *Journal of Medical Ethics*, *Vogue*, *Qualitative Sociology*, *Ms.*, *Mothering*, and *Gender and Society*, of which she is associate editor.

JEANNE MAGER STELLMAN, PH.D.

Jeanne Mager Stellman is Executive Director of the Women's Occupational Health Resource Center and an Associate Professor at the Columbia University School of Public Health. She has served as a visiting staff member of the Occupational Safety and Health Branch of the International Labour Office in Geneva. Dr. Stellman taught at the University of Pennsylvania School of Medicine and the Rutgers University Labor Education Center. She is Editor of the journal *Women and Health* and, with Mary Sue Henifin, coauthor of *Office Work Can be Dangerous to Your Health.*

NADINE TAUB, L.L.B.

Nadine Taub has been the Director of the Project on Reproductive Laws for the 1990s since its origin in 1985. As a Professor of Law and Director of the Women's Rights Litigation Clinic at Rutgers University School of Law-Newark, she has worked extensively in the area of reproductive rights. Her litigation has involved the right to abortion, problems of sterilization abuse, and surrogacy. Among her writings are "Reproductive Rights" in *Women and the Law*, coauthored with Judith Levin, and the forthcoming book *The Law of Sex Discrimination*, coauthored by J. Ralph Lindgren. She is a member of the American Public Health Association's Legal Committee, the American Civil Liberties Union's Reproductive Freedom Project Advisory Committee, and the Planned Parenthood Federation of America's Volunteer Lawyers Panel.

TABLE OF CASES

INDEX